Electric Power Systems: Analysis and Design

.

Electric Power Systems: Analysis and Design

Edited by Marko Silver

CLANRYE INTERNATIONAL
www.clanryeinternational.com

Clanrye International,
750 Third Avenue, 9th Floor,
New York, NY 10017, USA

ISBN: 978-1-63240-600-2

Cataloging-in-Publication Data

Electric power systems : analysis and design / edited by Marko Silver.
    p. cm.
Includes bibliographical references and index.
ISBN 978-1-63240-600-2
1. Electric power systems. 2. Electrical engineering. 3. Electric power. 4. Electric power production.
5. Renewable energy sources. I. Silver, Marko.
TK1001 .E44 2017
621.042--dc23

For information on all Clanrye International publications
visit our website at www.clanryeinternational.com

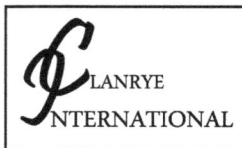

$\mathcal{CL}$ANRYE
INTERNATIONAL

Printed in the United States of America.

Contents

Preface

This book has been an outcome of determined endeavour from a group of educationists in the field. The primary objective was to involve a broad spectrum of professionals from diverse cultural background involved in the field for developing new researches. The book not only targets students but also scholars pursuing higher research for further enhancement of the theoretical and practical applications of the subject.

Electric power systems are highly effective ways to transmit electrical energy for public and private use. The grid is the most popular form of electric power system which can be divided into generators, distribution system and transmission system. The various studies that are constantly contributing towards advancing technologies and evolution of this field are examined in detail. The various advancements in electric power systems are glanced at and their applications as well as ramifications are discussed herein. The book is appropriate for students seeking detailed information in this area as well as for experts. It will help the readers in keeping pace with the rapid changes in the field of electrical engineering.

It was an honour to edit such a profound book and also a challenging task to compile and examine all the relevant data for accuracy and originality. I wish to acknowledge the efforts of the contributors for submitting such brilliant and diverse chapters in the field and for endlessly working for the completion of the book. Last, but not the least; I thank my family for being a constant source of support in all my research endeavours.

Editor

Electrical network reduction for load flow and short-circuit calculations using powerfactory software

Funso K. Ariyo

[1]Department of Electronic and Electrical Engineering, Ile-Ife, Nigeria
[2]Obafemi Awolowo University, Ile-Ife, Nigeria

Email address:
funsoariyo@yahoo.com (F. K. Ariyo)

Abstract: The primary purpose in constructing equivalents is to represent a portion of a network containing many buses but having only a few "boundary buses" by a reduced network containing only the boundary buses and, perhaps, a few selected buses from within the original sub-network. The equivalent constructed gives an exact reproduction of the self and transfer impedances of the external system as seen from its boundary buses. PowerFactory's network reduction algorithm produces an equivalent representation of the reduced part of the network and calculates its parameters. This equivalent representation is valid for both load flow and short-circuit calculations, including asymmetrical faults (that is, single-phase faults).

Keywords: Boundary Buses, Reduced Network, Powerfactory, Load Flow, Short-Circuit Calculation

1. Introduction

A typical application of network reduction is a project where a specific network has to be analyzed, but this network cannot be studied independent of a neighboring network of the same or a higher/lower voltage level. In such cases, one option is to model both networks in detail for calculation purposes. There may, however, be situations when it is not desirable to perform studies with the complete model. Such situations may include, for example, cases when the calculation times would increase significantly or when the data of the neighboring network is confidential and cannot be published in detail. In these cases it is common practice to provide a representation of the neighboring network, which contains the interface nodes (connection points) that may be connected by equivalent impedances and voltage sources [1].

Network reduction for load flow is an algorithm based on sensitivity matrices. The basic idea is that the sensitivities of the equivalent grid, measured at the connection points in the retained grid, must be equal to the sensitivities of the grid that has been reduced. This means that for a given (virtual) set of ΔP and ΔQ (active an reactive power variations) injections in the branches, from the retained grid to the grid to be reduced, the resulting Δu and Δφ (voltage magnitude and voltage phase angle variations) in the boundary nodes must be the same for the equivalent grid, as those that would have been obtained for the original grid (within a user defined tolerance).

Network reduction for short-circuit is an algorithm based on nodal impedance/nodal admittance matrices. The basic idea is that the impedance matrix of the equivalent grid, measured at the connection points in the retained grid, must be equal to the impedance matrix of the grid to be reduced (for the rows and columns that correspond to the boundary nodes). This means that for a given (virtual) additional ΔI injection (variation of current phasor) in the boundary branches, from the retained grid to the grid to be reduced, the resulting Δu in the boundary nodes must be the same for the equivalent grid, as those that would have been obtained for the original grid (within a user defined tolerance). This must be valid for positive sequence, negative sequence, and zero sequence cases, if these are to be considered in the calculation (unbalanced short-circuit equivalent) [2-7].

2. Methodology

Power system network matrices [8]:

$$I_{bus} = Y_{bus} V_{bus} \qquad (1)$$

where

I_{bus} is the vector of the injected bus currents;

V_{bus} is the vector of bus voltages measured from the reference node.

Y_{bus} is known as the bus admittance matrix.

In a power system, the current injection is always zero at buses where there are no external loads or generators connected; such nodes may be eliminated. Then for an n-bus system, the reduced admittance matrix by eliminating node k is:

$$I_k = 0 \qquad (2)$$

where node k has zero current injection, therefore,

$$I_k = 0$$

The new bus admittance matrix is:

$$Y_{ij}^{(new)} = Y_{ij} - \frac{Y_{ij} Y_{kj}}{Y_{kk}} \qquad (3)$$

where i, j = 1, 2, ..., n. $i, j \neq k$

In stability studies, all nodes are eliminated except for the internal generator nodes and obtain the Y matrix for the reduced network:

$$I = YV \qquad (4)$$

where

$$I = \begin{bmatrix} I_n \\ 0 \end{bmatrix} = \begin{bmatrix} Y_{nn} & Y_{nr} \\ Y_{rn} & Y_{rr} \end{bmatrix} \begin{bmatrix} V_n \\ V_r \end{bmatrix} \qquad (5)$$

The matrix in equation (5) is the reduced matrix Y. It has dimensions (n x n), where n is the number of generators.

Network reduction can be applied only to those nodes that have zero injection current. An equivalent represents a network, which contains many buses but only a few boundary buses, by a reduced network that contains only the boundary buses and a few of the original buses. Equivalents are used in two circumstances: both to allow larger areas of major interconnected systems to be represented in studies and also to achieve improved computational speed in simulations by removing buses and branches that influence system behavior.

A boundary cuts a set of tie-lines between areas or otherwise identifiable sections of a network but passes through no buses. A boundary bus is part of only one area. An equivalent makes more efficient use of storage when the ratio of branches to buses in the equivalent is reduced. As a general rule, however, reducing a system into a number of small equivalents is more efficient than reducing of a large system in one step to produce a single equivalent [9-11].

3. Network Modeling in PowerFactory

Figure 1 shows IEEE nine-bus system with three synchronous generators. Generator (G_1) was selected the swing generator. The transformer, load and generators data are given in Table 1, 2 and 3 respectively. The network model contains the electrical and graphical information for the grid. To further enhance manageability, this information is split into two subfolders: diagrams and network data. An additional subfolder, Variations, contains all expansion stages for planning purposes. The network model folder contains the all graphical and electrical data which defines the networks and the single line diagrams of the power system under study. This set of data is referred as the network data model.

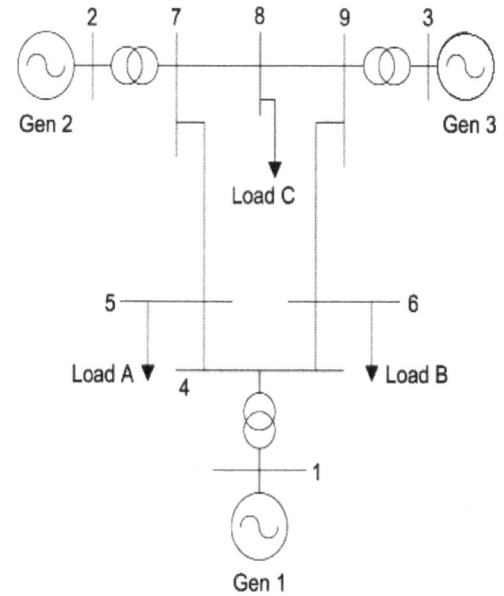

Figure 1. IEEE three-machine, nine-bus test system [12].

Table 1. Transformers data.

	T_{r1}	T_{r2}	T_{r3}
MVA	250	200	150
HV (kV)	230	230	230
LV (Kv)	16.5	18	13.8

Table 2. Loads data.

	Load$_A$	Load$_B$	Load$_C$
MV	125	90	100
Mvar	50	30	35

Table 3. *Generators data.*

Parameters	G_1 (swing)	G_2	G_3
Apparent power (MVA)	247.5	192	128
Voltage (kV)	16.5	18	13.8
Power factor	1	0.85	0.85
Active power (MW)	-	150	85
Reactive power (Mvar)	-	6.7	-10.9
Voltage (p.u.)	1.04	1.025	1.025
Rotor type	salient	round	round
Syn. reactances, x_d	0.36	1.72	1.68
Syn. reactances, x_q	0.24	1.66	1.61
Sub. transient (x''_d)	0.2	0.2	0.2
Sub. transient (x''_q)	0.1	0.12	0.2
transient (x'_d)	0.15	0.23	0.23
transient (x'_q)	0.00	0.378	0.32
Rotor resistances	0.00	0.005	0.00
Time constant T'_d	3.73	0.8	0.806
Time constant T''_d	0.05	0.05	0.05
Accel. Time constant	9.55	4.162	2.765
Leakage reactances x_l	0.083	0.141	0.0949

4. Network Reduction in Power Factory

The Network Reduction function, by default, retains the original network data. The boundary, necessary to split the grid into the part to be reduced, was defined and the part to retain its detailed representation. To ensure that the boundary splits the network into two regions, Check Split button in the ElmBoundary dialogue could be used.

The distribution network in Fig. 1 is fed by two busbars, 'Bus 5' and 'Bus 6', in the center of the transmission system. It is represented by Load A and Load B and the corresponding two transformers. If the distribution system is to be studied in details, while the transmission system is to be represented only by its equivalent model. The transmission system has a connection to another transmission system, which is represented by the External Grid and the connecting line. This second transmission system shall remain as it is, the aim of the paper is to reduce only one transmission system. The "Boundary" (shown by dotted line in Figure 1) defines which part shall be reduced [1]. After this, the network is reduced using PowerFactory. The network reduction procedure will automatically create a new variation/system stage to represent the original grid.

In the new system stage(s), the part of the grid which was not reduced retains its full representation, whereas the part that was reduced is erased and the new simplified grid representation is added, with the connections to the other part of the grid (that is, to the part which is not reduced). The new system stage(s) will therefore represent the combined grid (retained grid and reduced equivalent). The new system stages will be automatically activated in the active study case. To calculate short-circuit equivalent, this option is used to specify whether the short-circuit equivalent shall be calculated (option enabled) or not (option disabled). PowerFactory currently supports only the complete short-circuit calculation method [1]. Figure 2 shows elements in the defined boundary. These are to be reduced to an equivalent three AC voltage sources (ElmVac) and three common impedances (ElmZpu) connecting the remaining busbars.

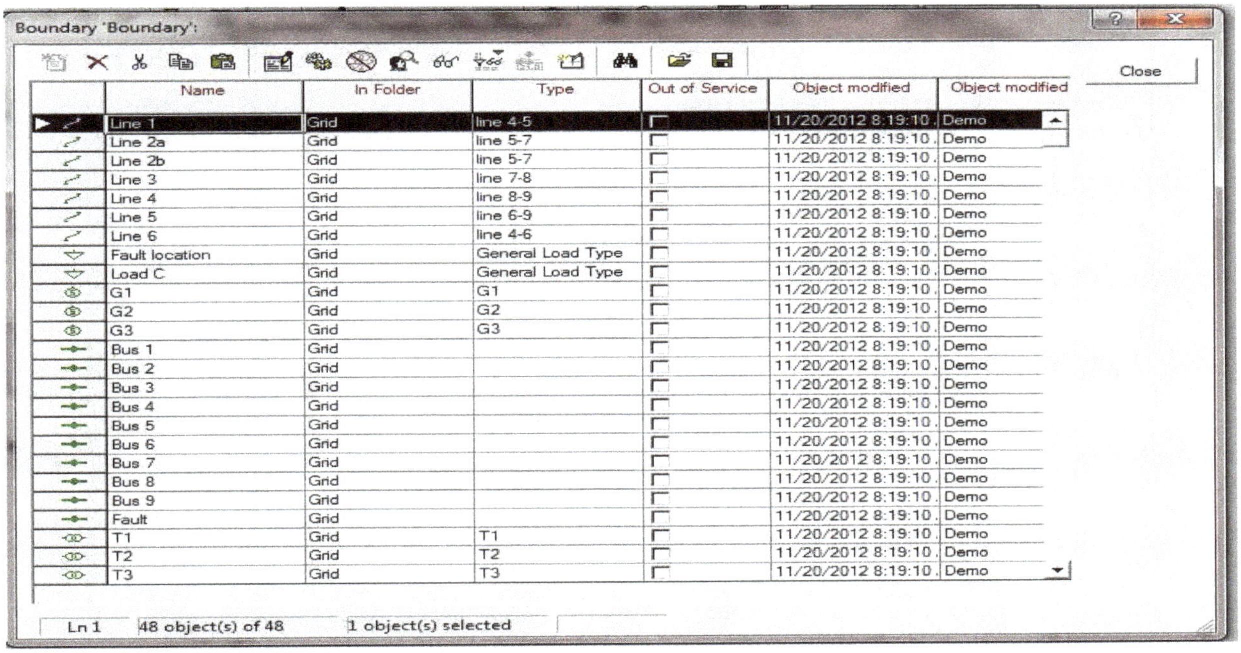

Name	In Folder	Type	Out of Service	Object modified	Object modified
Line 1	Grid	line 4-5		11/20/2012 8:19:10	Demo
Line 2a	Grid	line 5-7		11/20/2012 8:19:10	Demo
Line 2b	Grid	line 5-7		11/20/2012 8:19:10	Demo
Line 3	Grid	line 7-8		11/20/2012 8:19:10	Demo
Line 4	Grid	line 8-9		11/20/2012 8:19:10	Demo
Line 5	Grid	line 6-9		11/20/2012 8:19:10	Demo
Line 6	Grid	line 4-6		11/20/2012 8:19:10	Demo
Fault location	Grid	General Load Type		11/20/2012 8:19:10	Demo
Load C	Grid	General Load Type		11/20/2012 8:19:10	Demo
G1	Grid	G1		11/20/2012 8:19:10	Demo
G2	Grid	G2		11/20/2012 8:19:10	Demo
G3	Grid	G3		11/20/2012 8:19:10	Demo
Bus 1	Grid			11/20/2012 8:19:10	Demo
Bus 2	Grid			11/20/2012 8:19:10	Demo
Bus 3	Grid			11/20/2012 8:19:10	Demo
Bus 4	Grid			11/20/2012 8:19:10	Demo
Bus 5	Grid			11/20/2012 8:19:10	Demo
Bus 6	Grid			11/20/2012 8:19:10	Demo
Bus 7	Grid			11/20/2012 8:19:10	Demo
Bus 8	Grid			11/20/2012 8:19:10	Demo
Bus 9	Grid			11/20/2012 8:19:10	Demo
Fault	Grid			11/20/2012 8:19:10	Demo
T1	Grid	T1		11/20/2012 8:19:10	Demo
T2	Grid	T2		11/20/2012 8:19:10	Demo
T3	Grid	T3		11/20/2012 8:19:10	Demo

Figure 2. *Elements in the defined boundary.*

5. Results and Discussion

The network reduction was carried out with PowerFactory by DIgSILENT version 14.1 [1]. The equivalent model for power injection is extended ward equivalent method.

The short-circuit method used is complete method with 0.1 seconds as break time, while the fault clearing time is 1 second.

Figures 3 and 4 show PowerFactory command dialogue for AC voltage source and common impedance respectively.

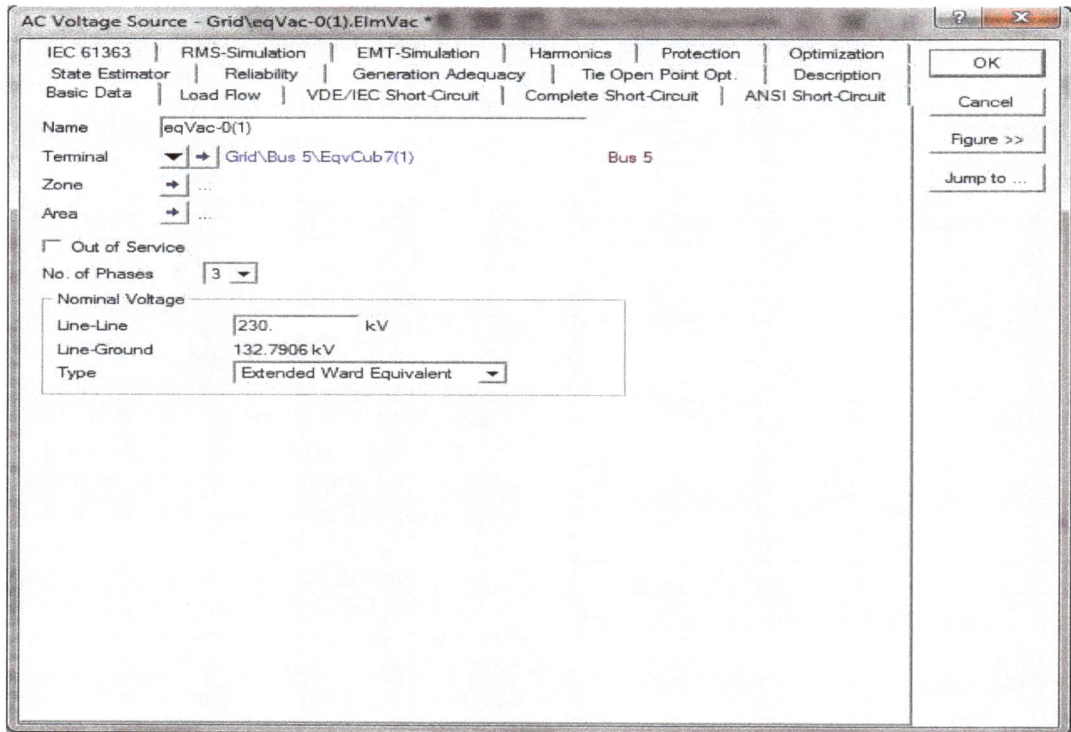

Figure 3. *Command dialogue for AC voltage source.*

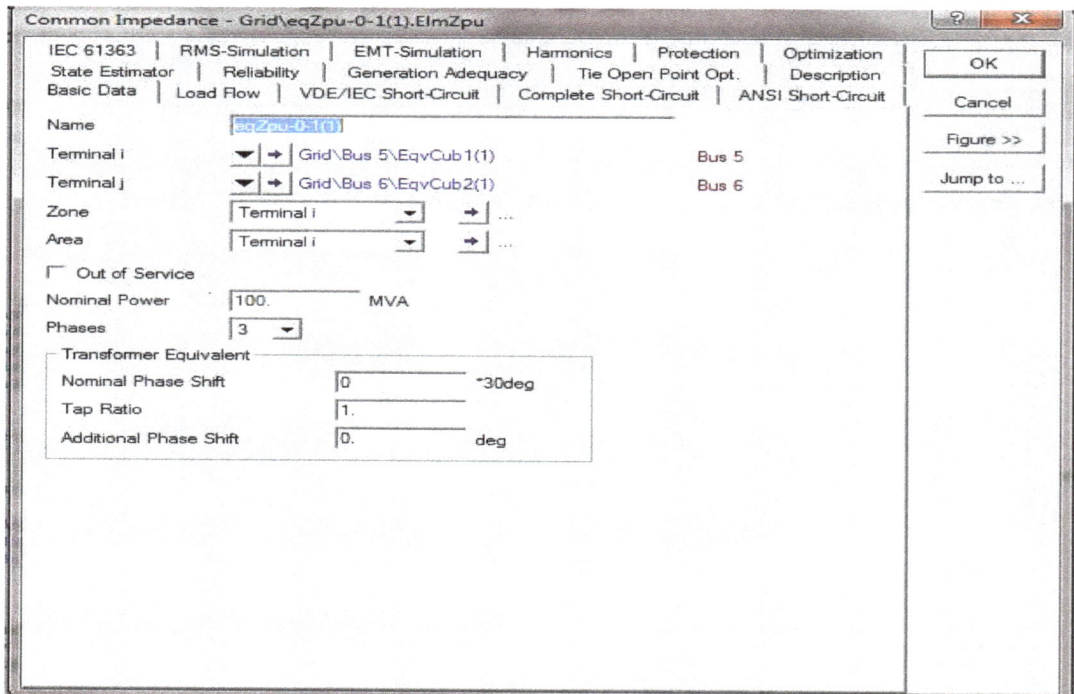

Figure 4. *Command dialogue for common impedance.*

The transmission network was reduced to an equivalent representation by three AC voltage sources (eqVac-0 at bus 6; eqVac-1 at bus 5 and eqVac-2 at bus 4) and three common impedances (eqZpu-0-1 between buses 6 and 5; eqZpu-0-2 between buses 6 and 4; eqZpu-1-2 between buses 5 and 4) connecting the remaining busbars. The resulting load flow and method short-circuit parameters are given in Table 4 - 6.

Table 4. *Parameters of the AC voltage sources.*

Parameters	eqVac-0	eqVac-1	eqVac-2
Terminal voltage (kV)	230	230	230
Active power (MW)	28.211	78.667	-0.00068
Reactive power (Mvar)	27.89	21.389	17.6732
V_{mag} (p.u.)	1.0143	1.0026	1.02871
r_{ward} (ohm)	0.00	0.00	0.00
X_{ward} (ohm)	312.85	114.68	-1583.6
r_{short} (ohm)	48.318	13.530	-13.298
X_{short} (ohm)	247.67	132.06	-13171.2

Table 5. *Parameters of the common impedances (load-flow).*

Components	Load flow			
	positive sequence		negative	zero
	(real)	(imag.)	(re, im)	(re, im)
eqZpu-0-1	0.1151325	0.4091812	0	0
eqZpu-0-2	0.0170001	0.0920005	0	0
eqZpu-1-2	0.01	0.0850005	0	0

Table 6. *Parameters of the common impedances (complete short-circuit).*

Components	complete short-circuit			
	positive sequence		negative	zero
	(real)	(imag.)	(re, im.)	(re, im)
eqZpu-0-1	0.150755	0.150755	0	0
eqZpu-0-2	0.0170326	0.0919994	0	0
eqZpu-1-2	0.0100325	0.0850005	0	0

The load flow calculation in the reduced network gives the same results for the distribution network as for the original (non-reduced) network as shown in Table 7 and 8. In Table 7, the buses 1, 4, 5 and 6 busbars report for the reduced network is the same as the busbars report for buses 1, 4, 5 and 6 for non-reduced network as shown in Table 8.

Table 7. *Load flow (busbars) report for non-reduced network.*

Busbars	rated Voltage	Bus - Voltage		deg.
	(kV)	(kV)	(p.u.)	
Bus 1	16.5	17.16	1.04	0
Bus 2	18	18.45	1.025	4.81
Bus 8	230	232.6	1.011	147.17
Bus 9	230	237.14	1.031	149.23
Bus 3	13.8	14.14	1.025	1.93
Bus 6	230	233.29	1.014	144.99
Bus 4	230	236.61	1.029	147.31
Bus 5	230	230.6	1.003	145.6
Bus 7	230	234.51	1.02	149.66

Table 8. *oad flow (busbars) report for reduced network.*

Busbars	rated Voltage	Bus - Voltage		deg.
	(kV)	(kV)	(p.u.)	
Bus 1	16.5	17.16	1.04	0
Bus 6	230	233.29	1.014	144.99
Bus 4	230	236.61	1.029	147.31
Bus 5	230	230.6	1.003	145.6

6. Conclusion

PowerFactory's network reduction algorithm produces an equivalent representation of the reduced part of the network and calculates its parameters. This equivalent representation is valid for both load flow and short-circuits calculations, including asymmetrical faults (that is, single-phase faults). By using small, equivalent networks, the computational requirements can be significantly reduced. Network reduction is usually performed by computing impedances and by eliminating unnecessary elements.

This reduction usually results in a highly dense impedance matrix; therefore, using the reduced network may not significantly increase efficiency. Equivalent networks have been used for load flow and short circuit studies because they can reproduce the same voltages and currents of the remaining buses as the original systems do [13]. A load flow calculation or a short-circuit calculation in the reduced network gives the same results for the distribution network as for the original (non-reduced) network.

Nomenclature

p.u.: per unit;
HV: high voltage;
LV: low voltage;

MW: megawatts;

Mvar: mega volt ampere reactive;

kV: kilo voltage;

V_{vag}: voltage magnitude for extended ward method;

R_{ward}: resistance for the extended ward;

X_{ward}: reactance for the extended ward;

r_{short}: positive-sequence complete short-circuit resistance;

x_{short}: positive-sequence complete short-circuit reactance;

References

[1]   DIgSILENT PowerFactory Version 14.1 Tutorial, DIgSI-LENT GmbH Heinrich-Hertz-StraBe 9, 72810 Gomaringen, Germany, May, 2011.

[2]   Buygi M. O., H. M. Shanechi, G. Balzer, and M. Shahideh-pour, Transmission Planning Approaches in Restructured Power Systems, IEEE Powertech, June 2006.

[3]   Ward J. B.. Equivalent circuits for power flow studies. AIEE Trans. Power App. Syst., 8:373-380, February 1949.

[4]   Singh H.K. and Srivastava S.C.. A sensitivity based network reduction technique for power transfer assessment in dere-gulated electricity environment. Transmission and Distribu-tion Conference and Exhibition: Asia Pacific, IEEE/PES, October 2002.

[5]   Tinney W. F. and J. M. Bright, "Adaptive reductions for power flow equivalents," IEEE Trans. Power Syst., vol. 6, no. 2, pp. 613–621, May 1991.

[6]   Cheng X. and T. J. Overbye, "PTDF-based power system equivalents," IEEE Trans. Power Syst., vol. 20, no. 4, pp. 1868–1876, Nov. 2005.

[7]   Enns M. K. and J. J. Quada, "Sparsity-enhanced network reduction forfault studies," IEEE Trans. Power Syst., vol. 6, no. 2, pp. 613–621, May 1991.

[8]   Saadat, H. 'Power System Analysis'. McGraw- Hill Interna-tional Editions, 1999.

[9]   P.M. Anderson & A.A. Fouad, 'Power System Control and Stability', 2nd edition, IEEE Press Power Engineering Se-ries, Wiley-Interscience, 2003.

[10] Arthur R. Bergen & Vijay Vittal, 'Power System Analysis', 2nd edition, Prentice Hall, Inc., 2000.

[11] Yao-nan Yu, 'Electric Power System Dynamics', Academic Press, Inc., 1983.

[12] www.sciencedirect.com.

[13] Hyungseon Oh, 'A New Network Reduction Methodology for Power System Planning Studies', IEEE Transactions on Power Systems, Vol. 25, No. 2, May 2010.

Analysis of smart grid with 132/33 KV sub-transmission line in rural power system of Bangladesh

A. S. M. Monjurul Hasan[1, *], Md. Habibullah[1], A. S. M. Muhaiminul Hasan[2]

[1]Department of Electrical & Electronic Engineering, IUT, Dhaka, Bangladesh
[2]Department of Electrical & Electronic Engineering, AUST, Dhaka, Bangladesh

Email address:
a.s.m.monjurul.hasan@hotmail.com(A. S. M. M. Hasan), habibullahiut@gmail.com(Md. Habibullah)

Abstract: "Smart Grid" is a modern concept which refers to the conversion of the mainstream or typical electric power grid to a modern power grid. This new conversion is a foreseeable solution to the power system problems of the modern century. Rejuvenation of the current electric power distribution system is an important step to implement the Smart Grid technology. So, distribution system engineers should be acquainted with the knowledge of Smart Distribution System. Also the customers should acknowledge the benefits that they will be enjoying from this modernized power system. There are power crisis everywhere in the world, besides there is system loss in the existing power system. It is happening in Bangladesh also. To reduce power crisis renewable source of energy like solar energy, wind energy, raw coal energy may be used. But the quantity of electricity produced by this renewable energy source is low and several kilowatts range. This electricity is utilized by smart grid which is hard for usual power grid system. For this reason we need hybrid ac/dc smart control grid system. In this report models using Wind-mil Software is proposed for reducing system loss and also incorporate smart metering so that the power flow can reach easily to the consumers. Smart grid also helps sole proprietor and individual business man to sell their little amount of electricity to the grid which they produce by their own entrepreneurship through smart meter. Line loss and regulated voltage, regulator and capacitor are also inserted here to reduce the loss and make the effective and efficient power supply to the consumer. This metering is also centrally controlled and works over a huge area. At present developing country should develop this system.

Keywords: Controllable Loads, Wind Turbine Generator, Smart Grid, Smart Metering, Windmill Software

1. Introduction

Energy generation is one of the key factors in driving the socio-economic growth of any country. In Bangladesh, increasing demands for energy has already exceeded the capacity from existing plants from conventional sources of energy. Thus access to electricity is very limited where Per capita energy consumption is about 237 KOE. There are still lots of area where there is no supply of electricity. Attention is being focused on renewable energy sources and to harness electricity from them to meet the national energy demand. In Bangladesh solar photovoltaic (PV) systems are being widely deployed in rural areas and large scale coverage in rural areas with renewable energy sources is being actively considered with mini-grid structure. Such a grid system can implement smart grid techniques by the efficient management of the power grid systems in many countries around the globe. A Smart Grid is a form of electricity network using digital technology. A smart grid delivers electricity from suppliers to consumers using two-way digital communications to control appliances at consumers' homes; this could save energy, reduce costs and increase reliability and transparency if the risks inherent in executing massive information technology projects are avoided. The "Smart Grid" is envisioned to overlay the ordinary electrical grid with information and net metering system, this system includes smart meters. Smart grids are being promoted by many governments as a way of addressing energy independence, global warming and emergency resilience issues. The function of an Electrical grid is not a single entity but an aggregate of multiple networks and multiple power generation companies with multiple operators employing varying levels of communication and coordination, most of which is manually controlled. Smart grids increase the connectivity, automation and coordination between these suppliers, consumers and networks that perform either long distance transmission or local distribution task.

Transmissions networks move electricity in bulk over medium to long distances, are actively managed, and generally operate from 345kV to 800kV over AC and DC lines. Local networks traditionally moved power in one direction, "distributing" the bulk power to consumers and businesses via lines operating at 132kV and lower. This paradigm is changing as businesses and homes begin generating more wind and solar electricity, enabling them to sell surplus energy back to their utilities. Modernization is necessary for energy consumption efficiency, real time management of power flows and to provide the bi-directional metering needed to compensate local producers of power. Power demand in isolated islands has been increasing rapidly. Diesel generators fueled by fossil fuels mostly supply the power for this power demand. For greenhouse gas reduction and oil substitution, introduction of renewable energies such as photovoltaic and wind energy is important. Renewable power resources are safe, clean, and abundant in nature. However, due to the power fluctuation of renewable energy sources, voltage and frequency deviations are occurred in island power systems whose ability to maintain stable supply–demand balance is low. Therefore, it is necessary to control the system frequency and voltage at the supply side. At the supply side, installation of storage equipment and pitch angle control of a wind generator has been proposed for control of the distribution power system. However, the installation of storage equipment that needs large storage capacity and the cost of maintenance for battery degradation are not expected. Hence, in case of using the renewable energy plants connected to power system, supply-side control has limitations. Therefore, mutual cooperation control with the demand side is required because it is difficult to maintain the power quality by only the supply-side control. From this viewpoint, a smart grid, which maintains stable supply–demand balance by monitoring the power information of the demand side, is necessary. Smart grids provide an excellent opportunity to manage power quality better and reduce harmonic distortions of the power networks. The impact on power quality is taken into consideration from the viewpoint of a generator side, a grid side, and a demand side management. It is expected that smart technologies will lead to reduced investment in primary equipment and it will increase higher availability of power supply.

2. Present Scenario of Transmission and Distribution System of Bangladesh Electrical Power

Because of major reforms, restructuring and corporatization process of Bangladesh power sector, a number of distribution entities were formed with the

objective of bringing commercial environment including increase of efficiency, accountability and dynamism with the aim of reaching electricity to all citizens by 2021. In order to increase and improve power generation and customer service with an aim to bring a greater mass under electrification, major integrated power distribution programs have been undertaken. Presently the following five organizations are responsible for the distribution of power:

1. Bangladesh Power Development Board (BPDB)
2. Rural Electrification Board (REB)
3. Dhaka Power Distribution Company (DPDC)
4. Dhaka Electric Supply Company (DESCO)
5. West Zone Power Distribution Company (WZPDC)

Table 1. *Overall scenario of distribution line in Bangladesh*

Total Distribution Lines	278,000 KM
Total Consumers	12.5 million
Irrigation Consumer	2.77 Lac
Access to electricity	50%
Distribution Loss	12.75%
Accounts Receivable	2.22 Equivalent months

At present the government has taken measures to reduce system loss and increase customer satisfaction. Under this project 409 interface meters have been installed at all generating stations throughout the country and Dhaka distribution zone and transmission network at 230KV, 132KV and 33KV level. All the meters are connected with the main server which is located at Bidyut Bhaban (13th floor), Dhaka, Bangladesh. Energy inflow/outflow, demand, voltage, current, power factor, meter tempering etc. may be known from the main server. BPDB, REB, DPDC, DESCO and WZPDC have individual workstation and can read data at some level. The interface meters have been used as billing meters. This will be extremely beneficial for the energy auditing system. Operator's performance will also be enhanced significantly and accountability and transparency will be established in the energy auditing system.

The government has crossed some prominent hurdles in this challenging field. But Bangladesh still faces power shortages and that is the reason the GoB has set a target for providing electricity to all citizens by 2021. This electrification target is unlikely to be met by grid expansion alone, as rest of the populations live in remote areas which are far away from existing grid line and sometimes isolated from the main land. Considering this over arching goal, the government has identified private sector participation as an important requirement. Since power system development is highly capital-intensive, the government encourages private sector investment to implement RAPSS. Under the RAPSS concept, private investor will be given an area (the RAPSS Area) for the development, operation and maintenance of the electricity distribution and retail supply system, including generation as a utility for a period of 20 years. The government has taken initiatives to establish solar mini-grid for remote off-grid area under RAPSS where grid expansion is not planned for the next 15 to 20 years.

3. Smart Grid platform for Bangladesh

The power system in Bangladesh is very complex and quite aged with lots of lacking. But, there are many scopes to convert the power grid of Bangladesh to the smart grid. To address the power crisis and other problems, it is the high time to initiate the plans to form grids which are more smart, receptive and flexible than present power grids. In Bangladesh, not only by integrated communication techniques but also by increasing the usage of renewable resources the implementation of smart grid technology can be achieved. In prospective to the socio-economic condition of Bangladesh; smart grid will enable consumer empowerment to manage their energy usage and financial savings. In recent days, an interest is increasing rapidly about the small-scaled grid system based on several tens of Photovoltaic power generation. Such a grid system, which is called as micro grid, has advantages to increase an operational efficiency and economics when it is connected to grid or supply a secured electric power at islands, mountains and remote areas without connecting grid. The micro grid is divided into ac micro grid and dc micro grid, which is classified by whether, distributed sources and loads are connected on the basis of ac or dc grid. ac micro grid has a benefit to utilize existing ac grid technologies, protections and standards but stability and requirement of reactive power are the inherent demerits of it. On the other hand, dc micro grid has no such demerits of ac micro grid and assures reliable implementation of environment-friendly distributed generation sources.

4. Proposed Smart Grid

Our proposed smart grid model consists of a smart model of the Khulna region power system. The model includes different power stations there. It also included the simulated the output of this virtual designed smart power system. In our proposed system we tried to apply the local power system in our design and then simulated and also debugged the design that we find a case. We tried to find good efficiency and reduce the power loss. We also given the practical data (from goalpar110 MW power plant, PGCB Khulna South and also from nearest substations) as input.

4.1. System Configuration

In our main system configuration of figure 1, we have drawn a model of small area smart grid model. We put all the transformer, transmission lines, substations, generation power plant also and we simulated our design with actual data and we can increase performance and efficiency Power system of least developed country is very poor in performance and classical, we tried to figure out something better than existing model. The WTG (Wind Turbine

Generator) has no gear. It is rated several like 1 or 2-MW generator. Permanent Magnet Synchronous Generator (PMSG) generally used in WTG. Its structure is not complex and performs well with high efficiency.

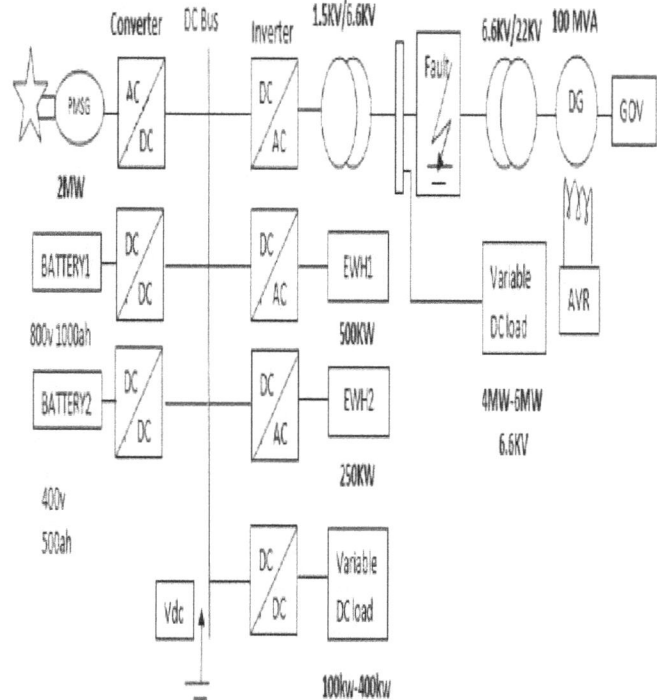

Figure 1. DC smart grid system

It is predicted that the future wind turbine generator will be PMSG based. Generator side used a convertor in this dc model of smart grid, also a grid-side inverter used here. Various loads like chargeable batteries and electric heaters are connected to this grid. Total system is attached to a diesel generator and variable ac load via the grid side inverter and the transformer. The output power of Wind turbine is given to the PMSG. To get optimal power a pulse width modulated converter is used as a controller of PMSG. DC distribution system is used to supply the PMSG's output power to consumer. Rest of the power is sent to the ac load in assistant with a grid-side inverter.

Rotational side converter effects on the rotational speed of the PMSG effects The PMSG model is explained in. A generator side converter effects on the rotational speed of the PMSG. This is for getting variation of speed. Also for maximum power operation and maximum power point tracking (MPPT) control. This makes the system torque under control. For observing the speed control we the speed control of the PMSG are sensed on a dynamic frame. Here the rotational speed error is given as input of the controller of speed. From the controller the axis stator current is tuned automatically. Usually, the salient pole type synchronous machine is controls the d-axis stator current and the reference are expressed by the following equation [1]:

$$i_{1d}^* = \frac{\phi_f}{2(L_d - L_q)} - \sqrt{\frac{\phi_f^2}{4(L_d - L_q)^2} + i_{1q}^2} \qquad 1$$

4.2. Grid-Side Inverter & Controllable Load

For frequency and voltage controls grid-side inverter is preferable. The control system the grid-side inverter is mainly done by controlling axis current. Current and voltage can be controlled by the axis current. We set the frequency to 60 Hz and the voltage to 6.6 kV. The dc bus voltage is controlled by the voltage control of the controllable loads. EWH and battery are used as controllable loads. The decentralized EWH model for each house connected to the dc grid. The EWH is modeled as a current source and each EWH is controlled to consume power within the rated range. The temperature of accumulated water of EWH is controlled by feedback control with an integral of the power consumption. In this control system, the dc bus voltage fluctuation is suppressed by the droop control, the power consumption. Command is decided by the droop coefficient and the controlled variable produced by controller of water temperature. The design of the droop coefficient is explained in the next section.

4.3. System Design

In this section we have some interface that is user operated. There have commands like add device, change feed, back feed, split, disc (disconnect), connect, rotate, move, devices, zoom to fit, zoom in, zoom out, calculate analysis etc. with this command one has to work with Wind-mil software. In the system design part we have designed a smart grid system with respect to our country and we have taken as a reference Khulna Central power grid. In this design we included a source, three transformer, overhead lines, and five nodes which is different substation of the Khulna Central Grid. As we worked in student version we had to use limited components. If we get the main version then we will do the more sophisticated smart grid design. For grid setting we double clicked on the source of the design and then a circuit element editor appeared and then we clicked to the navigator input the data and at last push on apply change tab. By clicking the impedance code min & max in the editor in the right box, we selected Khulna Central from the Equipment list and clicking ok we came out of the editor

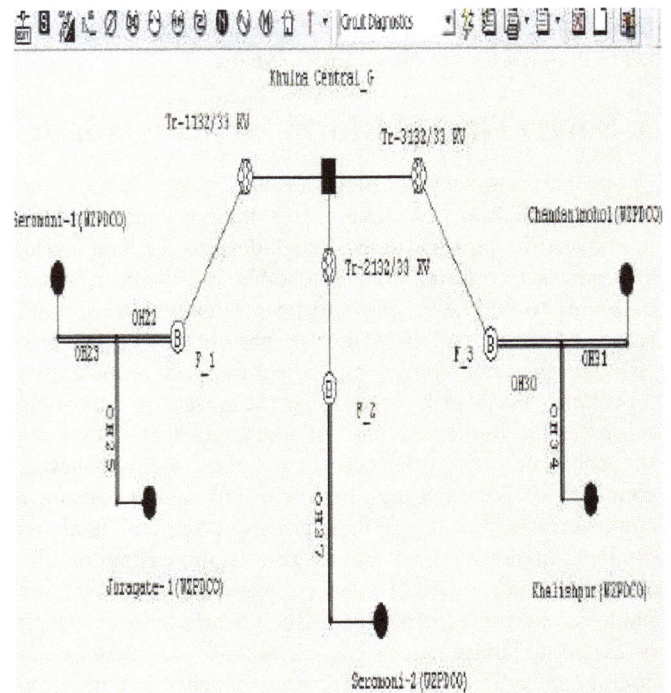

Figure 2. *Smart grid system design for Khulna local sub-station area, Bangladesh*

The analysis starts with Circuit Diagnosis. At first, we put circuit diagnosis and then recalculate analysis command is pressed. Then we found a new window showing errors and warnings.

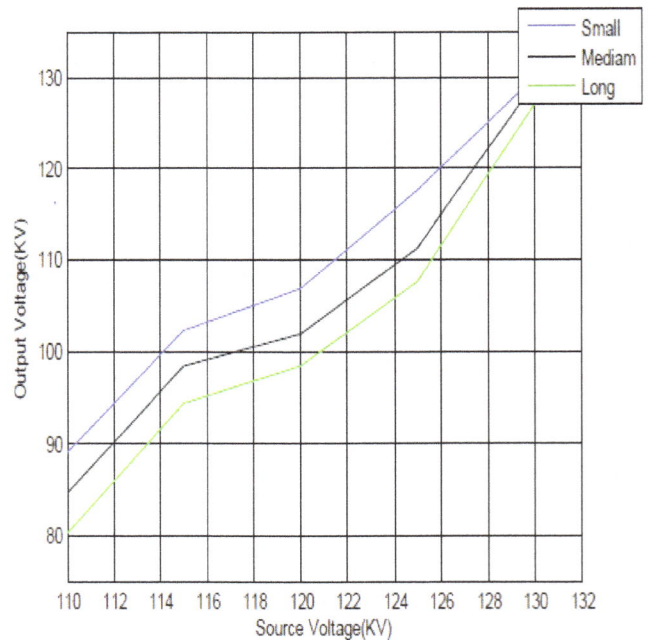

Figure 3. *Source Voltage Vs Output Voltage in our smart grid system. (All voltage values are assumed comparing with the data of PGCB Khulna South substation).*

Table 2. Output voltage of substation with connection of capacitor and without connection of capacitor

Source Voltage(KV)	Output Voltage (KV) for S-6000KW, M-6000 KW, L-8000KW					
	Without Capacitor			With Capacitor		
	S	M	L	S	M	L
132	130.1	130	124	130	131	125.7
125	116.59	123.01	118	116	123.9	119
120	107.2	118	109.7	107	119	110.16
115	97.99	112.5	100	98.3	114	100
110	89.05	101.9	90	89.87	103.05	90.34

5. Some Aspects of DC Micro Grid over AC Micro Grid for Bangladesh

Since Bangladesh is a densely populated country; a high quality of power is required for such agricultural dependable society and dc grid can assure it. For example, if a blackout or voltage sag occurs in a bulk dc and ac hybrid power system, most inverters might be tripped. So, it is difficult for ac micro grids to keep a super high quality power supplying continuously in islanding operation. A significant part of the system is transferred to the consumer's premises; reducing the initial cost of the power company and overall maintenance of the system. Since the cost of the battery and inverter is taken away from the power company, the overall maintenance is simplified so as the cost of energy (can be as much as 40%). In addition to this small scale rural industries like irrigation, rice husk can use the high voltage (240 V dc) directly without using dc-dc converter. The ac micro grid has some inherent problems such as synchronization, stability, need for reactive power while dc grid system is immune to those. Moreover, it is to be mentioned that the dc micro grids can be turned into a viable proposition due to its low cost over ac micro grids since the energy cost comparison between an ac and dc grid has been shown in previous research work.

6. Conclusion

At present smart grid is being developed according to practical hypothesis. This will make the world more economic and people can get the best output through it. The extra energy of the total system will also be managed by this grid system. Here we tried to design a small interconnected smart grid. Country like Bangladesh should implement smart grid in their transmission and distribution system to utilize the generated power. As people of Bangladesh are suffering for power shortage, government should take immediate steps to generate more electricity as well take necessary steps for proper distribution of the generated power.

Acknowledgements

Authors are grateful to Prof. Dr. A. K. M. Sadrul Islam, Department of Mechanical & Chemical Engineering, & Assistant Professor Syeed Shihav, Department of Electrical and Electronic Engineering, IUT for helping us in different aspects

References

[1] Duic N. & Carvalho M. G. (2004) "Increasing renewable energy sources in island energy supply: case study portosanto," *RENEWABLE SUSTAINABLE ENERGY*, VOL. 8, PP. 383–399.

[2] J. Clerk Maxwell, A Treatise on Electricity and Magnetism, 3rd ed., vol. 2. Oxford: Clarendon, 1892, pp.68–73.

[3] Senjyu T., Sakamoto R., Kaneko T., Yona A. & Funabashi T. (2009) "OUTPUT POWER LEVELING OF WIND FARM USING PITCH-ANGLE CONTROL WITH FUZZY NEURAL NETWORK," *ELECTRIC POWER COMPON. SYST.*, VOL. 36, NO. 10, PP. 1048–1066.

[4] Y. Yorozu, M. Hirano, K. Oka, and Y. Tagawa, "Electron spectroscopy studies on magneto-optical media and plastic substrate interface," IEEE Transl. J. Magn. Japan, vol. 2, pp. 740–741, August 1987 [Digests 9th Annual Conf. Magnetics Japan, p. 301, 1982].

[5] Johnson A. P. (2010), "The History of the Smart grid evolution at southern California edision," Presented at the IEEE pes innovative smart grid technologies conf., GAITHERSBURG, MD. SOOD V. K., FISCHER D., EKLUND J. M., AND BROWN T. (2009) "DEVELOPING A COMMUNICATION INFRASTRUCTURE FOR THE SMART GRID," IN *PROC. ELECTR. POWER ENERGY CONF.*, PP. 1–7.

[6] http://www.powerdivision.gov.bd/user/brec1/43/1

3

Application of STATCOM to increase transient stability of wind farm

Bouhadouza Boubekeur, Ahmed Gherbi, Hacene Mellah

Department of Electrical Engineering, Sétif-1 University, Algeria

Email address:

bouhadouza_b@yahoo.fr (B. Bouhadouza), gherbi_a@yahoo.fr (A. Gherbi), has.mel@gmail.com (H. Mellah)

Abstract: In this paper we interested to the study the necessary of Facts to increase the transient stability on the presence of faults and the integration of new renewable source, like wind energy, these lasts make the electrical grid operate in a new conditions, the STATCOM is one of the important Facts element, It provides the desired reactive-power generation and absorption entirely by means of electronic processing of the voltage and current waveforms in a voltage source converter (VSC). This function is identical to the synchronous condenser with rotating mass. In present work we propose a transient stability improvement using STATCOM under faults, in the first time we study the transient stability with and without STATCOM for clearly his advantages. In the second time we know the relation between the reactive power injecting by a STATCOM and the critical clearing time, some simulation results are given, commented and discussed.

Keywords: Transient Stability, Reactive Power, FACTS, STATCOM, Wind Power, CCT

1. Introduction

There is now general acceptance that the burning of fossil fuels is having a significant influence on the global climate. Effective mitigation of climate change will require deep reductions in greenhouse gas emissions, with UK estimates of a 60–80% cut being necessary by 2050 [1], Still purer with the nuclear power, this last leaves behind dangerous wastes for thousands of years and risks contamination of land, air, and water; the catastrophe of Japan is not far[2], to avoid the problems of the pollution, the energy policy decision states that the objective is to facilitate a change to an ecologically sustainable energy production system such as wind power [3], but the major problem is how associate the wind power stations to the grid with assure the linking conditions[4]. In addition, now a day's power transmission and distribution systems face increasing demands for more power, better quality and higher reliability at lower cost, as well as low environmental effect. Under these conditions, transmission networks are called upon to operate at high transmission levels, and thus power engineers have had to confront some major operating problems such as transient stability, damping of oscillations and voltage regulation etc [5], in this work we interest to the transient stability, this last indicates the capability of the power system to maintain synchronism when subjected to a severe transient disturbances such as fault on heavily loaded lines, loss of a large load etc [6].Generator excitation controller with only excitation control can improve transient stability for minor faults but it is not sufficient to maintain stability of system for large faults occur near to generator terminals [6]. Researchers worked on other solution and found that flexible AC transmission systems (FACTS) are one of the most prominent solution [7], [8].

The objective principal to use FACTS technology for the operators of the electric power is to have an opportunity for the control of the power flow and by increasing the capacities usable of these lines under the normal conditions. The parameter which controls the operation of transmission of energy in a line such as the impedances series and shunts, running, tension and phase angle is controlled by utilizing FACTS controllers. FACTS devices increases power handling capacity of the line and improve transient stability as well as damping performance of the power system [7], [8].

According to the specialized literature we find several types of FACTS [6-11], in our work we are limited to the study a great disturbance, so the FACTS element used for reactive power compensation both assuring the low cost and high efficiency is STATCOM.

The static synchronous compensators (STATCOM) consist of shunt connected voltage source converter through coupling transformer with the transmission line. STAT-

COM can control voltage magnitude and, to a small extent, the phase angle in a very short time and therefore, has ability to improve the system [7], [8].

2. Wind Turbine Model

2.1. Squirrel Cage Induction Generator

The fixed speed wind generator systems have been used with a multiple-stage gearbox and a SCIG directly connected to the grid through a transformer [11].

The well-known advantages of SCIG are it is robust, easy and relatively cheap for mass production [11], electrically fairly simple devices consisting of an aerodynamic rotor driving a low-speed shaft, a gearbox, a high-speed shaft and an induction generator [12].

The gearbox is needed, because the optimal rotor and generator speed ranges are different, we find also a pole-changeable SCIG has been used in some commercial wind turbines; it does not provide continuous speed variations [11]. The generator is directly grid coupled. Therefore, rotor speed variations are very small, because the only speed variations that can occur are changes in the rotor slip[13], because the operating slip variation is generally less than 1%, this type of wind generation is normally referred to as fixed speed [12].

A SCIG consumes reactive power. Therefore, in case of large wind turbines and/or weak grids, often capacitors are added to generate the induction generator magnetizing current, thus improving the power factor of the system as a whole [13].

The power extracted from the wind needs to be limited, because otherwise the generator could be overloaded or the pullout torque could be exceeded, leading to rotor speed instability. In this concept, this is often done by using the stall effect. This means that the rotor geometry is designed in such a way that its aerodynamic properties make the rotor efficiency decrease in high wind speeds, thus limiting the power extracted from the wind and preventing the generator from being damaged and the rotor speed from becoming unstable [13], so the operating condition of a squirrel-cage induction generator, used in fixed-speed turbines, is dictated by the mechanical input power and the voltage at the generator terminals. This type of generator cannot control bus bar voltages by itself controlling the reactive power exchange with the network. Additional reactive power compensation equipment, often fixed shunt-connected capacitors, is normally fitted [12]; this system concept is also known as the 'Danish concept' and is depicted in Fig 1 [13].

The slip is generally considered positive in the motor operation mode and negative in the generator mode. In both operation modes, higher rotor slips result in higher current in the rotor and higher electromechanical power conversion. If the machine is operated at slips greater than unity by turning it backwards, it absorbs power without delivering anything out i.e. it works as a brake. The power in this case

is converted into I heat loss in the rotor conductor that needs to be dissipated [14].

Fig. 1 shows the torque-slip characteristic of the induction machine in the generating mode. If the generator is loaded at constant load torque T_L only $P1$ is stable. The loading limit of the generator i.e. the maximum torque it can support is called the breakdown torque and represented in the Fig.1 as T_{max} If the generator is loaded under a constant torque above T_{max}, it will become unstable and stall, draw excessive current and destroy itself thermally if not properly protected [14].

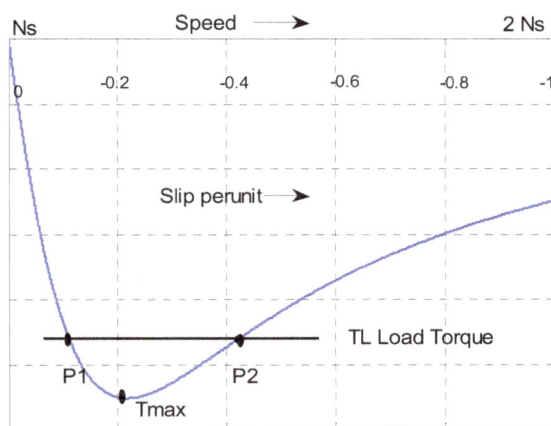

Figure 1. *Torque versus slip characteristic of an induction generator [14].*

2.2. Modeling for Fixed - Speed Wind Turbines

The modeling of wind turbine plays an important role in the building of stability concept. Every research recently uses grid model, wind turbine model and wind speed model as a foundation. The specific simulation approach used to study the dynamics of large power systems is reduced-order modeling of wind turbine. This model uses several assumptions and gives the models the various subsystems of each of the recent wind turbine types as presents at the Fig.2 [14].

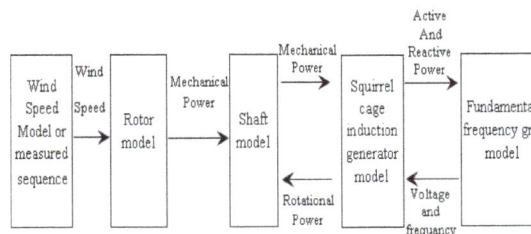

Figure 2. *Generator structure of fixed-speed wind turbine model [6].*

We use Matlab to modeling the wind turbine system in two main blocks: rotor model and generator model.

2.2.1. Rotor Model

The traditional rotor model in wind turbine simulation is base on the well known equation which gives the relationship between the power extracted from wind and wind speed [14]:

$$P_{wt} = C_p P_v = C_p(\lambda, \beta) . \frac{\rho A_{wt} v_w^3}{2} \quad (1)$$

Where C_p is the power coefficient of wind turbine (C_p is the function of the blade pitch angle β and the tip-speed ratio); ρ is the air density; A_{wt} is the swept area; v is the wind speed. The tip-speed ratio λ.is defined as:

$$\lambda = \frac{w_{wt} . R}{v} \quad (2)$$

Where w_{wt} is mechanical angular velocity of wind turbine blades; R is radius of wind turbine blades. The numerical method of C_p is in Ref [15].

$$C_p(\lambda, \beta) = 0.5176 \left(\frac{116}{\lambda_i} - 0.4\beta - 5 \right) e^{\frac{-21}{\lambda_i}} + 0.0068\lambda \quad (3)$$

Where

$$\lambda_i = \left[\frac{1}{\lambda + 0.08\beta} - \frac{0.035}{\beta^3 + 1} \right]^{-1} \quad (4)$$

There is always an optimum tip speed ratio λ_{opt} corresponding to the maximum power coefficient of wind turbine C_{pmax} for any pitch angleβ. The $\beta = 0$ without considering wind turbine status at extreme wind speed.

The output torque of wind turbine is [4]:

$$T_m = \frac{P_m}{w_{wt}} \quad (5)$$

The relation betweenC_p, β and λ is shown in Fig .3.

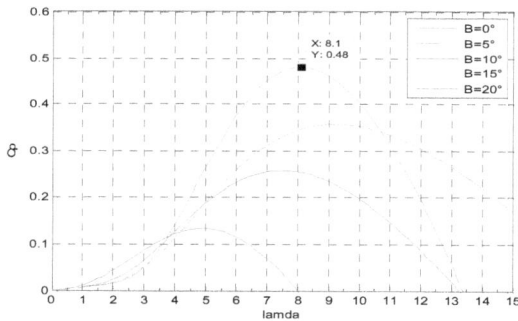

Figure 3. *Aerodynamic power coefficient variationC_p against tip speed ratio λ and pitch angleβ.*

The maximum value of C_p ($C_{pmax} = 0.48$) is achieved for $\beta = 0$ degree and for$\lambda_{opt} = 8.1$.

To extract the maximum power generated, we must fix the advance report λ_{opt} is the maximum power cient C_{pmax}.

2.2.2.Generator Model

In real wind power market, three types of wind power system for large WTs exist. The first type is fixed-speed wind power (SCIG), directly connected to the grid. The second one is a variable speed wind system using a DFIG or SCIG. The third type is also a variable speed WT, PMSG [16].

The IG space vector model is generally composed of three sets of equations: voltage equations, flux linkage equations, and motion equation. The voltage equations for the stator and rotor of the generator in the arbitrary reference frame are given by [17]:

$$\begin{cases} v_{ds} = \overline{R}_s \overline{i}_{ds} - \overline{\varphi}_{qs} + \frac{1}{w_s} \frac{d}{dt} \overline{\varphi}_{ds} \\ v_{qs} = \overline{R}_s \overline{i}_{qs} + \overline{\varphi}_{ds} + \frac{1}{w_s} \frac{d}{dt} \overline{\varphi}_{qs} \end{cases} \quad (6)$$

$$\begin{cases} v_{dr} = \overline{R}_r \overline{i}_{dr} - s.\overline{\varphi}_{qr} + \frac{1}{w_s} \frac{d}{dt} \overline{\varphi}_{dr} = 0 \\ v_{qr} = \overline{R}_r \overline{i}_{qr} + s.\overline{\varphi}_{dr} + \frac{1}{w_s} \frac{d}{dt} \overline{\varphi}_{qr} = 0 \end{cases} \quad (7)$$

The electrical torque is given by this equation after several converted steps:

$$T_e = \overline{L}_m \left(\overline{i}_{dr} . \overline{i}_{qs} + \overline{i}_{qr} . \overline{i}_{ds} \right) \quad (8)$$

$$\frac{dw_r}{dt} = \frac{T_m - T_e}{J} \quad (9)$$

The power flow studies in the IG are represented in Fig .4 [14].

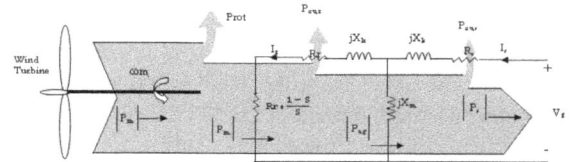

Figure 4. *Power flow and losses in an IG.*

3. Statcom

A STATCOM is a controlled reactive-power source. It provides the desired reactive-power generation and absorption entirely by means of electronic processing of the voltage and current waveforms in a voltage source converter (VSC). This function is identical to the synchronous condenser with rotating mass, but its response time is extremely faster than of the synchronous condenser. This rapidity is very effective to increase transient stability, to enhance voltage support, and to damp low frequency oscillation for the transmission system [5].

The schematic representation of the STATCOM and its equivalent circuit are shown in Fig 5.

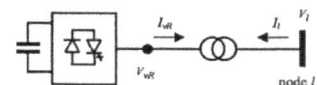

Figure 5. *STATCOM, VSC connected to the AC network via a shunt transformer.*

The STATCOM has the ability to either generate or absorb reactive power by suitable control of the inverted voltage $|V_{vR}| < \theta_{vR}$, with respect to the AC voltage on the high-voltage side of the STATCOM transformer, say node $l, |v_l| < \theta_l$.

In an ideal STATCOM, with no active power loss involved, the following reactive power equation yields useful insight into how the reactive power exchange with the AC system is achieved.

$$Q_{vR} = \frac{|v_l|^2}{x_{vR}} - \frac{|v_l||v_{vR}|}{x_{vR}}\cos(\theta_l - \theta_{vR})$$

$$= \frac{|v_l|^2 - |v_l||v_{vR}|}{x_{vR}}$$

Where $\theta_l = \theta_{vR}$ for the case of a lossless STATCOM;

If $|v_l| > |v_{vR}|$ then Q_{vR} becomes positive and the STATCOM absorbs reactive power. On the other hand, Q_{vR} becomes negative if $|v_l| < |v_{vR}|$ and the STATCOM generates reactive power.

In power flow studies the STATCOM may be represented in the same way as a synchronous condenser, which in most cases is the model of a synchronous generator with zero active power generation. It is adjusts the voltage source magnitude and phase angle using Newton's algorithm to satisfy a specified voltage magnitude at the point of connection with the AC network as presents at the Fig .5.

$$v_{vR} = |v_{vR}|(\cos\theta_{vR} + j * \sin\theta_{vR})$$

It should be pointed out that maximum and minimum limits will exist for $|v_{vR}|$ which are a function of the STATCOM. Capacitor rating. On the other hand, θ_{vR} can take any value between 0 and 2π radians but in practice it will keep close to θ_l [18].

STATCOM is capable of providing capacitive reactive power for network with a very low voltage level near 0.15pu. It also is able to generate its maximum capacitive power independent of network voltage. This capability will be very beneficial in time of a fault or voltage collapse or other restrictive phenomena, as presents at the Fig 6 [10].

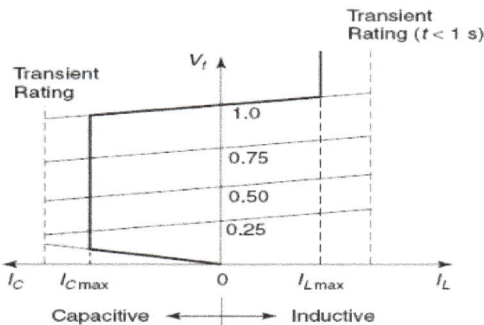

Figure 6. *Voltage current characteristic of STATCOM.*

4. Simulation Results

The proposed test system has a wind farm connected to a network of 6 bus bars; the type of generators is an IG. Under normal operating conditions, the wind farm provide 9MW, the bank condenser used for offer a reactive power to the IG ,as presents at the following Fig .7.

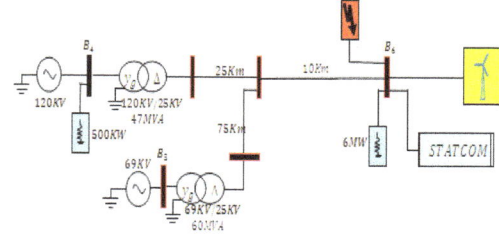

Figure 7. *Test system.*

The first objective of this paper is to evaluate the specific needs of the system to restore to its initial state as quickly as possible after fault clearing.

4.1. Without STATCOM

The effect of a three phase short circuit fault at the load bus is studied. The ground fault is initiated at $t = 15s$ and cleared at $t = 16s$. The system is studied under different conditions at the load bus as chosen below.

Fig 8 and Fig 9 shows the active and reactive power at the load bus, we can see the active power curve reached 8.7MW in transient state operation and return near to zero in the steady state mode even with the presence of the fault, however we find a peak in the reactive power curve at the time of the application fault and stabilized at -1Mvar.

Fig 10 and 11 shows the active and reactive power of each wind turbine.

It is clear according to these results that the active and reactive power of wind farm are disconnected before the appearance of fault, because the insufficient of the excitation condenser of generator, and the wind farm protection systems, however the reactive power gives a negative value because the presence of the condenser.

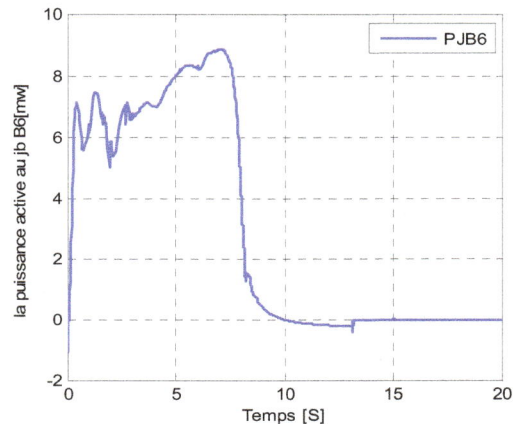

Figure 8. *Active power at bus 6.*

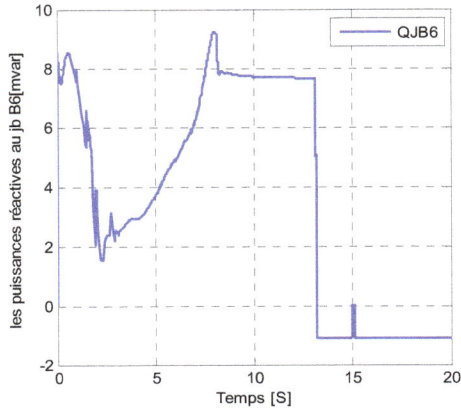

Figure 9. Reactive power at bus 6.

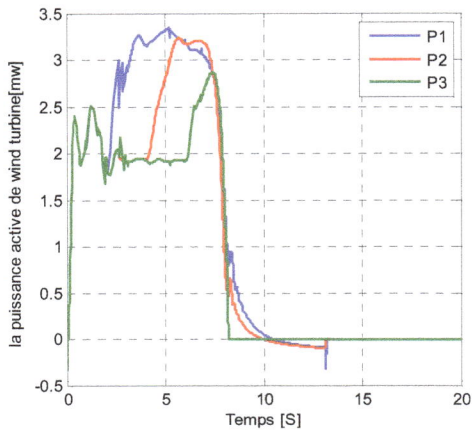

Figure 10. Active power of wind farm.

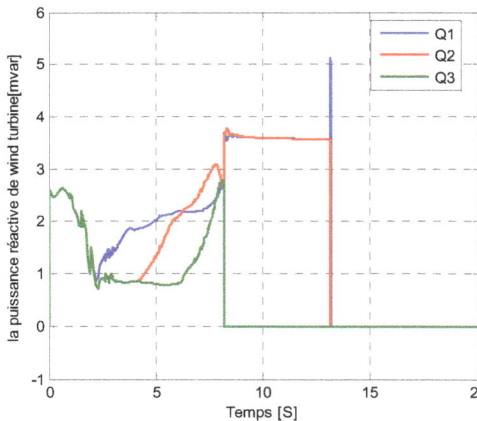

Figure 11. Reactive power of wind farm.

4.2. With STATACOM

According to the previous simulation results, we added the STATACOM at bus 6 for view the STATACOM effects.

Fig 12 and 13 shows the active and reactive power at the load bus, we can note that in the both curves the two powers also stabilized faster with less oscillation compared with the preceding case in the transient state and even after the fault, however fig 14 and 15 shows the active and reactive power for each wind turbine.

According to the simulation results, the curves presented above shows the importance of the compensation when the wind farm recovers its operation after the fault and takes its stability with some oscillation by the intervention of STATCOM at bus bar 6.

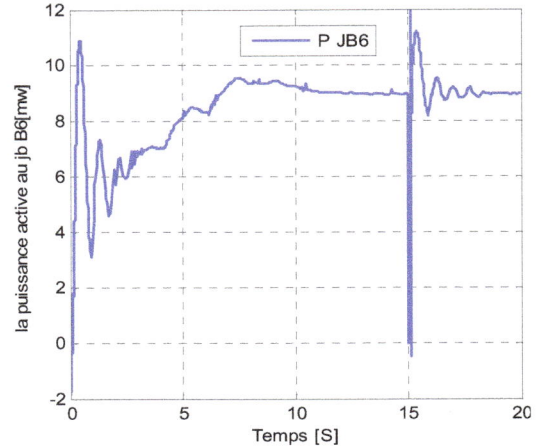

Figure 12. Active power at bus 6.

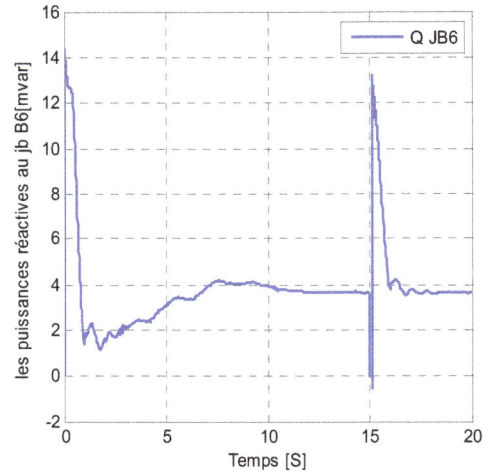

Figure 12. Active power at bus 6.

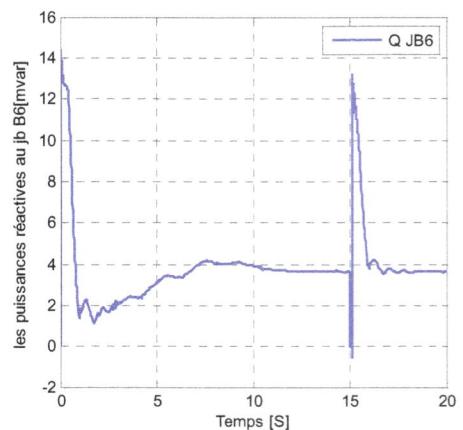

Figure 13. Reactive power at bus 6.

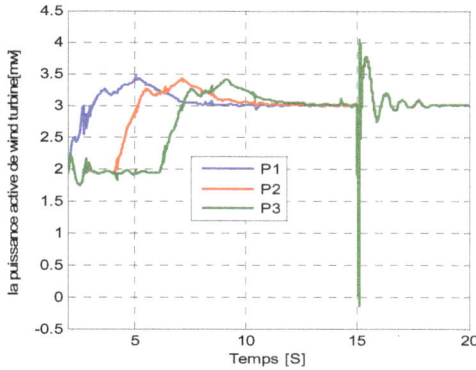

Figure 14. *Active power of Wind Farm.*

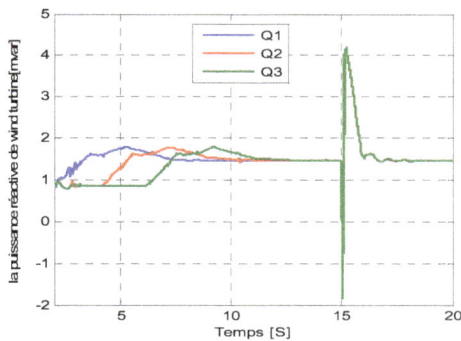

Figure 15. *Reactive power of Wind farm.*

4.3. Transient Stability

In this section, the following evaluation index is used to show the impact of grid-connected wind farms of IG type on the transient stability test system.

Critical clearance time (CCT) of faults is generally considered as the best measurement of severity of a contingency and thus widely used for ranking contingencies in accordance with their severity; in addition CCT is defined as the longest allowed fault clearance time without losing stability [4]. In our studies, the CCT is employed as a transient stability index to evaluate the test system; we use a different value of reactive power injecting by a STATCOM for controller the CCT.

Fig 16 shows CCT for several values of STATCOM, we illustrate that the relation between the reactive power injecting by a STATCOM and the CCT is nearly a linear.

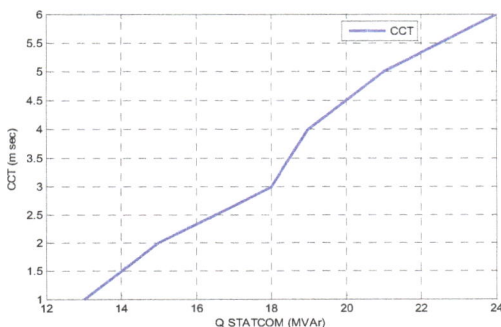

Figure 16. *Critical time for several values of STATCOM.*

5. Conclusions

The increasing penetration of renewable energy sources in the grid, high demands, caused destabilized the electrical network, so the researchers must be finding and master a new techniques for produced more power, better quality and higher reliability at lower cost. In first section a global description of system was presented, for each its component a brief presentation are given, modeled and simulated.

In the second section, the dynamics of the grid-connected wind farm is compared with and without the presence of STATCOM under fault, our test network contain three wind farm each wind farm has two equal wind turbine, according to the simulation results, it clearly illustrates the need of STATCOM improvement when the wind farm recovers its operation after the fault and takes its stability and do not leave the wind farm disconnect in the insufficient of the excitation condenser case. In the last section, a several successive simulation are executed for understand the relation between the STATCOM dimension and the CCT.

References

[1] AL. Olimpo, J. Nick, E. Janaka, C. Phill and H Mike, "Wind Energy Generation Modelling and Control," John Wiley & Sons, Ltd 2009.

[2] H. Mellah and K. E. Hemsas, "Design and Analysis of an External-Rotor Internal-Stator Doubly Fed Induction Generator for Small Wind Turbine Application by Fem," International Journal of Renewable and Sustainable Energy. vol. 2, no. 1, pp. 1-11, 2013.

[3] A. Petersson, "Analysis Modeling and Control of Doubly-Fed Induction Generators for Wind Turbines," PhD thesis, Chalmers university of technology, GOteborg, Sweden 2005.

[4] B .BOUHADOUZA, "Amélioration de la Stabilité Transitoire des Fermes Eoliennes par l'utilisation des STATCOM," magister theses of Setif university, Algeria, 2011.

[5] A. Pahade and N. Saxena, "Transient stability improvement by using shunt FACT device (STATCOM) with Reference Voltage Compensation (RVC) control scheme," International Journal of Electrical, Electronics and Computer Engineering, vol. 2, pp.7-12, 2013.

[6] S. Chauhan, V. Chopra, S. Singh, "Transient Stability Improvement of Two Machine System using Fuzzy Controlled STATCOM," International Journal of Innovative Technology and Exploring Engineering, vol. 2,pp. 52-56, March 2013.

[7] V. K. Chandrakar, A. G. Kothari, "Comparison of RBFN based STATCOM, SSSC and UPFC Controllers for Transient Stability improvement," Power Systems Conference and Exposition, PSCE '06, pp. 784 – 791, IEEE PES 2006.

[8] M. Mohammad Hussaini, R. Anita, "The Study of Dynamic Performance of Wind Farms with the Application Trends in Engineering," International Journal of Recent Trends in Engineering, vol. 2, pp. 158-160, 2009

[9] P. Kumkratug, "The Effect of STATCOM on Inter-Area

Power System Stability Improvement," IEEE computer society conference, Second UKSIM European Symposium on Computer Modeling and Simulation, pp. 359 – 363, 2008.

[10] X. Zhang, C. Rehtanz and B. Pal, "Flexible AC Transmission Systems: Modelling and Control," Publisher: Springer, 2006.

[11] H. Li, Z. Chen, "Overview of different wind generator systems and their comparisons," IET, Renewable Power Generation, vol. 2, pp. 123–138, 2008.

[12] A. Olimpo, J. Nick, E. Janaka, C. Phill and H Mike, "Wind Energy Generation Modelling and Control, John Wiley & Sons, Ltd 2009.

[13] J.G. Slootweg, S. de Haan, H. Polinder, W. L. Kling, "Modeling wind turbines in power system dynamics simulations," Power Engineering Society Summer Meeting, Vol. 1, pp. 22 – 26, IEEE, 2001.

[14] M. Huong Nguyen, T. K. Saha, "Dynamic simulation for wind farm in a large power system," Power Engineering Conference, AUPEC '08. Australasian Universities, pp. 1 – 6, IEEE, 2008.

[15] L. Shenghu , L. Zhengkai , H. Xinjie , J. Shusen, "Dynamic equivalence to induction generators and wind turbines for power system stability analysis," 2nd IEEE International Symposium on Power Electronics for Distributed Generation Systems (PEDG), pp. 887 – 892, IEEE 2010.

[16] H. Mellah, K. E. Hemsas, "Simulations Analysis with Comparative Study of a PMSG Performances for Small WT Application by FEM," International Journal of Energy Engineering, vol.3, no 2, pp. 55-64, 2013.

[17] B. wu, Y. Lang, N. Zargari, S. Kouro, "power conversion and control of wind energy systems," IEEE press, wiley, Canada 2010.

[18] E.Acha, V.G.Agelidis, O.Anaya-lara, T.J.E.Miller, "power Electronic Control in Electrical Systems", Newnes. A division of reed educational and professional publishing ltd, 2002.

[19] M. A. Kamarposhti, M. Alinezhad, "Comparison of SVC and STATCOM in Static Voltage Stability Margin Enhancement", International Journal of Electrical Power and Energy Systems Engineering, 2010.

4

Alleviation of harmonics for the self excited induction generator (SEIG) using shunt active power filter

A. M Bouzid[1,2], A. Cheriti[1], M. Bouhamida[2], M. Benghanem[2]

[1]Departmentof Electrical and Computer Engineering, University of Quebec at TroisRivieres UQTR, QC, CANADA
[2]Department of electrical engineering, University USTO MB, Oran, ALGERIA

Email address:

Allal.El.Moubarek.Bouzid@uqtr.ca(A. M. Bouzid), Ahmed.Cheriti@uqtr.ca(A. Cheriti)

Abstract: The Self Excited Induction Generator (SEIG) is an isolated power source, whose terminal voltage and frequency are controlled by the excitation of capacitance or the load impedance. A new strategy based on an active power filter (APF) for controlling the current and power quality of the self-excited induction generator (SEIG) is also presented in this paper. The proposed active filter proved to play an important role and give good dynamic response and robust behavior upon changes in load parameters. This investigation demonstrated that power average control strategy can facilitate the improvement of the power quality. This control method extracts fundamental (reference) components of the source current for the shunt active power line conditioners for nonlinear loads and unbalanced loads. The shunt APF in conjunction with the proposed controller perform perfectly under different steady state and transient conditions. The simulation results with nonlinear loads and unbalanced loads have showed the effectiveness of the proposed scheme for harmonic reduction in Wind based Power Generation.

Keywords: SEIG, Induction Generator, Harmonics, Shunt Active Filter, Power Electronics

1. Introduction

In recent years, the Self-Excited Induction Generator (SEIG) has emerged as the best electromechanical energy converter to replace the conventional synchronous generator in isolated power generators driven by renewable energy resources: biogas, micro-hydroelectric, wind etc. The main advantages of the SEIG are: low cost, ruggedness, absence of a separate DC source for excitation, brushless rotor construction and ease of maintenance. The fundamental problem with using the SEIG is its inability to control the terminal voltage and frequency under varying load conditions. The analysis of the SEIG under steady-state conditions and imposed speed is already known[1]-[2]. Active Power Filters (APF) are often used in applications where low current harmonics are desirable and/or improvement of quality of energy taken from the power grid are needed with the use of APF, it is possible to draw near perfect sinusoidal currents and voltages from the grid or renewable distributed power sources, where the shape of currents and voltages should be very close to sinusoidal. Another possibility is to balance load currents in different phases which is important in stand-alone power generation like wind turbines. Unsymmetrical load currents e.g. could lead to torque pulsation in generator's shaft and decrease reliability. The currents taken by office consumers have high harmonic contents. It is related to increasing number of loads with rectifier and capacitor, where the current is drawn at the peak of voltage sinusoid. The APF can be used to prevent any kind of harmonic generation. The benefits of using APF could be summarized as: reduction of harmonic content in the grid, reduction of peak value of the current drawn from the grid, reduction of the inrush current taken from the grid, compensation of neutral line current, active power factor correction and transformers are not necessary[3].

2. Description of the Proposed Control

A schematic diagram of the proposed system is shown in "Fig. 1". It consists of a three phase star-connected induction generator driven by an uncontrolled micro hydroelectric turbine. The generator is operated as an SEIG by connecting a fixed terminal capacitor with a value so as to result in rated terminal voltage at full load[4]. When SEIG supplies a non-linear load, the load draws a fundamental component of current and a harmonic current from the generating system which have to be properly controlled. The shunt APF can compensate the harmonic current by continuously tracking the changes in harmonic content. This APF consists of a voltage fed converter with a PWM current controller and an active filter controller that implements an almost instantaneous control algorithm as shown in "Fig.1". As the input power is nearly constant, the output power of the SEIG must be held constant at all consumer loads. Any decrease in load may accelerate the machine and raise the voltage and frequency levels to prohibitively high values, resulting in large stresses on other connected loads.

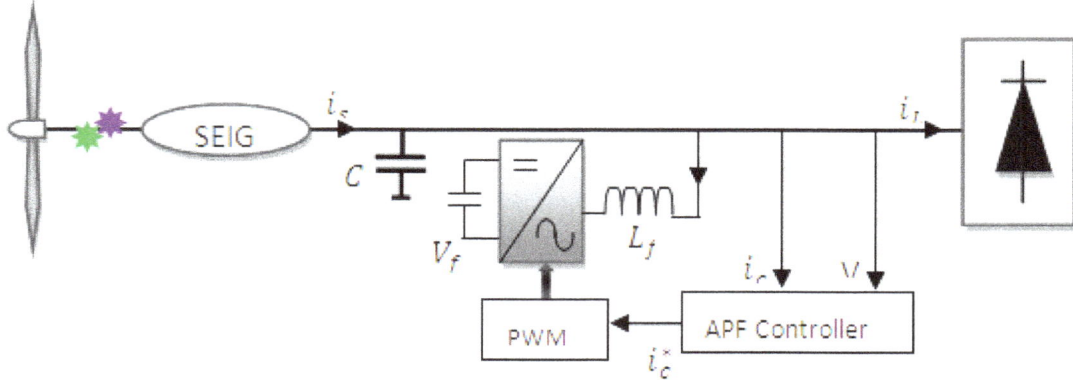

Figure 1. Block diagram of the APF with SEIG.

3. Mathematical Model of the Self Excited Induction Generator

A classical matrix formulation using d-q axes modeling is used to represent the dynamics of conventional induction machine operating as a generator. The representation includes the self and mutual inductances as coefficients widely used in machine theory. Using such a matrix representation, one can obtain the instantaneous voltages and currents during the self-excitation process, as well as during load variation. The dynamic model of the three-phase squirrel cage induction generator is developed by using stationary d-q axes references frame and the relevant volt-ampere equations are as[5]-[6]:

$$[V] = [R][I] + [L]\frac{d}{dt}[I] + \omega_r[G][I] \quad (1)$$

From which, the current derivative can be expressed as:

$$\frac{d}{dt}[I] = -[L]^{-1}\{[R][I] + \omega_r[G][I] - [V]\} \quad (2)$$

Where $[V]$, $[I]$, $[R]$, $[L]$ and $[G]$ defined below:

$$L = \begin{bmatrix} L_s & 0 & L_m & 0 \\ 0 & L_s & 0 & L_m \\ L_m & 0 & L_r & 0 \\ 0 & L_m & 0 & L_r \end{bmatrix} \quad G = \begin{bmatrix} 0 & 0 & 0 & 0 \\ 0 & 0 & 0 & 0 \\ 0 & L_m & 0 & L_r \\ -L_m & 0 & -L_r & 0 \end{bmatrix}$$

$$L_s = L_{ls} + L_m, L_r = L_{lr} + L_m$$

$$[V] = [V_{ds} V_{qs} V_{dr} V_{qr}]^T, [I] = [I_{ds} I_{qs} I_{dr} I_{qr}]^T$$

$$[R] = diag[R_s \ R_s \ R_s \ R_s] \quad and \quad K = 1/(L_m^2 - L_s L_r)$$

3.1. Magnetizing Inductance

The SEIG operates in the saturation region and its magnetizing characteristics are non-liner in nature. Magnetizing current should be calculated in every step of integration in terms of stator and rotor d-q currents as:

$$I_m = \sqrt{(I_{ds} + I_{dr})^2 + (I_{qs} + I_{qr})^2} \quad (3)$$

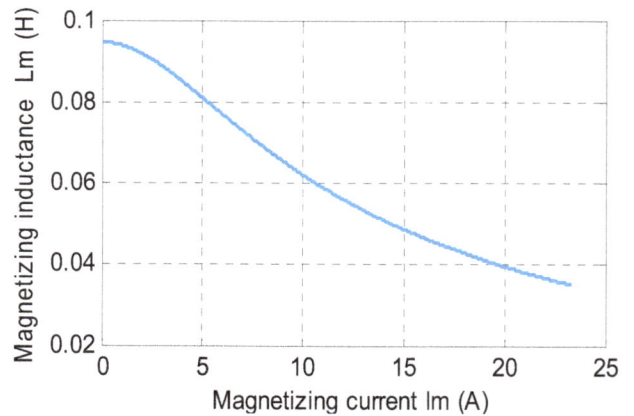

Figure 2. Variation of magnetizing inductance as a function of magnetizing current.

Magnetizing inductance is calculated from the magnetizing characteristics which is obtained by synchronous speed test for the machine under test and defined as:

$$L_m = 0.63 atan(0.15 I_m)/I_m \quad (4)$$

3.2. Electromagnetic Torque

Developed electromagnetic torque of the SEIG is:

$$T_e = (3P/4)L_m(I_{qs}I_{dr} - I_{ds}I_{qr}) \quad (5)$$

4. Reference Current Generation Using Average Power Method

The average power method gives accurate results even if the current is distorted. A PLL based unit vector template is used to obtain fundamental component of mains voltage. To get unit vector templates of voltage, the input voltage is sensed and multiplied by a gain equal to $1/vpk$ where vpk is the peak amplitude of fundamental supply voltage. These unit vectors are then passed through a PLL for synchronization of signals. Three phase fundamental components are multiplied by vpk to get fundamental mains voltage. The Power average method needs reduced calculation, since it works directly with $abc - phase$ voltage and line currents. The elimination of the Clark transformation makes this control strategy simple[7][8]. The Power average method presents a minimum rms value to draw the same three phase average active power from the source as the original load current. The control strategy principle for the shunt active power filter based on three-level inverter is illustrated in "Fig 3".

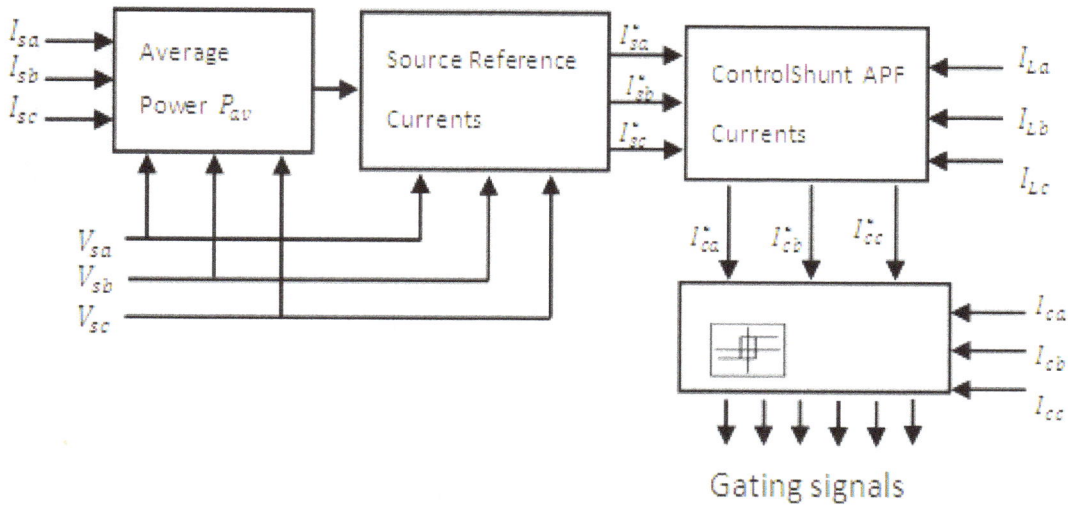

Figure 3. *Block diagram of the proposed shunt active power filter control scheme.6. Analysis and Modeling of the Active Power Filter*

The three phase instantaneous source current can be written as

$$i_s(t) = i_L(t) - i_c(t) \qquad (6)$$

The instantaneous source voltage is given by

$$V_s(t) = V_m \sin \omega t \qquad (7)$$

If a nonlinear load is applied, then the load current will have a fundamental component and harmonic components, which can be written as:

$$i_L(t) = \sum_{n=1}^{\infty} I_n \sin(n\omega t + \varphi_n) = I_1 \sin(\omega t + \varphi_1) + \left(\sum_{n=2}^{\infty} I_n \sin(n\omega t + \varphi_n)\right) \qquad (8)$$

The reduction of current harmonics in the load current is achieved by injecting equal but opposite current harmonic components at the point of common coupling, thereby cancelling the original distortion and improving the power quality[9].

5.1. Computation of the Average Power

The sensed load currents (iLa, iLb, iLc) and bus voltages (Va, Vb, Vc) through PLL are used to derive the instantaneous power p_{ave} as given by:

$$P_{ave}(t) = V_{sa}(t)i_{La}(t) + V_{sb}(t)i_{Lb}(t) + V_{sc}(t)i_{Lc}(t) \qquad (9)$$

The three phase instantaneous reactive power in each phase becomes[8]:

$$q_{La} = V_b I_{Lc} - V_c I_{Lb}$$

$$q_{Lb} = V_c I_{La} - V_a I_{Lc} \qquad (10)$$

$$q_{Lc} = V_a I_{Lb} - V_b I_{La}$$

The instantaneous active and reactive power delivered to a nonlinear load must satisfy (10) and (11).

$$p_L = p_s + p_c = p_{L1} + p_{Lh} \qquad (11)$$

$$q_{fk} = q_{LK}, k = a, b, c \qquad (12)$$

Where p_s- Instantaneous active power supplied by the source

p_f- Instantaneous active power supplied by the APF

p_{L1}- Instantaneous active fundamental power of the load

p_{Lh}- Instantaneous harmonic power of the load

q_{LK}- Instantaneous reactive power generated by the APF at phase k.

In order to ensure that the fundamental active power is supplied to the load from the source, the instantaneous reactive power and harmonic power must be compensated by the APF. When considering the compensation of both harmonic and reactive power, P_f is expressed as:

$$P_f(t) = V_{sa}(t)i_{ca}(t) + V_{sb}(t)i_{cb}(t) + V_{sc}(t)i_{cc}(t) \qquad (13)$$

5.2. Computation of the Average Power

From (12) and (13), the reference compensating currents are determined as:

$$\begin{cases} I_{sa}^* = I_{La} - \dfrac{p_{L1}}{v_{sa}^2 + v_{sb}^2 + v_{sc}^2} V_a \\ I_{sb}^* = I_{Lb} - \dfrac{p_{L1}}{v_{sa}^2 + v_{sb}^2 + v_{sc}^2} V_b. \\ I_{sc}^* = I_{Lc} - \dfrac{p_{L1}}{v_{sa}^2 + v_{sb}^2 + v_{sc}^2} V_c \end{cases} \quad (14)$$

Finally the desired 3-phase references of the APF currents$(I_{ca}^*, I_{cb}^*, I_{cc}^*)$ are computed by taking the difference between the three phase instantaneous reference source currents $(I_{sa}^*, I_{sb}^*, I_{sc}^*)$ and the actual source currents(I_{La}, I_{Lb}, I_{Lc}) as below:

$$I_{ca}^* = I_{sa}^* - I_{La}$$

$$I_{cb}^* = I_{sb}^* - I_{Lb} \quad (15)$$

$$I_{cc}^* = I_{sc}^* - I_{Lc}$$

6. Results and Discussion

The performance of the proposed control strategy is evaluated through simulation using SIMULINK toolbox in the MATLAB. The parameters of SEIG are shown in Table 2. The system parameters values are: source impedance ofR_S, L_S is 0.1 Ω and 1 mH respectively; filter impedance of Rc, Lc is 1Ω and 20mH respectively; diode rectifier R_L, L_L load in steady state: 300 Ω and 100 mH and unbalanced load Rl1, Ll1 :150 Ω and 100 mH, Rl2, Ll2 :75 Ω and 100 mH , Rl3, Ll3: 50 Ω and 10 mH respectively; DC voltage (VDC) is 500V; C_{dc} = 1100µF; Power devices used are IGBT/Diode.

6.1. Performance of Self Excited Induction Generator

6.1.1. Excitation with and without Saturation

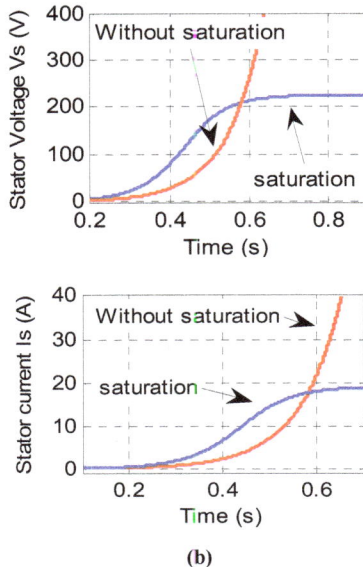

(b)

Figure 4. *Simulation of SEIG with/without saturation*

6.1.2. Excitation with Saturation and no Load

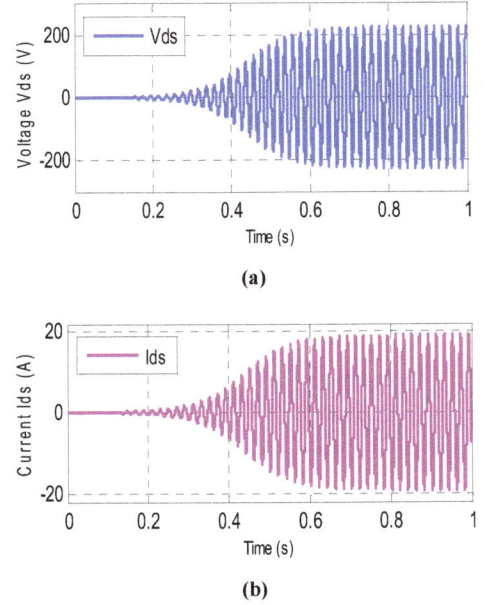

(a)

(b)

Figure 5. *Simulation of SEIG with saturation (a) Stator voltage; (b) Stator current*

When SEIG is excited with capacitance value of C=270µF and rotor speed w_r=1500 rpm, the generated voltage and current attain their steady state values of 380 Volts and 19 A in 0.8 sec as shown in "Fig. 4".

6.1.3. Variation of Speed

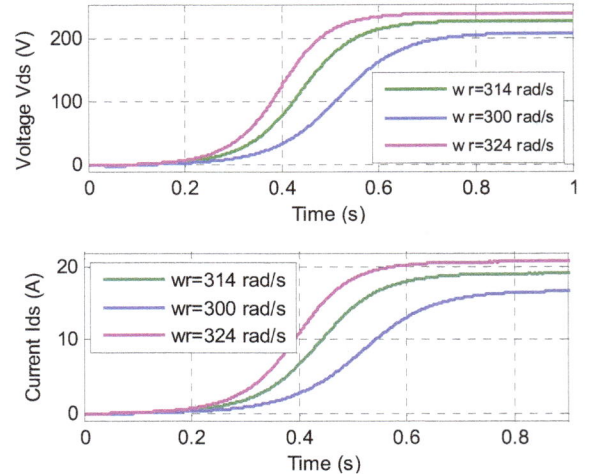

Figure 6. *Simulation of the self-excitation at C = 270 µF with variation of speed (graphs: 1. Stator voltage 2.Stator currents)*

The speed has a direct influence on the voltage for the same magnetizing current, the relation$E' = E\, n'/n$, shows that when the speed of rotation is proportional to the voltage. This is illustrated in Figure 6. And it is not limited by the saturation as in the case of the capacitor.

The speed change also affects the frequency of the voltage, otherwise say if the speed increases with increasing frequency $f_s = n.p/60\ (g \cong 0)$ .In the case of autonomous operation, the speed of the SEIG must be fixed in a restricted range.

6.1.4. Variations of Speed after Full-Excitation

The simulation results presented in Figure 7 shows:

- The transition from speed 314 rad / s to 280rad / s, causes a decrease in the stator voltageand a decrease in the frequencyand stator current delivered by the machine.

- The transition from speed 314 rad/s to330rad/s, causes anincrease in the stator voltageand anincrease in the frequencyand stator current delivered by the machine.

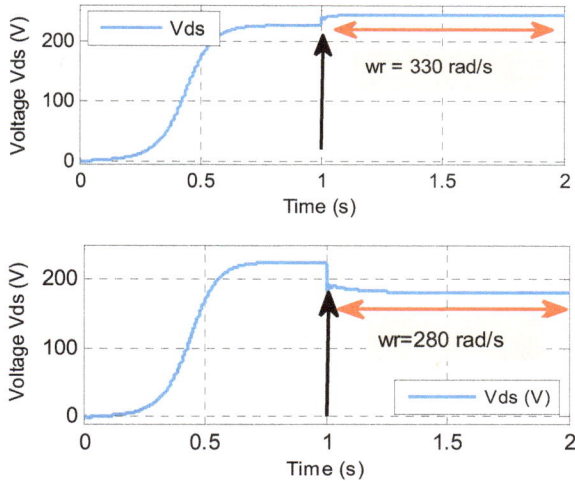

Figure 7. *Simulation of the self-excitation with variation of speed increase/reduction*

7.2. Shunt Active Power System Performance

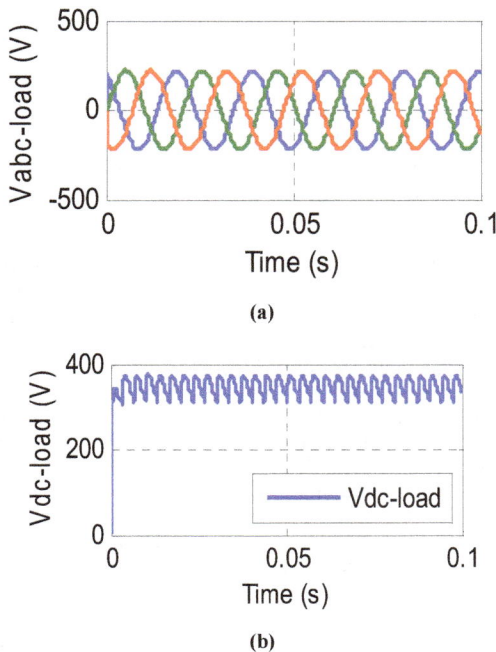

Figure 8 *(a) Unbalanced load voltage; (b) Nonlinear load voltage*

The unbalanced load RL voltage before compensation is shown in "Fig 8(a)" and the six-pulse diode rectifier RL load voltage before compensation is shown in "Fig 8(b)".

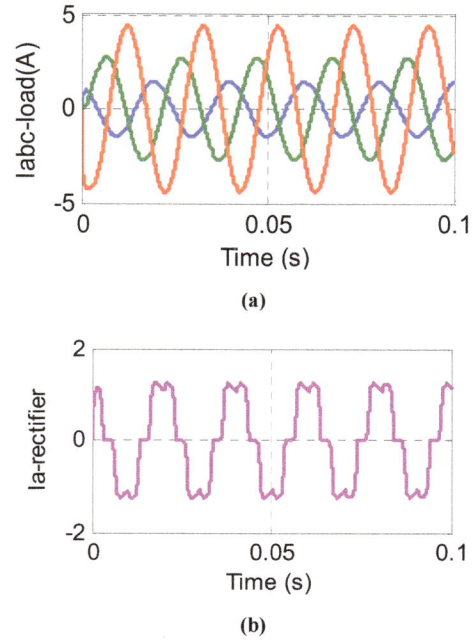

Figure 9 *(a) Unbalanced load currents; (b) Nonlinear load current.*

The computer simulation results are provided to verify the effectiveness of the proposed control scheme. The unbalanced load RL current before compensation is shown in "Fig 9(a)" and the six-pulse diode rectifier RL load current or source current before compensation is shown in "Fig 9(b)".

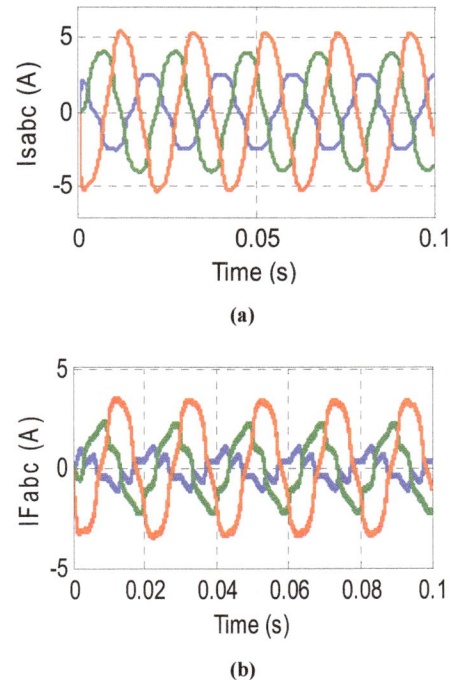

Figure 10 *(a) Source current before compensation ;(b) Reference current before APF.*

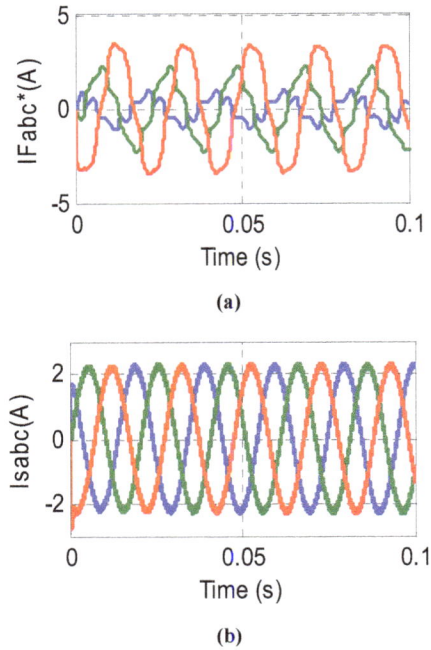

Figure 11 (a) Source current after compensation ;(b) Reference current after APF.

"Fig. 10 (a)" shows the simulated results of the load currents. The harmonic currents of a nonlinear load and unbalanced load are compensated by the shunt active power filter. The actual reference currents for the three phases are shown in "Fig. 10(b)". This waveform is obtained from the proposed average power controller. The source current after compensation is illustrated in "Fig. 11(b)" which indicates that the current becomes sinusoidal. After active filter operation, the AC-source current only supplies the active fundamental current to the load. The shunt APLC supplies the compensating current that is shown in "Fig. 11(a)". The current after compensation shown in Fig. 11(b) would have taken a shape as shown in "Fig. 10(b)" without APF. It is clearly visible that this waveform is sinusoidal with some high frequency ripples.

The total harmonic distortion is measured using the source current waveform and presented in Table 1 with and without APLC.

Table 1. Total harmonic distortion (THD %) of source current

ConditionTHD	(Is)Without APLC	(Is)With APLC
Steady state	23.08%	2.01%

The FFT analysis that was carried out confirms that the active filter brings the THD of the source current down to less than 5% which is in compliance with IEEE-519 standards for harmonics.

8. Conclusion

This paper has presented the implementation of a cage-rotor IG system completely isolated from the utility grid, in order to supply rural sites or isolated areas. In this paper we also discussed the problem of terminal current stabilization of the self-excited induction generator (SEIG) in standalone mode from which a new method of stabilization of the current is used to improve the performance characteristics of the SEIG. This investigation demonstrated also that the generalized Power average control strategy can facilitate the improvement of the power quality. Simulation results are included in order to validate the proposed control technique. It has been shown that the Power average approach additionally maintains the voltage of the capacitor (of the PWM inverter) nearly constant without any external control circuit. Different types of linear and nonlinear loads for reactive power and current harmonics compensation have been connected to the APF to analyze the steady-state and transient performance of the system. The APF has been proved to remarkably eliminate the harmonic and reactive components of load current resulting in sinusoidal and unity power-factor source currents.

Appendix

Table 2. Parameters of SEIG

Rated Power	3.5KW
Rated Line to Line Voltage	380 V
Rated line to line Current	14 A
Rated Frequency	50 Hz
Number of poles, P	4
Rated Rotor speed Nn	1410 rpm
Stator Resistance, Rs	0.76 Ω
Stator Leakage inductanceLls	0.003mH
Rotor Resistance, Rr	0.74 Ω

References

[1] D. Joshi, K. S. Sandhu, and M. K. Soni',"Performance Analysis of Self-Excited Induction Generator Using Artificial Neural Network",Iranian Journal of electrical and computer engineering, vol. 5, no. 1,pp 57-62, winter-spring 2006.

[2] Avinash Kishore, G. Satish Kumar, "Dynamic modeling and analysis of three phase self-excited induction generator using generalized state-space approach",IEEE International Symposium on Power Electronics, Electrical Drives, Automation and Motion, SPEEDAM, pp 52-59, 2006.

[3] ArkadiuszKulka, "Digital Control of Power Electronics for Reliable Distributed Power Generation", PhD Projects 2006 at Dep. of Electrical Power. Eng. University of Science and Technology. Norwegian, Jan 2006.

[4] Li Wang, Member, 'Transient Performance of an isolated induction generator under unbalanced excitation capacitor', IEEE Transaction on Energy conversion,Vol 14,no 4, pp 887-893, Dec. 1999.

[5] B. Singh, S.S. Murthy and S. Gupta,"Analysis and implementation of an electronic load controller for a self-excited induction generator", IEE Proc. C, Gener.

Transm. Distrib, vol. 151, pp. 51-60, Jan. 2004.

[6] Khan.P.K.S, Chatterjee.J.K, Salam. M.A. and Ahmad.H, "Transient Performance of Unregulated Prime Mover Driven Stand Alone Self-Excited Induction Generator With Solid-state Lead-Lag Var Compensator, " IEEE TENCON 2000, vol.1, pp. 235 – 239, sep 2000.

[7] Bhim Singh, Kamal Al-Haddad, and Ambrish Chandra: "A New Control Approach to Three-phase Active Filter for Harmonics and Reactive Power Compensation" ,IEEE Transactions on Power Systems, Vol. 13, No. 1,pp. 133-138, Feb. 1998.

[8] Youssef, K.H. Wahba, M. Yousef, H. Sebakhy, O, "A new method for voltage and frequency control of stand-alone self-excited induction generator using PWM converter with variable DC link voltage",IEEETrans.American Control Conference, pp. 2486-2491, Jun. 2008.

[9] A. Eid, M. Abdel-Salam, H. El-Kishky and T. El-Mohandes, "Active power filters for harmonic cancellation in conventional and advanced aircraft electric power systems", Elsevier, Electric Power Systems Research 79, pp80–88, 2009.

5

Designing of dynamic voltage restorer (DVR) to improve the power quality for restructured power systems

M. Kavitha[1], T. Chandrasekhar[2], D. Mohan Reddy[3]

[1]Dept of Computational Engineering, APIIIT, RGUKT, Nuzvid, A.P,India
[2]Dept of CSE, APIIIT, RGUKT, Nuzvid, A.P,India
[3]Dept of EEE, SVIET,Pedana (M), Krishna (Dt),A.P,India

Email address:

merugukavitha2013@gmail.com(M. Kavitha), Chandra.indra@gmail.com(T. Chandrasekhar), usmohanus@gmail.com(D. M. Reddy)

Abstract: In Restructured power systems, Power quality is one of the major concerns in the present era. The problem of harmonics, voltage sags and swells and its major impact on sensitive loads are well known. To solve this problem, custom power devices are used. One of those devices is the Dynamic Voltage Restorer (DVR), which is one of the most efficient and effective modern custom power devices used in power distribution networks for the power quality improvement. Control of power quality problems involves cooperation between network operator (utility), customer and equipment manufacturer. A Dynamic Voltage Restorer (DVR) is a distribution voltage DC-to-AC solid-state switching converter that injects three single phase AC output voltages in series with the distribution feeder and in synchronism with the voltages of the distribution system. A DVR is interface equipment between utility and customer connected in series between the supply and load to mitigate the three major power quality problems, namely the harmonics, voltage sags, and swells etc. This paper concentrates on the designing of Dynamic Voltage Restorer (DVR) for the harmonics compensation so that it can improve the Power Quality (PQ).

Keywords: Dynamic Voltage Restorer (DVR), Harmonics,Voltage Swell And Power Quality

1. Introduction

Power quality may be defined as any power problems manifested in voltage, current or frequency deviations that result in failure or mis -operation of customers equipment.Power Quality (PQ) is an important measure of an electrical power system. The term PQ means to maintain purely sinusoidal current wave form in phase with a purely sinusoidal voltage wave form. The power generated at the generating station is purely sinusoidal in nature. The deteriorating quality of electric power is mainly because of current and voltage harmonics due to wide spread application of static power electronics converters, zero and negative sequence components originated by the use of single phase and unbalanced loads, reactive power, voltage sag, voltage swell, flicker, voltage interruption etc. To improve the power quality traditional compensation methods such as passive filters, synchronous capacitors, phase advancers, etc. were employed. However traditional controllers include many disadvantages such as fixed compensation, bulkiness, electromagnetic interference, possible resonance etc.. These disadvantages urged power system and power electronic engineers to develop adjustable and dynamic solutions using custom power devices. Custom power devices are power conditioning equipments using static power electronic converts to improve the Power Quality (PQ) of distribution system customers.

2. Dynamic Voltage Restorer (DVR)

The schematic diagram of Dynamic Voltage Restorer (DVR) is shown below.

DVR (Dynamic Voltage Restorer) is a static var device that has seen applications in a variety of transmission and distribution systems. It is a series compensation device, which protects sensitive electric load from power quality problems such as voltage sags, swells, unbalance and distortion through power electronic controllers that use voltage source converters (VSC).

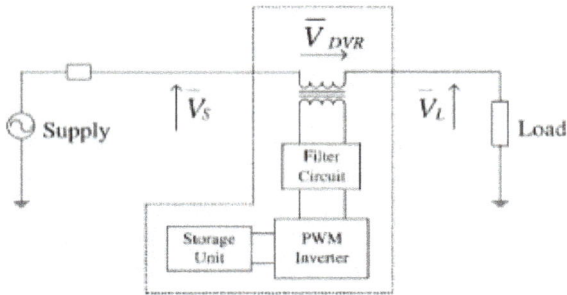

Figure 1. *Schematic diagram of a Dynamic Voltage Restorer*

A DVR, Dynamic Voltage Restorer is a distribution voltage DC-to-AC solid-state switching converter that injects three single phase AC output voltages in series with the distribution feeder and in synchronism with the voltages of the distribution system. The series voltage controller is connected in series with the protected load. Usually the connection is made via a transformer, but configurations with direct connection via power electronics also exist. The resulting voltage at the load bus bar equals the sum of the grid voltage and the injected voltage from the DVR. The converter generates the reactive power needed while the active power is taken from the energy storage. The energy storage can be different depending on the needs of compensation.

DVR can compensate voltage at both transmission and distribution sides. Usually a DVR is installed on a critical load feeder. During the normal operating condition (without sag condition) DVR operates in a low loss standby mode during this condition the DVR is said to be in steady state. When a disturbance occurs (abnormal condition) and supply voltage deviates from nominal value, DVR supplies voltage for compensation of sag and is said to be in transient state.

3. DVR Components

The major components that are used in Dynamic Voltage Restorer (DVR) are explained below.

3.1. Energy Storage Unit

The required energy for compensation of harmonics, load voltage during sag can be taken either from an external energy storage unit (batteries) or from the supply line feeder.

3.2. Inverter Circuit

Since the vast majority of harmonic disturbances and voltage sags seen on utility systems are unbalanced, mostly due to single-phase events, the VSC will often be required to operate with unbalanced switching functions for the three phases, and must therefore treat each phase independently.

Mitigation of harmonics and Voltage Sags in a refinery with Induction Motors Using Dynamic Voltage Restorer

(DVR), a sag on one phase may result in a swell on another phase, so the VSC must be capable of handling both sags and swells simultaneously. The variable output voltage of the inverter is achieved using PWM scheme.

3.3. Filter Unit

The nonlinear characteristics of semiconductor devices cause distorted waveforms associated with high frequency harmonics at the inverter output. To overcome this problem and provide high quality energy supply, a harmonic filtering unit is used. This can cause voltage drop and phase shift in the fundamental component of the inverter output, and has to be accounted for in the compensation voltage.

3.4. Series Injection Transformers

Three single-phase injection transformers are used to inject the missing voltage to the system at the load bus. To integrate the injection transformer correctly into the DVR, the MVA rating, the primary winding voltage and current ratings, the turn-ratio and the short-circuit impedance values of transformers are required. The existence of the transformers allow for the design of the DVR in a lower voltage level, depending upon the stepping up ratio. In such case, the limiting factor will be the ability of the inverter switches to withstand higher currents.

3.4. Controller and Auxiliary Circuits

By-Pass switches, breakers, measuring and protection relays are some auxiliaries to the DVR block, in addition to the controller of the DVR.

3.6. Carrier Phase Shifting Pulse Width Modulation (PWM)

Carrier phase-shift sinusoidal pulse width modulation (PS-SPWM) switching scheme is proposed to operate the switches in the system. Optimum harmonic cancellation is achieved by phase shifting each carrier by

$$(k-1)\,\pi/n \tag{1}$$

Where k is the kth inverter, n is the number of series-connected single phase inverters.

$$n = (L-1)/2 \tag{2}$$

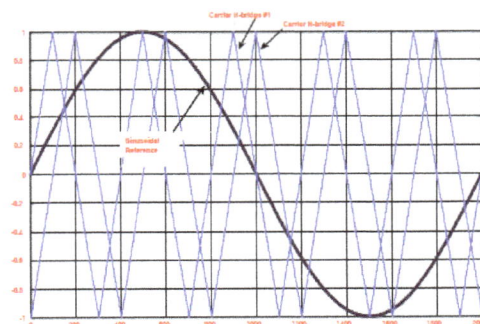

Figure 2. *Phase Shifted Carrier PWM*

Fig-2 shows the PSCPWM. In general, a multilevel inverter with m voltage levels requires (m–1) triangular carriers. In the PSCPWM, all the triangular carriers have the same frequency and the same peak-to-peak amplitude, but there is a phase shift between any two adjacent carrier waves, given by φcr=3600/ (m–1). The modulating signal is usually a three-phase sinusoidal wave with adjustable amplitude and frequency. The gate signals are generated by comparing the modulating wave with the carrier waves. It means for five-level inverter, four triangular carriers are needed with a 90° phase displacement between any two adjacent carriers. In this case the phase displacement of Vcr1 = 0°, Vcr2 = 90°, Vcr1- = 180° and Vcr2- = 270°.

4. Operation

In normal conditions, the dynamic voltage restorer operates in stand-by mode. However, during disturbances, nominal system voltage will be compared to the voltage variation. This is to get the differential voltage that should be injected by the DVR in order to maintain supply voltage to the load within limits.

The amplitude and phase angle of the injected voltages are variable, thereby allowing control of the real and reactive power exchange between the dynamic voltage restorer and the distribution system. The DC input terminal of a DVR is connected to an energy storage device of appropriate capacity. As mentioned, the reactive power exchange between the DVR and the distribution system is internally generated by the DVR without AC passive reactive components. The real power exchanged at the DVR output AC terminals is provided by the DVR input DC terminal by an external energy source or energy storage system.

5. Basic Principle

The basic idea of a DVR is to inject the missing voltage cycles into the system through series injection transformer whenever voltage sags are present in the system supply voltage. As a consequence, sag is unseen by the loads. During normal operation, the capacitor receives energy from the main supply source.

When voltage dip or sags are detected, the capacitor delivers dc supply to the inverter. The inverter ensures that only the missing voltage is injected to the transformer. A relatively small capacitor is present on dc side of the PWM solid state inverter and the voltage over this capacitor is kept constant, by exchanging energy with the energy storage reservoir. The required output voltage is obtained by using pulse-width modulation switching pattern. As the controller will have to supply active as well as reactive power, some kind of energy storage is needed. In the DVRs that are commercially available now large capacitors are used as a source of energy.

6. Caution

DVRs may provide good solutions for end-users subject to unwanted power quality disturbances. However, there is a caution regarding their application in systems that are subject to prolonged reactive power deficiencies (resulting in low voltage conditions) and in systems that are vulnerable to voltage collapse.

In many cases, the main protection of the power system against voltage collapse is the natural response of load to decrease demand when system voltage drops. The application of DVRs would tend to maintain demand even when incipient voltage conditions are present. As a result, this reduces the innate ability to prevent a collapse and increases the chance of cascading interruptions.

In addition, from the transmission viewpoint, the dynamic voltage restorer would extend the voltage range if the load is a constant power type. The combination of direct-connected DVRs, voltage-switched capacitor banks and on-load tap-changing distribution transformers, leads to more current drawn from the transmission system during periods of reactive deficiency and low voltages.

Therefore, when applying DVRs, it is vital to consider the nature of the load whose voltage supply is being secured, as well as the transmission system which must tolerate the change in voltage-response of the load. It may be necessary to provide local fast reactive supply sources in order to protect the system, with the DVR added, from voltage collapse and cascading interruptions. A comprehensive simulation study, which includes the transmission system, is highly recommended.

7.Matlab Simulink Circuit

The matlab simulink circuit diagram of Dynamic Voltage Restorer (DVR) is shown below. This will give the voltage output waveforms by using with and without DVR.

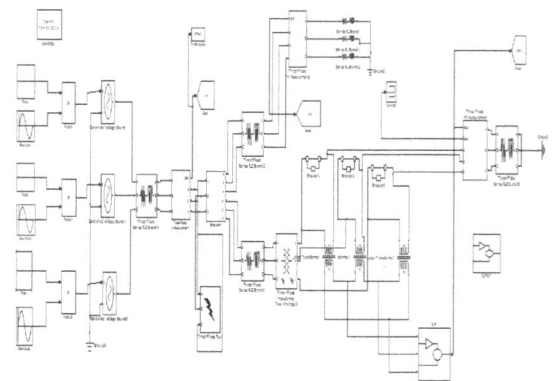

Figure 3. Matlab simulink circuit diagram of Dynamic Voltage Restorer (DVR)

8. Results

The input and outputs of without and with Dynamic Voltage Restorer (DVR) for the above circuit are as shown below.

The three phase ac supply as input for the designed circuit in the form of waveform is as shown below.

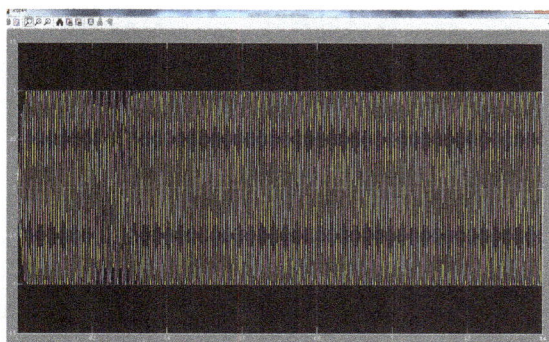

Figure 4. Input voltage waveform

8.1. Without DVR

The output waveform of the circuit without using Dynamic Voltage Restorer (DVR) is as shown below. In this waveform we can easily observe the harmonics effect which shows an effect on power quality. This can problem can be solved by using Dynamic Voltage Restorer (DVR).

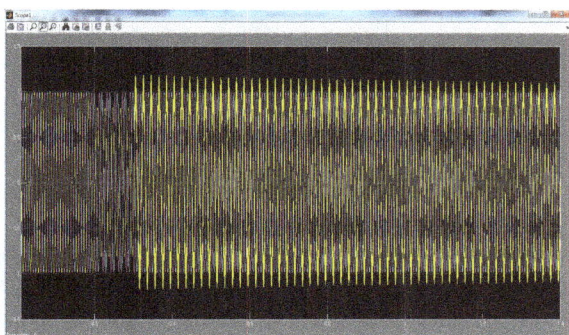

Figure 5. Output voltage waveform without DVR

8.2. With DVR

The output waveform of the circuit with Dynamic Voltage Restorer (DVR) is as shown below. In this waveform we can easily observe the elimination of harmonics effect from the above waveform. With this Dynamic Voltage Restorer (DVR) circuit we can maintain the power quality better than without using the DVR.

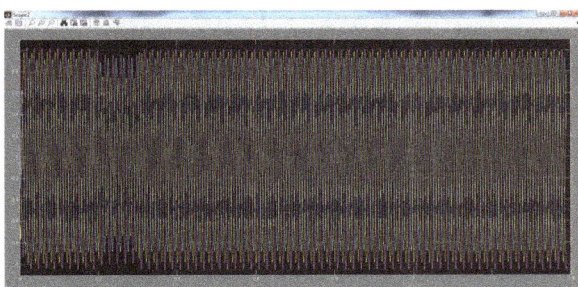

Figure 6. Output voltage waveform with DVR

9. Conclusion

The Dynamic Voltage Restorer (DVR) is a promising and effective device for power quality enhancement due to its quick response and high reliability. The conclusion is that the DVR is an effective apparatus to protect sensitive loads from short duration harmonics and voltage swells and sags. The DVR can be inserted both at the low voltage level and at medium voltage level. The series connection with the existing supply voltages makes it effective at locations where voltage dips are the primary problem. However, the series connection makes the protection equipment more complex as well as the continuous conduction losses and voltage drop. The role of a DVR in mitigating the power quality problems in terms of harmonics, voltage sag, swell and is explained.

References

[1] S. Chen, G. Joos, L. Lopes, and W. Guo." A nonlinear control method of dynamic voltage restorer" , IEEE 33rd Annual Power Electronics Specialists Conference. pp. 88-93.Sept2002.

[2] Ghosh and G. Ledwich. " Power Quality Enhancement Using Custom Power Devices", 2002. Kluwer Academic Publishers.

[3] Math H.J. Bollen, "Understanding power quality problems: voltage sags and interruptions", IEEE Press, New York, 2000.

[4] Sasitharan S., Mahesh K. Mishra, Member, IEEE, B.Kalyan Kumar, and Jayashankar V., member, IEEE, " Rating and Design Issues of DVR Injection Transformer", IEEE Press., New York, 2008.

[5] Madrigal, E.Acha., "Modelling of Custom Power Equipment Using Harmonic Domain Techniques", IEEE 2000.

[6] R.Mienski,R.Pawelek and I.Wasiak., "Shunt Compensation for Power Quality Improvement Using a STATCOM controller Modelling and Simulation", IEEE Proceedings., Vol.151, No.2. Jan. 2006.

[7] Bollen, M.H.J.," Voltage sags in three-phase systems" Power Engineering Review, IEEE, Vol. 21, Issue: 9, pp: 8 - 11, 15, Sept. 2001.

[8] Anaya-Lara O, Acha E., "Modeling and analysis of custom power systems by PSCAD/EMTDC", IEEE Transactions on Power Delivery, Vol.17, Issue: 1, Pages:266 – 272, Jan. 2002.

[9] G. Yaleinkaya, M.H.J. Bollen, P.A. Crossley, "Characterization of voltage sags in industrial distribution systems", IEEE transactions on industry applications, vol.34, no. 4, July/August, pp. 682-688, 1999.

[10] Haque, M.H., "Compensation of distribution system voltage sag by DVR and D-STATCOM", Power Tech Proceedings, 2001 IEEE Porto, vol.1, pp.10-13, Sept. 2001.

6

The current status of wind and tidal in-stream electric energy resources

Hamed H. H. Aly, M. E. El-Hawary

Department of Electrical and Computer Engineering, Dalhousie University, Halifax, Nova Scotia, Canada, B3H 4R2

Email address:

hamed.aly@dal.ca (H. H. H. Aly), elhawary@dal.ca (M. E. El-Hawary)

Abstract: Renewable energy is an effective and clean source of supplying electrical loads especially in remote and rural areas. In this paper we discuss offshore wind and tidal in-stream energy as they rely on similar technologies for generating electricity at offshore sites. In particular, we survey the impacts of offshore wind and tidal current integration into the grid, various types of generators and their dynamic modeling, fault ride-through techniques used to improve generator and grid integration performance, the aggregated wind turbines modeling and finally put the light on the stability and control problems.

Keywords: Wind Power, Tidal In-Stream Power, Doubly Fed Induction Generators (DFIG), Direct Drive Permanent Magnet Synchronous Generator (DDPMSG), Power System Dynamic Stability, Power System Modeling

1. Introduction

Wind energy is the energy produced from the simple air in motion and this motion is caused by the uneven heating of the earth's surface by the sun. The air over the sea absorbs the heats faster than the land and so the air moves from the sea to the land causing the wind but in the night the air motion is changed from the land to the sea because the air over the sea cools faster than the air over the land. This wind is hardly predictable source of energy. Tidal energy is due to the gravitational influence of the moon and the sun on the earth due to the rotation of the earth relative to the moon and the sun which produces two high and two low waters each day (12.4 h cycle). This rotation makes the rise and fall of the tides and these tides are predictable. These tides run approximately six hours in one direction and then reverse for another six hour in the opposite direction [1, 2]. Tidal in-stream energy has various advantages such as high energy density, hence cheap rotors for power output, predictable energy; hence its integration is easy, low environmental impact and low bird disturbance [3]. The main difference between the wind and tidal is the high density of the seawater (800 times greater than air) if it is compared to wind and this helps to use a smaller system for tidal and obtaining the same energy [4, 5].

While wind turbines have negative issues such as audio noise, visual impacts, erosion, birds and bats killed and radio interference, it is useful in rural area applications where access to transmission facilities is limited. Moreover, wind energy helps to reduce the environmental damage (Green house Gas emissions) and climate change due to fossil fuel replacement [6]. The wind power resource is intermittent and challenging to predict, and requires using some form of storage to integrate it in the electric grid. New control techniques and improved forecasting methods help establish operating practices which will increase reliability of wind energy supply to the grid.

The problems of wind and tidal-stream may be overcome by the advanced technology in the near future and will become from the preferred ways for obtaining electrical energy.

2. Renewable Energy from Canadian and Nova Scotian Perspective

Various types of renewable energy are used at present. For instance, solar energy is used directly, usually via solar panels, to heat and power homes. Similarly, the heat of the sun drives the winds to produce wind energy. The wind and the sun cause water evaporation, which turns into rain and snow and contributes to rivers and waterfalls, whose energy can be captured through hydro power turbines. The sunlight and rain cause plants to grow, and these can eventually be harvested for biomass energy. Other renewable energy sources are geothermal energy, which is generated and stored in the earth, and marine energy, on which this research is based.

Canada is one of the world leaders in the use of marine renewable energy due to its unique geography, abundant resources, and expertise in ocean engineering and offshore operations. Billions of tonnes of seawater ebb and flow every day along Canadian shorelines. Indeed, developing marine energy has become an integral part of government energy and economic strategy, according to one government minister who stated that "The Marine Renewable Energy Technology Roadmap demonstrates how government, industry and academics are working together to advance the commercialization of marine energy technologies in Canada while sharpening our global competitiveness" [10].

Being almost completely surrounded by seawater, the province of Nova Scotia has abundant marine renewable energy resources from offshore wind, waves and tides. The Bay of Fundy, located on the province's western shore, has a 100 billion tonnes of seawater flowing into it each day, delivering a commercial potential of approximately 2,400 megawatts of power. This massive inflow of seawater exceeds the daily combined flow of the world's freshwater rivers. The energy potential is so huge compared to other countries that one industry expert has dubbed the Bay of Fundy the "Saudi Arabia" of marine renewable energy. The United States (U.S.)-based Electric Power Research Institute (EPRI) has also identified the Bay of Fundy as a prime site for potential tidal power generations. Ocean energy presents a significant opportunity for generating electrical energy, and tidal current and wave energy technologies are at the investigative stage. The development of renewable energy in Nova Scotia will help to contribute to the long-term renewable electricity mix, reducing greenhouse gases and other air pollutants, decreasing dependence on fossil

fuels, reducing emissions, providing a diverse and more secure mix of energy, producing clean, green energy, and creating employment opportunities that build wealth and exports [11].

Despite the rosy picture being painted by industry and government investors, there are several issues that must be taken into consideration when dealing with marine renewable energy as a new source of energy. These issues include the following:

1. The protection of the marine ecosystem.
2. Health, safety and environmental protection.
3. The conservation of natural resources (not economic gain) as a top priority.
4. Sustainable industry development.

As well, there are environmental impacts that must be investigated and properly handled, such as [11, 12]:

1. The sediments, substrates and disruption of the currents and waves.
2. Electric and magnetic field effects.
3. Noise due to turbine blade rotation.
4. Navigation impacts and water quality changes.
5. Impacts on sea and land animal migration.

Renewable energy is needed to reduce dependence on imported fossil fuels, make Nova Scotia less susceptible to fluctuating market prices, and diversify the energy mix to bring stability to electricity rates. The total amount of renewable electricity in Nova Scotia based on 12,000 GWh/yr of total provincial electricity sales was 1100 GWh/yr (9%) pre-2001, 1300 GWh/yr (11%) at the end of 2009, and 1700 GWh/yr (14%) at 2011. It is expected to be 2300 GWh/yr (19%) by 2013, 3000 GWh/yr (25%) by 2015, and 4800 GWh/yr (40%) by 2020. Figure (1) shows the percentage of renewable energy compared to other sources in Nova Scotia for 2001 and 2009 and the expected percentage for 2015 and 2020 [12, 13].

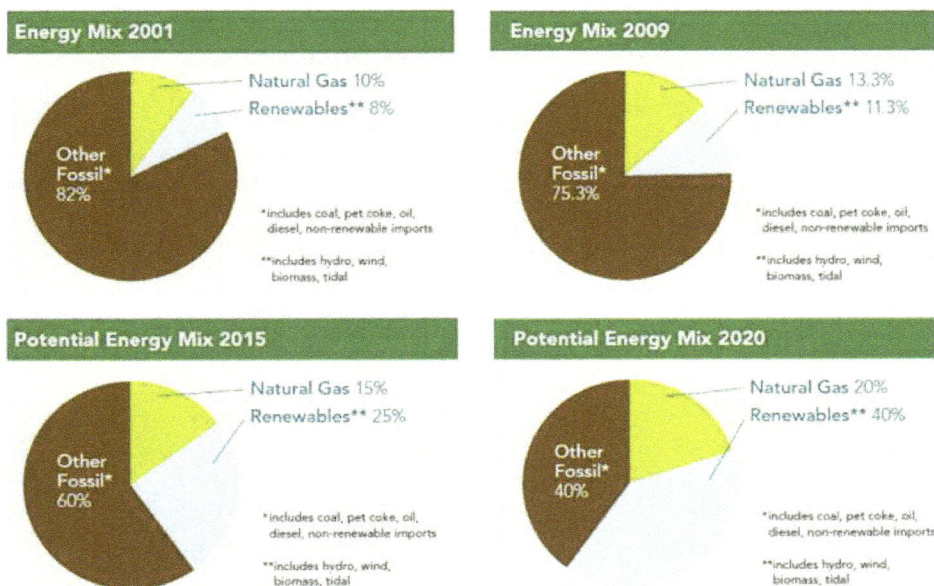

Figure 1. Potential energy mix in 2001 and 2009 and the expected potential energy mix in 2015 and 2020 [13].

Ocean waves produced by winds passing over the surface of the water are converted to electricity. There are approximately 50 competing designs being tested around the world in search of commercially viable wave-energy technology. The cost of wave energy is estimated by the International Energy Agency (IEA) to be in the range of $0.20 to $0.75 per kWh, depending on location. Nova Scotia's best wave resources are far from land, and so the overall cost will be relatively high. Hence, other parts of the world with better waves closer to shore have a competitive advantage and wave technology is therefore considered to be a lower priority for Nova Scotia [12].

Wind power is another attractive renewable energy source, but the best winds (offshore winds) are expensive to harness. The cost of onshore wind power in Nova Scotia is in the range of $0.07 to $0.09 per kWh, but onshore winds are not as constant as offshore winds. A 2011 report by the International Energy Agency put the current costs of offshore winds between $0.17 and $0.35 per kWh, making it far less economical than onshore winds due to the significantly higher construction and maintenance costs associated with these projects [11, 12].

Marine renewable energy is a new trend for generating electricity. It is expected to create jobs and grow the economy in the near future. Tidal current energy is an ideal renewable energy source because it is more predictable than wind and solar power, and this will reduce the backup capacity and improve reliability, which potentially could be reflected in the cost.

The US-based Electric Power Research Institute estimated that underwater turbines could safely extract 300 megawatts of energy from the Minas Channel alone. In July 2011, the government of Nova Scotia announced a plan to create 'winning conditions' for the development of an in-stream tidal energy sector that will generate 65 MW by 2015 and an additional 300 MW in 5 to 10 years to replace approximately 10% of Nova Scotia's current power supply.

This amount would be more or less equivalent to Nova Scotia's existing coal-fired generation. The province has recently introduced the Renewable Electricity Plan, which sets out a detailed program to move Nova Scotia away from carbon-based electricity generation towards greener, more local sources. Power from tidal current is expected to start contributing electricity around the middle of this decade and could make a significant contribution to electricity generation by 2020. The government currently provides support for tidal energy through FORCE (Fundy Ocean Research Centre for Energy). However, the cost of electricity from marine renewable energy resources is still so high that it is not yet competitive with other sources. This is because the technology is still in its infancy stage and many technical challenges remain to be resolved before large-scale commercial development can be implemented. Nevertheless, as the technology develops, the cost is expected to become competitive [14, 15].

Until now, there has been limited experience in assessing the costs associated with large-scale tidal energy. As mentioned, current costs are generally high, averaging $0.44 to $0.51 per kWh for initial deployments. Costs are even higher for smaller projects (in the range of $0.652 per kWh). Despite this financial hurdle, rising oil and coal prices along with the growing demand for clean and safe energy, are important issues bolstering the attractiveness of renewable energy. Moreover, the range of benefits and impacts created by the generation of marine renewable energy will differ depending on project location and technology used [14, 15]. Figure (2) shows a map of the mean power that can be easily extracted from tidal current passages around Nova Scotia while reducing the volume of water flowing through the passages by 5%. These values are calculated using simulation programs for tidal currents. In Cape Breton, the values are calculated using the characteristics of the flow and power extraction theory [12].

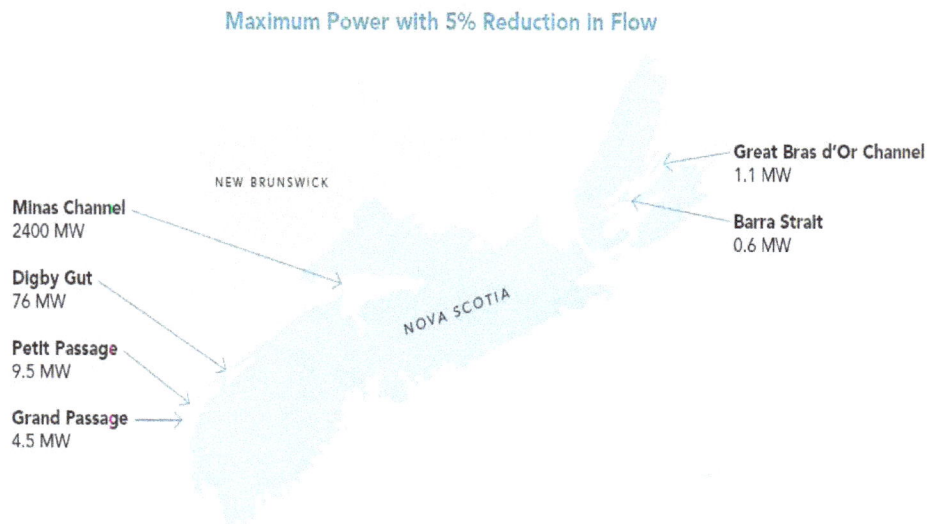

Figure 2. Map for the mean power that can be easily extracted from tidal current passages around Nova Scotia [12].

The federal and provincial governments have spent more than $75 million in support of marine renewable energy development projects over the past five years. An additional $100 million will be invested in phase 1 of FORCE, and the installation of technology arrays will involve a $500-million investment in the coming five years. From these projects, 75 MW will be installed by 2016, 250 MW will be installed by 2020, and 2,000 MW will be installed by 2030. It is worth noting that more than 50% of marine energy projects around the world use Canadian technology or expertise [16, 17].

The European Union (EU) member states have a target of deploying 1.95GW of marine energy by 2020. Figure (3) shows the marine targets for EU member states, most of which are situated along the Atlantic coast (UK, Ireland, France, Spain and Portugal) [18].

The speed varies in a very small range (may be considered as fixed speed) according to the generated power at variable wind changes depending on the mechanical system, hence during the fault it may lead to voltage instability, especially at low voltage. Some of the drawbacks of this approach are size, maintenance requirements, noise, lower reliability and efficiency. With the induction generator connected directly to the grid, the voltage level at the grid cannot be controlled, also blade rotation causes power variations and, this will affect the voltage and cause the frequency to vary from 1 to 2 Hz in the grid. The FSIG consumes reactive power and this may cause voltage issues after the fault is cleared. Compensating capacitors are used as the squirrel cage generator absorbs reactive power from the grid. The transient stability behavior of FSIG is poor and the machine must be switched off under fault conditions. [19, 20].

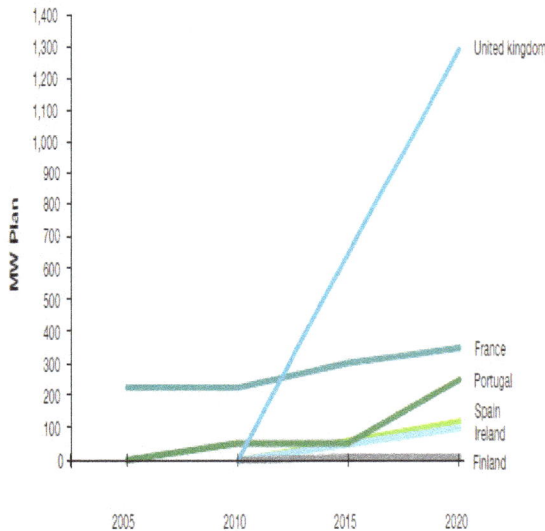

Figure 3. *Marine targets for EU member states [18].*

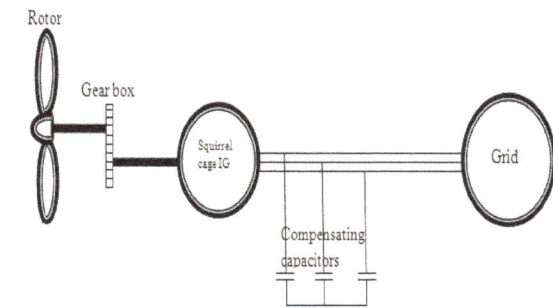

a. Fixed Speed Squirrel Cage Induction Generator

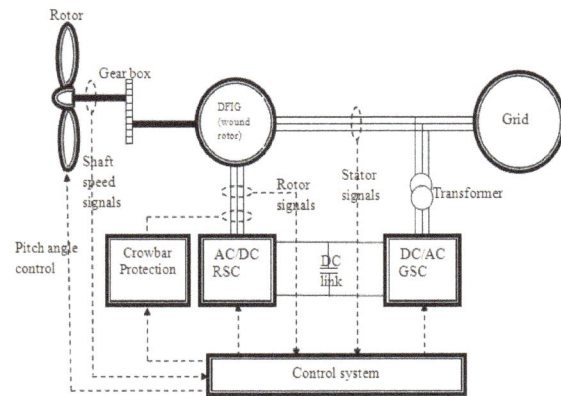

b. Doubly Fed Induction Generator

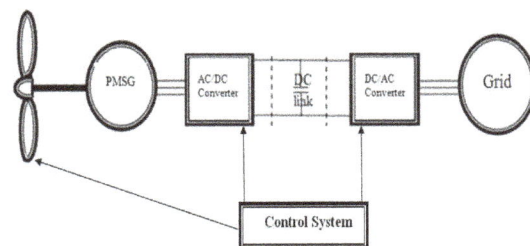

c. Direct Drive Synchronous Generator

Figure 4. *Types of wind turbine.*

3. Offshore Wind Turbines

This section consists of five subsections ordered as follows; the types of the wind turbines, the dynamic modeling of the wind turbine system, the integration of wind turbines in the power system grid, the stability problems of the wind turbines, its mitigation methods and finally the wind farm aggregation.

3.1. Wind Turbine Types

Most wind turbines are based on one of the three main wind turbine generator types as shown figure (4):

1) Fixed speed with directly grid-coupled squirrel cage induction generator (FSIG). The rotor side of the fixed speed squirrel cage induction generator (FSIG) is connected to the wind turbine via a gearbox and the other side of the generator (stator winding) is connected directly to the grid.

2) Variable speed with doubly fed induction generator

(DFIG).

In the DFIG, a converter is used to feed or take power from the rotor and gives a variable speed (a partial scale power converter used as about 20-30% of the power passes through the converter). The rotor side of the DFIG is connected to the grid via a back to back converter. The converter on the side connected to the grid is called the supply side converter (SSC) or grid side converter (GSC) while the converter connected to the rotor is the rotor side converter (RSC). The RSC operates in the stator flux reference frame. The direct axis component of the rotor current acts in the same way as the field current as in the synchronous generator and thus controls the reactive power change and the quadrature component of the rotor current is used to control the speed by controlling the torque and the active power change. Thus the RSC governs both the stator-side active and reactive powers in an independent manner.

The GSC operates in the stator voltage reference frame. The d-axis current of the GSC controls the DC link voltage to a constant level, and the q-axis current is used for reactive power control. The GSC is used to supply or draw power from the grid according to the speed of the machine. If the speed is higher than synchronous speed it supplies power, otherwise it draws power from the grid but its main objective is to keep the dc-link voltage constant regardless of the magnitude and direction of the rotor power [21]. DFIG is more stable as the rotor speeds of DFIGs is easy to be controlled by the generator side, also the active power and reactive power are controlled independently by using the converter on the rotor side, and the fluctuation of the voltage is minimized [22]. The dynamic behavior of the turbine is improved, the noise at low speed is reduced, the power production higher than FSIG and mechanical stresses reduced. The power quality in DFIG is improved but it is more complex than FSIG and the overall cost is increased due to the use of power electronic devices for control [22-25]. DFIG is the most commonly used one for wind integration due to its high efficiency, fast reaction and robustness during the fault. This machine is able to give a controlled reactive power to the grid [26].

3) Variable speed based on a direct drive synchronous generator. In the variable speed wind turbine *with* direct drive synchronous generator, the generator and the grid are connected by means of a converter, which is allowing variable speed operation (a full scale converter is used, 100% of the power pass through the converter) [22,23].

The gear box between the generator and the turbine has various demerits because of the size, the maintenance, the noise, the lower reliability and efficiency, so the new technologies connect the generator directly to the turbine by using numerous pairs of poles but at the same time use filters on the generator output because of high harmonic content caused by operation at low speeds [22]. Table (1) shows a comparison between three types of generators.

Table 1. A comparison between three types of generators.

Comparison	FSIG	DFIG	Synchronous
1-Speed	Fixed	Variable	Variable
2-Converter scale	Zero	20-30%	100%
3-Power supplied to the grid	Directly	Partially via stator and the converter	Totally via the converter
4-Control	Poor	Good	Very good
5-Active and reactive power control	Dependent on each other	Independent	Independent
6-Voltage fluctuation	High	Limited	Limited
6-Robustness	Small	High	Very high
7-Fault reaction	Slow	High	High
8-Efficiency	Poor	High	High
9-Cost	Low	High	Very high

3.2. Dynamic Modeling of the wind Turbine System

The dynamic modeling of the overall wind turbine system contains all subsystems such as wind speed model that is used for generating a wind speed signal applied to the rotor, the rotor model, the generator, the converter, the rotor speed controller, the pitch angle controller (for changing the blade pitch angle to control the amount of energy during high speeds) , the voltage controller (for controlling the voltage near the reference value), and the protection system (for the protection of wind turbine) [27-28]. Figure (5) shows the overall wind turbine system and their interaction.

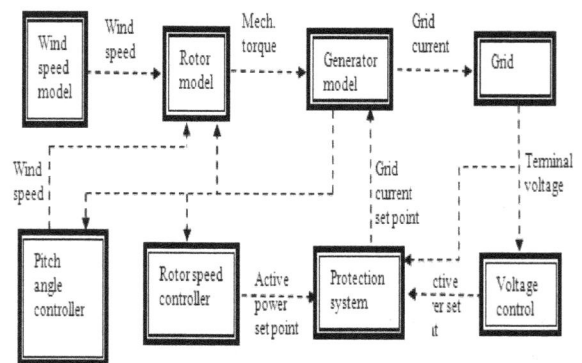

Figure 5. Overall wind turbine system and their interaction.

1) The Wind Speed Signals (v_w)

The wind speed signals model consists of four components, the mean wind speed (v_{mw}), the wind speed ramp (v_{rw}) (which is considered as the steady increase in the mean wind speed), the wind speed gust (v_{gw}), and the turbulence (v_{tw}). $v_w = v_{mw} + v_{gw} + v_{rw} + v_{tw}$. The mean wind speed is a constant speed; a simple ramp function will be used for ramp component (characterized by the amplitude of the wind speed ramp (A_r (m/s)), the starting time (T_{sr}), and the ending time (T_{er})). The wind speed gust component is characterized by the amplitude of the wind speed gust (A_g (m/s)), the starting time (T_{sg}), and the ending time (T_{eg}). The wind speed gust may be expressed as a sine, cosine wave function or a combination of both. One of the most used models is expressed as:

$$v_{gw} = A_g(1 - \cos(2\Pi(t/D_g - T_{sg}/D_g)))$$

$$T_{sg} \leq t \leq T_{eg}$$

$$v_{gw} = 0, \; t < T_{sg} \text{ or } \; t > T_{eg}$$

$$, D_g = T_{eg} - T_{sg}.$$

A triangle wave is used for the turbulence function which has adjusted frequency and amplitude [28, 29].

2) The Rotor Model

The rotor model is used for converting the kinetic energy to mechanical energy. The wind turbine is characterized by C_p (wind power coefficient), λ (tip speed ratio), and β (pitch angle). $\lambda = \omega_t R/v_w$, where R is the blade length in m, v_w is the wind speed in m/s, and ω_t is the wind turbine rotational speed in rad/sec. C_P-λ-β curves differ from manufacturer to another but there is an approximate relation for these curves as shown: $C_p = \frac{1}{2}(\frac{RC_f}{\lambda} - 0.026\beta - 2)e^{-0.295\frac{RC_f}{\lambda}}$

, C_f is the wind turbine blade design constant.

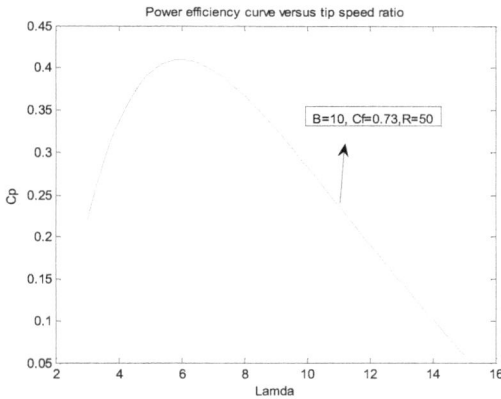

Figure 6. *The relation between e wind power efficiency and tip speed ratio.*

The rotor model may be represented by using the equation of the power extracted from the wind as follows:

$$P_w = 0.5\rho\Pi R^2 C_p v_w^3$$

And the mechanical torque applied to the turbine (T_m) can be expressed as: $T_m = 0.5\rho\Pi R^2 C_p v_w^3/v_m$

Where: v_m is the turbine speed at hub height upstream the rotor, ρ is the density of the air [29].

The shaft system may be represented by a two mass system (two mass model is represented, one for the turbine and the other for the generator) or by a single lumped mass system.

The two mass systems may be represented as:

$$2H_t\frac{d\omega_t}{dt} = T_t - K_s(\square_r - \square_t) - D_m(\omega_r - \omega_t)$$

$$2H_g\frac{d\omega_g}{dt} = -T_e + K_s(\square_g - \square_t) - D_m(\omega_g - \omega_t)$$

Where: D_t is the turbine self-damping, D_g is the generator self damping, D_m is the mutual damping, H_t and H_g are the turbine and generator inertia constants, respectively, K_s is the shaft stiffness, ω_t and ω_g the turbine and generator rotor speeds. $\square_t$ and $\square_g$ denote the turbine and generator rotor angles. There is a ratio of the torsion angles, damping and stiffness must be taken into account when one add the gear box as all above calculations must be referred to the generator and then calculated as follows[30].

$$a = \frac{\omega_g}{\omega_t} \; , \; \omega_t^{(t)} = \frac{\omega_t^{(g)}}{a} \; , \; {}_t^{(t)} = \frac{{}_t^{(g)}}{a}$$

$$K_s^{(t)} = a^2 K_s^{(g)} \; , \; D_m^{(t)} = a^2 D_m^{(g)}$$

One may express the shaft system as a single lumped mass as follows: $2H_m\frac{d\omega_m}{dt} = T_m - T_e - D_m\omega_m$, where, T_e is the electrical torque of the generator, H_m is the lumped inertia constant, ω_m is the rotational speed of the lumped mass system, and D_m is the damping of the lumped system.

3) The Generator Model

a) DFIG model: The modeling of the DFIG is developed by using a synchronously rotating d-q reference frame with the direct-axis oriented along the stator flux position. The reference frame is rotating with the same speed as the stator voltage. The stator and rotor active and reactive power can be described as [25, 31- 40]: $P_s = 3/2(v_{ds} \times i_{ds} + v_{qs} \times i_{qs})$,

$$P_r = 3/2(v_{dr} \times i_{dr} + v_{qr} \times i_{qr}),$$

$$P_g = P_s - P_r$$

$$Q_s = 3/2(v_{qs} \times i_{ds} - v_{ds} \times i_{qs}),$$

$$Q_r = 3/2(v_{qr} \times i_{dr} + v_{dr} \times i_{qr})$$

The model of the DFIG can be described as:

$$v_{ds} = -R_s \; i_{ds} - \omega_s \; \psi_{qs} + \frac{d}{dt}\psi_{ds}$$

$$v_{qs} = -R_s \; i_{qs} + \omega_s \; \psi_{ds} + \frac{d}{dt}\psi_{qs}$$

$$v_{dr} = -R_r \; i_{dr} - s \; \omega_s \; \psi_{qr} + \frac{d}{dt}\psi_{dr}$$

$$v_{qr} = -R_r \; i_{qr} + s \; \omega_s \; \psi_{dr} + \frac{d}{dt}\psi_{qr}$$

$$\Psi_{ds} = -L_{ss} \; i_{ds} - L_m \; i_{dr} \quad , \quad \Psi_{qs} = -L_{ss} \; i_{qs} - L_m \; i_{qr}$$

$$\Psi_{dr} = -L_{rr} \; i_{dr} - L_m \; i_{ds} \quad , \quad \Psi_{qr} = -L_{rr} \; i_{qr} - L_m \; i_{qs}$$

$$s = (\omega_s - \omega_r)/\omega_s$$

$$\frac{d\omega_r}{dt} = -\omega_s\frac{ds}{dt}$$

, where: d, and q are the indices of the direct and quadrature axis components, s, and r are the indices of the stator and the rotor, v is the voltage, R is the resistance , i is the current, ω_s, ω_r are the stator and rotor electrical angular velocities (rad/s), respectively, and ψ is the flux linkage. The converter in the grid side controls the dc-link voltage, regardless of the magnitude and direction of the rotor power and the converter in the rotor side controls the rotor currents. $L_{ss} = L_s + L_m$, $L_{rr} = L_r + L_m$, L_s, L_r and L_m are the stator leakage, rotor leakage and mutual inductances, respectively. For the DFIG the rotor is connected to the grid via a converter, hence: $v_{dr} = v_{dc}$, $v_{qr} = v_{qc}$. It is easy to find the state space representation of the induction machine as $\frac{d[i]}{dt} = [A][i] + [B][v]$, (by substituting ψ into the equation of v) or as $\frac{d[\psi]}{dt} = [A][\Psi] + [B][v]$.

$$
\begin{bmatrix} \frac{d\Psi_{ds}}{dt} \\ \frac{d\Psi_{qs}}{dt} \\ \frac{d\Psi_{dr}}{dt} \\ \frac{d\Psi_{qr}}{dt} \end{bmatrix} = \begin{bmatrix} \frac{-R_s}{\sigma L_{ss}} & \omega_s & \frac{-R_s}{\sigma L_{ss}} \times \frac{-L_m}{L_{rr}} & 0 \\ -\omega_s & \frac{-R_s}{\sigma L_{ss}} & 0 & \frac{-R_s}{\sigma L_{ss}} \times \frac{-L_m}{L_{rr}} \\ \frac{-R_r}{\sigma L_{rr}} \times \frac{-L_m}{L_{ss}} & 0 & \frac{-R_r}{\sigma L_{rr}} & s\omega_s \\ 0 & \frac{-R_r}{\sigma L_{rr}} \times \frac{-L_m}{L_{ss}} & -s\omega_s & \frac{-R_r}{\sigma L_{rr}} \end{bmatrix}
$$

$$
\times \begin{bmatrix} \Psi_{ds} \\ \Psi_{qs} \\ \Psi_{dr} \\ \Psi_{qr} \end{bmatrix} + \begin{bmatrix} V_{ds} \\ V_{qs} \\ V_{dr} \\ V_{qr} \end{bmatrix}, \; \sigma = \frac{L_{ss}L_{rr}-L_m^2}{L_{ss}L_{rr}}
$$

Under the steady state conditions one may write the stator and the rotor equations as follows:

$$ v_{ds} = -R_s \times i_{ds} - J\omega_s \times (-L_{ss} \times i_{qs} + L_m \times i_{qr}) + \frac{d}{dt}\psi_{ds} $$

$$ v_{qs} = -R_s \times i_{qs} + J\omega_s \times (-L_{ss} \times i_{ds} + L_m \times i_{dr}) + \frac{d}{dt}\psi_{qs} $$

$$ v_{dr} = R_r \times i_{dr} - s\times J\omega_s \times (L_{rr} \times i_{qr} - L_m \times i_{qs}) + \frac{d}{dt}\psi_{dr} $$

$$ v_{qr} = R_r \times i_{qr} + s\times J\omega_s \times (L_{rr} \times i_{dr} - L_m \times i_{ds}) + \frac{d}{dt}\psi_{qr} $$

The 5th order model may be reduced to a 3rd order by neglecting the stator transients and described it as follows:

$$ v_{ds} = -R_s \times i_{ds} + X\square \times i_{qs} + e_d $$

$$ v_{qs} = -R_s \times i_{qs} - X\square \times i_{ds} + e_q $$

$$ 2H_g \frac{d\omega_r}{dt} = T_e - T_m $$

Where

$$ \frac{de_d}{dt} = -\frac{1}{T_0}\left(e_d - (X-X) \times i_{qs}\right) + s \times \omega_s \times e_q - \omega_s \times \frac{L_m}{L_{rr}} \times v_{qr}, $$

$$ \frac{de_q}{dt} = -\frac{1}{T_0}\left(e_q - (X-X) \times i_{ds}\right) - s \times \omega_s \times e_d + \omega_s \times \frac{L_m}{L_{rr}} \times v_{dr}, $$

the components of the voltage behind the transient (the internal voltage components of the induction generator) are $e_d = -\frac{\omega_s L_m}{L_{rr}} \times \psi_{qr}$ and $e_q = \frac{\omega_s L_m}{L_{rr}} \times \psi_{dr}$. The stator reactance $X = \omega s\, L_{ss} = X_s + X_m$, and the stator transient reactance $X\square = \omega s\,(L_{ss} - L_m2/L_{rr}) = X_s + (X_r \times X_m)/(X_r$

$+ X_m)$. The transient open circuit time constant is To = $L_{rr}/R_r = (L_r + L_m)/R_r$, and the electrical torque is Te= (id-siqr-iqsidr)Xm/ω_s.

The 3rd order model doesn't give details about the transient but it is interesting in the dynamic studies of large power system as it reduces the computational time.

b) SCIG modeling: In the SCIG the rotor is short circuited and so, the stator voltages are the same as the DFIG:

$$ v_{ds} = -R_s \times i_{ds} - \omega_s \times \psi_{qs} + \frac{d}{dt}\psi_{ds} $$

$$ v_{qs} = -R_s \times i_{qs} + \omega_s \times \psi_{ds} + \frac{d}{dt}\psi_{qs} $$

The rotor voltages:

$$ 0 = R_r \times i_{dr} - s\times \omega_s \times \psi_{qr} + \frac{d}{dt}\psi_{dr} $$

$$ 0 = R_r \times i_{qr} + s\times \omega_s \times \psi_{dr} + \frac{d}{dt}\psi_{qr} $$

$$ P_r = 0; \; Q_r = 0 $$

Finally the equation of motion is $2H_g\frac{d\omega_r}{dt} = T_e - T_m$.

In the SCIG there are capacitors to provide the induction generator magnetizing current and for compensation. The current injected by a SCIG at the generated node is expressed as: $i_{dg} = ids + i_{dc}$, $i_{qg} = i_{qs} + i_{qc}$, $i_{dc} = v_{qc}/x$, $i_{qc} = v_{dc}/x$ [28].

c) The DDPMSG can be modeled as follows [42, 43]:

$$ v_{ds} = -R_s \times i_{ds} - \omega_s \times \psi_{qs} + \frac{d}{dt}\psi_{ds} $$

$$ v_{qs} = -R_s \times i_{qs} + \omega_s \times \psi_{ds} + \frac{d}{dt}\psi_{qs} $$

The flux linkages and the torque can be expressed as:

$$ \Psi_{ds} = -L_d \times i_{ds} + \psi_f $$

$$ \Psi_{qs} = -L_q \times i_{qs} $$

$$ T_e = (3/2)p\, i_{qs}((L_d-L_q)\, i_{ds} + \psi_f) $$

L_d, and L_q are the direct and quadrature inductances of the stator. Ψ_f is the excitation field linkage, and p is the number of pole pairs. Figure (4.8) shows the d-q axis component of the DDPMSG. In this work for simplicity we assume that $L_d = L_q = L_s$, because the difference is very small as we discussed in the introduction before, and so the generator model can be rewritten in a state space representation as:

$$ L_s\frac{d}{dt}i_{ds} = -v_{ds} - R_s \times i_{ds} + L_s \times \omega_s \times i_{qs} $$

$$ L_s\frac{d}{dt}i_{qs} = -v_{qs} - R_s \times i_{qs} - L_s \times \omega_s \times i_{ds} + \omega \times \psi_f $$

4) The rotor speed controller uses a power–speed curve to compute the reference power according to the actual speed.

5) The pitch angle controller model is active during the high wind speed to change the blade pitch angle to reduce Cp. The optimal pitch angle is zero below the nominal wind speed. The maximum rate of change is within 3 to 10 o/s,

depending on the size of the wind turbine. The pitch angle controller has a low frequency, of 1 to 3 Hz [28].

6) The voltage controller model is used to control the value of the terminal voltage v_{dr} by controlling the value of the reactive power as the reactive power is proportionally related to the terminal voltage [28]. The rotor side converter (RSC) is known as a controlled voltage source in which the q axis voltage vqr is controlling the rotor speed and the d axis voltage v_{dr} is controlling the reactive power but the grid side converter (GSC) is represented by a controlled current source, and provides the exchange of active power from the rotor circuit to the grid with unity power factor [41].

7) The protection system model consists of a two parts (that they switches off the wind turbine for the deviated voltage and for the deviated frequency) and a converter current limiter to protect the semiconductor switches (its boundaries are calculated depending on the maximum amount of the reactive power generated from the wind turbine and by using the nominal value of the active power and the voltage). The aims of the control system of the DFIG are to maximize the extracted power for a wide range of speed (power optimization), limit the output power to the rated for high speeds (power limitation), and adjust the active and the reactive powers (power regulation) to a specified value according to the power system operator [28].

3.3. Wind Integration and Mitigation Methods

Although, the conventional power plants are still necessary because of the low availability of wind turbines and their small contribution to reliability of the power system and tidal in-stream is still under development, the use of wind energy as a domestic energy resource is an important issue especially in rural areas and the growth of the wind is very fast. However, the use of wind power alone is not preferred [44]. Wind power is fluctuating, hardly predictable, and intermittent, so it is difficult to match the generation to the demand and there is impact of this power source on the transmission system (the medium and low voltage subsystems). Hence the influences of increasing wind energy generation weaken the system and render it vulnerable to power quality perturbations unless an extra reserve generation, storage batteries, new control techniques and forecasting methods are used to help for designing operating practice which will minimize these impacts and increasing reliability of wind energy on the grid [45-47]. The induction generator which is used in wind turbines may consume reactive power and if there is no compensation method this will lead to lagging power factor results. These generators are different from the conventional as the conventional used synchronous generators which are able to work during and after the fault remove and this also effects on the transmission capacity. Integration of the wind power turbines into a weak distribution grid requires an evaluation of the grid conditions by taking into account the wind turbine characteristics as this will affect on the quality of the system and its

stability [48].

3.3.1. Grid Integration Aspects

Successful integration of wind power into the grid requires solving some problems based on analytical studies to ensure system integrity, including [49]:

1. Power flow to ensure that lines and equipment are not overloaded or that their thermal limits are not exceeded.

2. Short circuit levels and values must be re-evaluated.

3. Transient stability margins need to be re-evaluated

4. Protection schemes need to be re-adjusted (constant speed turbines may require higher reactive power under fault conditions which may cause voltage collapse. This problem can be overcome by using dynamic reactive power sources. Variable speed wind turbines, however, restart normally following fault clearance and therefore do not have this problem. The problem with variable speed wind turbines is that they are disconnected when the fault occurs and this problem can be overcome by using specific ride-through arrangements.

5. A controllable energy source must be available in the system to compensate for the fluctuation of wind power.

6. The quality of power delivered to the system should be evaluated.

There are two main challenges for the wind integration into the power system, the intermittency and the grid reliability. The integration of small scale wind into the grid is not complicated as a large one [50].

3.3.2. Methods Used for Dealing with the Large Scale Wind Integration into The Grid

1) Revision of the methods for calculation of available transmission capacity (these methods depend on the country and the weather).

2) Transmission network reinforcement (by increasing the tensile stress of conductors, increasing the height of the towers, installing conductors with the higher load ability using capacitors, and using facts devices that have a great effect on the load flow).

3) Convert the power lines from HVAC to HVDC which will increase the rating of the power transmitted 2-3 times and also reduce the losses of the transmission lines or built a new transmission with high capacity but this method is time consuming.

4) Excess wind energy curtailments and excess wind energy storage in hydro reservoirs (this is can be done by making a coordination between the wind and the hydro energy stations).

For the wind farm not all wind turbine generators work at the same speed and so the maximum power production of the wind farm is less than the sum of the rated power production of each one. The peak of the wind turbine may not be at the same time of the peak of the transmission lines and the wind turbines works at a full rate power for a small time and this must be taken into account when one calculates the transmission lines capacity limits for wind integration [51].

There are some probabilistic indices that are useful for assessment of wind energy penetration from the point of view of reliability and the cost also, such as the expected wind energy supplied (EWES) which measures the use of wind energy instead of the conventional, the expected surplus wind energy (ESWE) which measures the reserve energy that is not used, and wind utilization factor (WUF) which is the ratio of EWES of the total wind energy. These indices are very important as the random fluctuation of the wind effects on the system stability and to solve this problem there might be some constraints on the penetration of wind energy by using previous indices [52].

3.4. Stability Problems of Wind turbines, Mitigation Methods and Some of the Used Control Schemes

The integration of wind energy effects on the stability of the grid, various problems are appearing, and some of them are described below:

1) As the penetration of the FSIG increases the transient stability decreases specially without using AVR and this may lead to instability conditions. On the contrary, the DFIG and the converter synchronous generators are more stable with or without AVR as there is the ability to control the active, reactive power and the terminal voltage. The full converter synchronous generator is more stable than the DFIG but it has a high cost. The use of AVR improved the stability margin in all types of generators. During the fault conditions as the number of the FSIG penetration increases the swing angle of the conventional synchronous generator increased because the increased FSIG increases the wanted reactive power and this will increase the amount of the wanted current, consequently the voltage drop increased. But after removing the fault the FSIG is still taking a huge amount of reactive power due to its inertia and this will reduce the voltage at the conventional generator also after removing the fault and so the increased penetration of the FSIG will affect on the stability. In contrary the DFIG penetration doesn't decrease the voltage profile for the conventional generators during and after removing the fault [53].

2) For the FSIG as the number of turbines increased the voltage at the PCC will decrease, also the torque speed characteristics will be in a small zone, hence the stability will be worse. On contrary for the DFIG as the number of turbines increased the stability will increase (because of the impedance change the DFIG will supply more reactive power around the synchronous slip) but at sub-synchronous speed (away from a zero slip) DFIG will require a higher voltage control [54].

3) In the FSIG the reactive power and the grid voltage level cannot be controlled, also the blade rotation causes power variations and, this will affect on the voltage. FSIG consumes reactive power through capacitors and this may cause voltage collapse after the fault cleared.

4) The transient stability behavior of FSIG is poor; this machine may loss the synchronism and must be switched off during the fault.

5) Fault in the power system may cause voltage sag at the connection point of the wind turbine and this will increase the current in the stator winding of DFIG, hence the current will also increase in the rotor due to the magnetic coupling, causing the destruction of the power electronic converters so in DFIG there is a protection system called crowbars which will disconnect the connection to the grid [25, 36].

6) Doubly feed induction generators (DFIGs) improve the transient stability margins if they are connected to low voltage ride through capability, reactive current boosting and fast voltage control. However, the wind source is connected to lower levels so the reactive losses are so high, hence the reactive contribution of wind energy is limited, as a result its integration have a negative impact on the transient stability. Their speed fluctuation is slow compared to time frame and it has not a direct effect on the transient stability but has an indirect effect as the wind energy is not predictable, hence it requires a higher spinning reserve and this adds inertia to the system [55].

7) The conversion control methods for wind energy using converters effect on its frequency specially frequencies between 2 and 8 Hz which may cause flicker in the grid and it is preferred to dampen these frequencies in the output power. There are various control methods which describe this phenomenon from which; optimal rotational speed control which concentrate on giving maximum power and the wind rotor at the rotor speed , torque control, average power control and stochastic dynamic optimization [56].

3.4.1. There are various methods used to mitigate the stability problem among

1) The wind generation with energy storage devices in the distribution systems may result in decreased of distribution losses and this depends on the generation relative to the local load. If this ratio is high the losses will increase. The use of storage energy is important to optimize the operation; it can be used to shift the generation to maintain the loading at optimal value [57].

2) The most effective way for improving the use of wind energy for security at steady state is to use energy storage. Energy storage is used to make the supply independent of time and this storage are used to deliver or accept energy from the grid if there is a shortage or a surplus and so at the peak load they deliver energy to the grid, hence it is the economical point also [58].

3) A DC link voltage boost scheme of insulated-gate bipolar transistors (IGBT) inverters for wind extraction is used to overcome the shortage in the voltage due to wind energy decreasing by adding a switch between one of the rectifier input legs and the middle point of the dc link reservoir capacitor and this switch turned on during the shortage and so double the value of the voltage and this method is called Patent pending. But this method makes an unsymmetrical operation, and this will bring a mechanical vibration on the wind turbine due to unsymmetrical and unbalanced operation, so it is preferred to use symmetrical

double voltage rectifier [59].

4) Dealing with Wind fluctuations: many methods are used to handle wind energy fluctuations. In terms of placement, the methods are divided into three categories:

a- At the wind turbine generator using two windings one of which is activated at high wind levels and the other is activated for low wind levels. Using shunt capacitors and/or another inverter to provide capacitive voltage support,

b- At the DC link (DC step up chopper or voltage boost rectifier), or

c- At the inverter output by using a step up transformer [60].

5) During the fault if all wind turbines disconnect from the grid this will affect on the overall stability, hence there must be a solution for this problem by decreasing the current through the rotor by using resistors. During longer voltage dips the rotor may feed a reactive power to the grid during the fault [25].

6) The wind farm stability is improved by increasing the shaft stiffness and/or moment of inertia of the generator rotor, reducing the impedance of the line between the wind farm and the network, and improving the operating power factor of the wind farm [61, 62].

From the above discussion one concludes that the fault in the power system is the most commonly problematic effects on the operation of the conventional generators and also the wind power generators unless there is a ride through. The impacts of the fault of the wind power plant are changed according to the location, the type of the fault, the setting of the protection relay, the wind generator type (if there is a ride through or not), the overall characteristic of the network power system, the load distance from the generator, the grid configuration (radial or ring), the method of compensation, and the control algorithm in the grid [23]. During the fault the characteristics of the DFIG are changed. The rotor current is increased (the current may exceed 2-3 times the rated value), this will lead to increase the DC-link voltage (it may reach 2-3 times the rated value), the GSC tries to stabilize the DC-voltage and this will lead to increase the GSC current (may reach up to 57% of the rated value), and finally the turbine will be exposed to oscillating torque and this will reduce the turbine life time. The separation of the wind turbine from the grid during the fault is not preferred as it may lead to a voltage collapse, so there must be a fault ride through to overcome these problems. The chopper module which will be connected in parallel to the dc-link may be used for increasing the normal range of the DFIG and smoothing the linked voltage during the imbalance conditions [23, 63].

3.4.2. Some of the Used Control Schemes

There are a various control schemes used nowadays for DFIG focusing on the active and reactive power. The control variables are the rotor voltage or current and the blade pitch angle. One of these control systems depends on the quadrature and direct components of the rotor current and blade pitch angle for controlling the speed, reactive and active power respectively. The second control system depends on the quadrature and direct components of the rotor voltage and blade pitch angle for controlling the activity, reactive power and the speed, respectively. The third control system is considered as a variant of the second control system in which there are two modes for the operation and so the speed is limited to its rated value by acting on the pitch angle. Control system schemes give the desired output but the third control system gives a higher reactive power when the wind farms maximizing the reactive power. The first and the second control system generate the reference power below the synchronous speed and so the rotor winding consumes active power, while the third control system generates power above the synchronous speed as a result it generates active power. At the power limitation, the output power for the first control system has a small variation, because of the pitch controller difficulty to control the output power. At down power regulation, the first and the second control systems give a similar performance, as they generate the reference power at the same rotational speed but, the third control system is different, because the turbine generates the reference power at the rated speed. Hence, the stator active power generated is less as compared with the other two controls [64].

Another voltage control strategy depends on both of converters on the grid and the rotor side to be in a coordinated manner in the DFIG to control the operation during the fault condition. The RSC is usually used as the main reactive power source but GSC is a supplementary one. During the fault conditions in case of DFIG the current in the stator increased rapidly as the stator is connected directly to the grid and this will increase the rotor current and voltage, as a result this increased power will increase the energized power in the dc link as there is a fault in the grid, as a result the rotor is protected by the crowbar resistance and in this case it is like a squirrel cage induction generator with an increased rotor resistance. In this case the GSC is used as a STATCOM and delivers a limited amount of reactive power to the grid but the RSG is connected to the impedances in the crowbar. These impedances improve the dynamic stability of the DFIG during the fault conditions but on the other hand will effect on the overall performance as the increased penetration of the wind farms [23].

Paper [65] described a new FMAC (flux magnitude angle controller) for the DFIG wind turbine to adjust the rotor voltage and the angle rotor for controlling the electric power. FMAC consists of AVR (for controlling the rotor voltage) and PSS (for controlling the rotor angle and used the stator electric power as its input signals). The use of the PSS shifted the eigenvalues to the left and this has a positive influence on the damping. For the rotor speeds close to the synchronization (for very small slip) DFIG control is limited, because the steady state rotor terminal voltage is affected by the slip value. At high value of the slip the rotor vector voltage is equal to the slip value times the internally generated vector voltage and as the magnitude of the internally generated voltage is approximately constant then the rotor voltage is proportional to the slip. At a low value of slip the magnitude of the rotor voltage is small and so the control for the DFIG is limited, the performance of the DFIG is near to the squirrel cage induction generator.

The voltage of the wind turbine during the fault and after the fault cleared improved by using a STATCOM; hence improve the system reliability and stability. The system voltage has no effect on the maximum compensating current; as a result the STATCOM is able to be operated at any capacity at low voltages and this enhances the flexibility of this device. The highest rating of a STATCOM the best recovery of the voltage during and after the fault removed but the overall cost will be increased[66, 67]. Papers [68, 69] describe a master control unit (MCU) to make a power schedule to reduce the effect of fluctuation of offshore wind energy and maintain the reliability of the power supply; this schedule may received by the transmission system operator (TSO) to use less wind power energy than it is available and so the control unit may be improved by making a good commitment and system management depending on the power forecasting. There two control units primary and overall (MCU). The primary is the single unit control (UCS) which is the control unit for each windmill.

3.5. Equivalent Wind Farm Model

The aggregation of the wind turbines will be easier if they receive the same wind and therefore generate the same output power. In this case the aggregating wind turbines, equal to the sum of the rated power of the individual wind turbines, and receive the same incoming wind. The equivalent wind turbine will present the same model of the individual wind turbines. However, the wind is not equal on all turbines. The wind farm may consist of many wind turbines arranged in rows (may be separated three or five times the rotor diameter at the same row) and column (may be separated five or three times the rotor diameter at the same column), hence the turbines at the same row may have the same wind but at the next row have a different wind from the previous because of the shadow and parking effect [70].

The aggregation of wind turbine was developed for both variable and fixed speed based on aggregating the power for each individual wind turbine (using 3th order model for simplicity)and neglecting the turbulence (stochastic) term for wind speed signal model due to the smoothing effect of a large number of turbines and using sum assumptions (using the electrical power in case of variable speed instead of the mechanical power as compared to fixed speed, Cp is assumed to be constant and replace by its maximum value, and the non-linear rotor speed versus control characteristic is replaced by a first order one) [71].

For different operating conditions of wind speed the aggregation of wind turbines may be easily calculated if the output mechanical torque of the individual turbine is used instead of the output power and by summing these torques one can find the total torque and used it as an input to the equivalent generator system, thus gives the equivalent output power at different wind conditions [70]. Figure (7) shows the block diagram of the equivalent wind turbine model. The equivalent impedance of the aggregated series or parallel wind farm calculated by using the expression proposed in [30]. The wind farm control system consist of two controllers one for the power controller and the other for the dispatching controller that distribute the wind farm generation between the wind turbines and design the active and relative power for each wind turbine depending on the system operator [49]. There are different techniques for calculating the equivalent impedance.

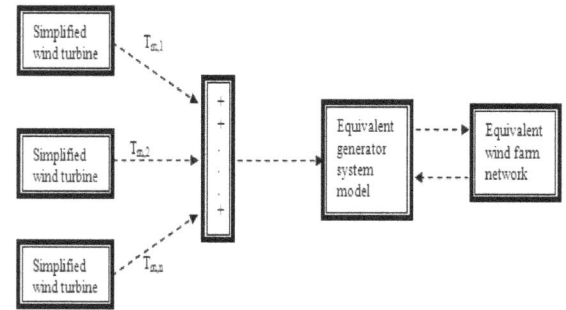

Figure 7. Block diagram of the equivalent wind turbine.

4. Tidal In-Stream Turbines

There are five types of ocean energy among tidal barrages and tidal in-stream. Barrages energy like hydro energy and it has an effect on plants and animals lived in the estuary, also affect on the quality of water and the fixed cost of making a barrage component is very high. Tidal in-stream is an attractive source of energy and its speed can be changed by changing the features of the place. One advantage is that the density of seawater is high (800 times that of air) compared to air resulting in higher energy density, therefore smaller size and cheaper rotors are needed for the same power output. Marine turbines are slow in moving, hence they cause a few disturbances in the underwater habitat compared to high speed wind turbines and have no discernible audio noise [72-79]. Marine turbines have larger capacity factor than wind turbines, hence they are more economical. Tidal in-stream turbines have smaller sizes and can use a wide variety of topologies. One of these topologies is the vertical axis turbine with a synchronous generator, the generator can be put above or below the turbine and so its size is not constrained [80]. The following sections discuss the tidal in-stream integration, the characteristics of the tidal in-stream, the technology used and marine turbine modeling.

4.1. Tidal In-stream Integration

Tidal flow results in predictable energy output patterns and forecasting marine currents depend on data gathered for short periods of time (predictable to within 98% accuracy,) while wind forecasting depends on data gathered over a longer period. The marine resource is easier to integrate in the electrical grid. The change of the weather doesn't affect on its production as it depends on the gravitational force of the molecules of the moon and the sun to the molecules of the earth and the rotation of the earth; this gravitational force may be described as F= K M m/d^2 , m (mass of the molecule of the earth), M (mass of the moon or sun), d (distance between the bodies) , and K (universal constant of gravitation). There are two types of tide; spring tide (the speed of the spring tides varies from 3.5 to 4m/s) which happens when the moon and the sun are in the same line and neap tide (the speed of the neap tides varies from 2 to 2.5m/s) and this happens when the moon and the sun at right angles as shown in figure (8) and so they pull water at the sea at different directions [80].

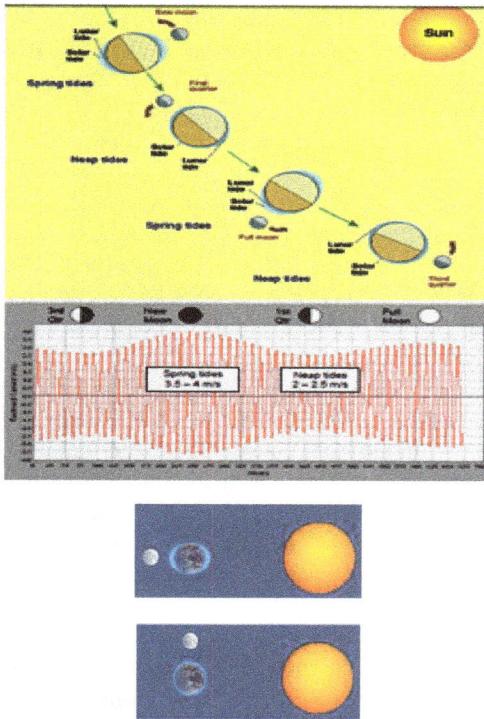

Figure 8. The power output during a spring and neap tide [83].

Some issues need to be considered in connecting marine generation into the grid such as intermittent, the effect of multi operation units, switch gear ratings during faults, plant size, generation mix, transmission line and cable thermal limits and quality of the power delivered including flickers, harmonics, and voltage sags. Reactive compensation may be required because using induction generators consume reactive power (53-51% at idle and 60% of rated) [81].

Wind and tidal generation fluctuate during the day. Without storage, this causes cycling (turning on and off) of conventional stations causing thermal stresses on the boiler, steam lines, turbines and auxiliary components which lead to component damage. As the diameter of the turbine blade and the depth of the turbine increase the obtained energy from the tidal increased [73].

The distribution density of the tidal current energy is

asymmetric everyday and so the electrical power from the tidal currents are unstable. Phase-locked-loop (PLL) as a control method which is a closed loop feedback control system is used to ensure that the grid-connected current and voltage have the same frequency and remain in phase with each other. The Phase Discriminator (PD) is used to monitor the phase difference between input and output signals. The Loop Filter (LF) is used as a filter for the noise and high frequency signals from PD, and then the signals go from LF to Voltage Controlled Oscillator (VCO) which is used to adjust the frequency. If this method of control based on digital signal processing (DSP), this will give a good result and this circuit will be named as a synchronous PLL (SPLL). The integration of tidal power into the grid is still under investigation [82].

4.2. Tidal In-stream Characteristics Compared To Off Shore Wind (The Advantages of Tidal In-stream Energy) [83, 84]

1. It is a predictable source of energy, hence its generated energy is more valuable than from random source (wind, wave and solar), thus enhance the reliability of this source and has a lower impact on the efficiency.

2. It has a higher energy intensity, hence a smaller rotor as it compared to wind turbines for the same power rating (see figure (7) from which it was found that for 1MW generation the blade diameter for offshore wind turbine is 66 meters, for tidal current is 18 meter diameter and this will affect on the overall cost).

3. It has low environmental impact as it compared to other sources of energy.

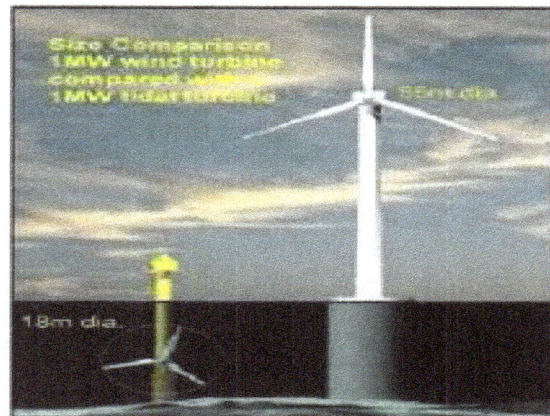

Figure 9. Wind turbine size versus tidal in-stream for the same rating (@MCT).

4. It has a turbine of much slower moving rotor and so will cause fewer disturbances for underwater survivals compared to the high speed for the wind turbines which will affect on the birds.

5. It causes very limited disturbances to shipping as its blade tip is immersed in the water at least 10 m.

6. It has roughly four times the energy intensity as it compared to a good wind site, hence the tidal turbines need

a quarter of the swept area of a wind turbine and this will reduce the cost of tidal turbines and also effects on the size of the place for putting this turbine.

7. It has no emissions of any polluted gas so it better for reducing the pollution.

8. It depends on data gathered in only one month with a great accuracy on the other hand wind depends on data gathered in two years.

9. It has greater capacity factor than the wind, so it is better for economics.

4.3. Tidal In-stream Technologies

Tidal in-stream turbines are under development, there are two types of these turbines depending on the axis of rotation horizontal axis (the axis of rotation is horizontal with respect to the ground and parallel to the flow direction) and vertical axis (the axis of rotation is perpendicular to the flow direction) [85].

There are various projects working on the development of tidal energy devices (TED) in the world and their progress is slow. From these projects MCT (Marine Current Turbines) which is ready for the use as shown in figure (7.a), apply the same technology as wind turbines and may use two turbines each has a rotor diameter between 15 and 22 m (the size depending on local site conditions), consists of twin axial flow rotors, each driving a generator connected to the turbine through a gear box and accommodated to work in bidirectional as their blades can be pitched through 180°. This technology is known as "SeaGen". The power units are easy to be raised above sea level for maintenance; this marine was tested in September 2005 and now they have a farm of turbines which may be used easily but still under development. The kinetic energy taken from water current depends on the square meters of flow cross-section, the water currents will drive the rotor at a speed of 10 and 22 revolutions per minute and this speed is slow to affect the lifetime of the blades [86].

1 MW prototype lunar energy turbine as shown figure (10.b) installed at the European Marine Energy Center in 2007 [87]. Figure (10c) shows another technology called an open hydro marine turbine technology [88]. Open hydro is one of the first energy technologies used in the world. The first test (6m) produces energy to supply 153 average European homes and save 473 tones emission of CO_2 each year.

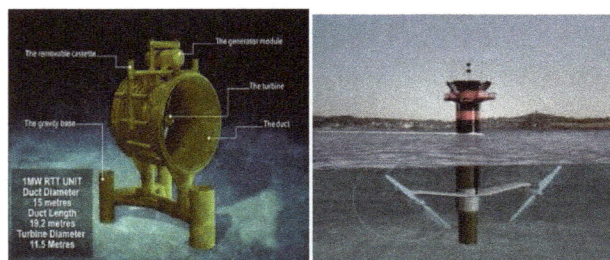

a. MCT   **b. Lunar energy**

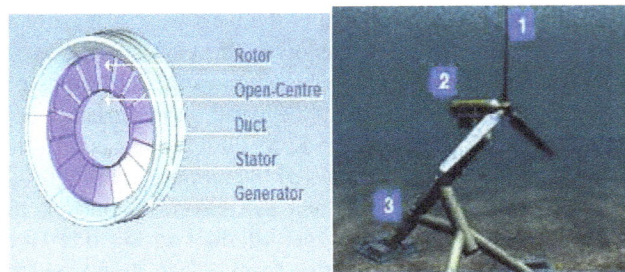

c. Open hydro   **d. Turbine components**

Figure 10. Some used technology (@MCT, @Openheydro).

The Engineering Business "Stingray" generator is another technology was developed in the United Kingdom and uses the oscillatory movement of hydroplanes driven by the water. The angle of the hydroplane is changed during the flowing of water (fall and rise) and the hydroplane is connected to a hydraulic cylinder. The movement will develop a pressure on the oil of a hydraulic cylinder; this pressure is used to drive a hydraulic motor which drives the electric generator.

The Hammerfest Stroem generator which has a horizontal axis prototype generator (similar to Seaflow project) was installed in Norwegian in 2003 to develop 310 kW. North America (US and Canada) developed some small prototype technologies such as the Verdant Power and Underwater Electric Kite (UEK) [89].

The Rotech Tidal Turbine (RTT) is a prototype system tested to extract power from tidal currents into electrical power in a unique and patented manner. This prototype used a symmetric duct and turbine blade sections to operate in both directions and this eliminate the used mechanics part in a reversing tide. The turbine is connected to a fixed displacement hydraulic pump. This pump converts the rotational energy into hydraulic flow and pressure, then this fluid energy is fed to a sealed pod which is used to drive two variable displacement hydraulic motors and these motors drive a synchronous generator. By using the swashplate the power extracted from the generator is adjusted at various tidal conditions. This prototype system is designed to produce 1 MW electrical power output at 11 KV [90].

It was concluded that to get positive benefits for the tidal generation for a case studied in Ireland the capital cost must be less than 664,000 euro per MW installed and this is very high as it is comparable to conventional energy till now [91]. Figure (10.d) shows the main components of the turbine 1) shows the blade system; 2) The nacelle which is called also the production system contains the generator, gearbox and control system, and 3) shows the subsea cables which are used to transmit the electrical power from the off shore to the shore system. The turbine is able to work in two directions according to the direction of the flow.

Various types of turbines are now available horizontal axis (this consists of straight and inclined axis. The straight axis contains two types, solid mooring and buoyant mooring. The buoyant mooring may be submerged or non-submerged) and vertical axis (this consists of four types

SC-Darrieus (Straight Blade), H-Darrieus (Straight Blade) , Darrieus (Curved Blade), Gorlov (Helical Blade) and Savonius (Straight/Skewed)). Table (2) shows a comparison between vertical and horizontal axis rotor used with tidal current turbines. The horizontal axis turbine is preferred due to its easy control and has a high starting torque [89].

Table 2. a comparison between vertical and horizontal axis rotor.

No	Type of comparison	Vertical	Horizontal
1	Design simplicity	Simple	Not simple
2	Cost	Less	High
3	Generator coupling	Placed in one end of the shaft and may be above the water surface	Using right angles gear coupling
4	Noise emission	Less	High
5	Floating and augmentation	Easy	Not easy
6	Skew flow	More suitable	Face problems
7	Starting torque	Poor	High (self starting)
8	Output torque	Has ripple	Hasn't
9	Efficiency	Low	High
10	Control	Not easy	Easy
11	Installation	Less hard	Hard
12	Knowing technology	Not well known	Well known as wind

4.4. Modeling of the Marine Turbines

As discussed in the modeling of the wind turbines, the marine turbine modeling consist of the speed signals (resource), the rotor, the generator, and the control model.

1) The speed signals (resource) model: the tidal speed may be expressed as a function of the spring tide speed, neap tide speed and tides coefficient (C), hence knowing tides coefficient, it is easy to derive a simple and practical model for tidal current speeds as follows:

$$V_{tide} = V_{nt} + \frac{(C-45)+(V_{st}-V_{nt})}{95-45}$$

, where: V_{nt} is the neap tide speed and V_{st} is the spring tide speed, C= 95 for spring, 45 in neap tide [77].

2) The Rotor model: The tidal current power (P_{ts}) may be found using the following equation: $P_{ts} = \frac{1}{2} \rho A (Vtide)3$,ρ is the density of the water (1044 kg/m3), A is the cross-sectional area perpendicular to the flow direction and V_{tide} is the velocity of the tidal in-stream. The marine turbines harness a fraction of this power, hence the power of the marine may be expressed as: $P_t = \frac{1}{2} \rho C_p A (V_{tide})^3$, Cp is the marine turbine blade design constant in the range of 0.35-0.5. The same model used for the offshore wind is used in tidal in-stream turbines; however, there is a number of differences in the design and operation of marine turbines due to the changes in force loadings, immersion depth, and different stall characteristics.

3) Dynamic modeling of the generator: The same dynamic modeling for the, DFIG used in the offshore wind [92, 93].

4) The control model: the tidal resource characteristics are nonlinear like turbulence and swell, also the inevitable uncertainties in DFIG based marine current turbines, and hence the need for nonlinear and robust control is of the requirements for marine turbines. One of the most popular nonlinear control methods is the higher order sliding mode (HOSM); this control method provides system dynamics with an invariant property to uncertainties [92].

5. Hybrid Offshore Wind and Tidal In-Stream Turbine (HOTT)

The simulations of the wind energy power and speed shows that wind power may lead to load problems at high levels of installed wind power if they used alone without any batteries and may destroy the stability of the grid.

Tidal in-stream and offshore wind generators use the same prime-moving principles, so a hybrid system of these two technologies may be very beneficial. Hybrid offshore wind and tidal-in-stream turbine systems (HOTT) are preferred because they offer flexible, stable, reliable and easy to use source of renewable energy. In some applications, the HOTT is connected electrically using a 6-pulse GTO rectifier to convert power to DC and then transmitted to shore using cables and again using a 6-pulse GTO inverter to convert the DC into AC and then connected to the grid using PWM control of the overall process [2, 5].

6. Conclusion

In this paper offshore wind turbines types were discussed, Different generator types were addressed, stability problems of wind turbines, mitigation methods, some control schemes and wind farm were shown as an introduction of the tidal current turbine. Then similarities and differences between tidal in-stream and offshore wind turbines, tidal in-stream technology and tidal in-stream turbine types were presented.

The wind is fluctuation and not easily predictable source of energy so the use of this source alone is not preferred. The hybrid of wind and other predictable or conventional source is more suitable. Since the techniques used in offshore wind turbines are approximately the same as the turbines used in the tidal in-stream (tidal current) so it is preferred to use a hybrid system of offshore wind and tidal current turbines and study its integration into the grid and its effect on the power system stability.

References

[1] José Luis Rodríguez, Santiago Arnalte, and Juan Carlos," Automatic Generation Control of a Wind Farm With Variable Speed Wind Turbines" IEEE Transactions on Energy

Conversion, Volume 17, No. 2, June 2002.

[2]   Hamed H. Aly, and M. E. El-Hawary "An Overview of Off-shore Wind Electrical Energy Systems" 23rd Annual Canadian IEEE Conference on Electrical and Computer Engineering, Calgary, Alberta, Canada, May 2-5, 2010.

[3]   "Tidal Energy" Available online (January 2010), http://www.gcktechnology.com/GCK/Images/ms0032%20final.pdf.

[4]   "Wave and Tidal Power" Available online (January 2010), http://www.fujitaresearch.com/reports/tidalpower.html.

[5]   Hamed H. H. Aly, and M. E. El-Hawary "State of the Art for Tidal Currents Electrical Energy Resources", 24th Annual Canadian IEEE Conference on Electrical and Computer Engineering, Niagara Falls, Ontario, Canada, 2011.

[6]   "Tidal Currents" Available online (April 2011), http://science.howstuffworks.com/environmental/earth/oceanography/ocean-current4.htm.

[7]   Beach, and Bathurst "Generation of Electricity from Tidal Currents "Available online (January 2010), http://www.ipenz.org.nz/conventionCD/Documents/Beach-Bathurst.pdf.

[8]   "Tidal Stream" Available online (January 2011), http://www.tidalstream.co.uk/html/background.html.

[9]   "Tidal Currents" Available online (April 2011), http://science.howstuffworks.com/environmental/earth/oceanography/ocean-current4.htm.

[10] "New Marine Energy Roadmap Launched in Montreal", Available online (August 2012), http://www.nrcan.gc.ca/media-room/news-release/2011/3195.

[11] "Nova Scotia Marine Renewable Energy Strategy ", Available online (August 2012), http://www.gov.ns.ca/energy/resources/publications/Nova-Scotia-Marine-Renewable-Energy-Strategy-May-2012.pdf.

[12] "Marine Renewable Energy Strategy" , Available online (August 2012), http://www.gov.ns.ca/energy/resources/publications/Marine-Renewable-Energy-FAQs.pdf.

[13] "Renewable Electricity Plan" , Available online (August 2012), http://www.gov.ns.ca/energy/resources/EM/renewable/renewable-electricity-plan.pdf.

[14] "Marine Renewable Energy Technology Roadmap" , Available online (August 2012), http://oreg.ca/index.php?p=1_58_Marine-Energy-TRM.

[15] "Charting the Course Canada's Marine Renewable Energy Technology Roadmap" , Available online (August 2012), http://oreg.ca/web_documents/mre_roadmap_e.pdf.

[16] "Renewable Energy", , Available online (August 2012), http://www.nspower.ca/en/home/environment/renewableenergy/default.aspx.

[17] "Marine Renewable Energy Legislation for Nova Scotia" , Available online (August 2012), http://www.oreg.ca/web_documents/ns-mre.pdf.

[18] "Wave and Tidal Energy in UK" , Available online (August 2012), http://www.bwea.com/pdf/marine/Wave_Tidal_energy_UK.pdf.

[19] Abreu, L.V.L.; Shahidehpour, M.; "Wind Energy and Power System Inertia", Power Engineering Society General Meeting, IEEE 2006.

[20] Roth, H.; Kuhn, P.; Wagner, U.; "Effects of Wind Energy on Thermal Power Plants" Clean Electrical Power, ICCEP '07. International Conference on 40-42 May 2007 Page(s):683 – 665. Summary: Increasing wind energy use affects both existing power plants as well as future investments in the electricity market. Only a small contribution to total system reliability is added by wind energy. Part load operation and lower efficiency caused by t.

[21] José Luis Rodríguez-Amenedo, Santiago Arnalte, and Juan Carlos Burgos," Automatic Generation Control of a Wind Farm With Variable Speed Wind Turbines" IEEE Transactions on Energy Conversion, Volume 17, No. 2, June 2002.

[22] http://www.unihildesheim.de/~irwin/inside_wind_turbines.html.

[23] Anca D. Hansen, Gabriele Michalke, Poul Sørensen and Torsten Lund, Florin Iov "Co-ordinated Voltage Control of DFIG Wind Turbines in Uninterrupted Operation during Grid Faults", wind energy 2007. Published online 10 August 2006 in Wiley Interscience.

[24] Li Lin, Yan Zhang, Yihan Yang " Transient Characteristics of the Grid-connected Wind Power Farm with DFIGs and SCIGs" Electric Utility Deregulation and Restructuring and Power Technologies. DRPT 2008. Nanjuing, China.

[25] Johan Morren, , and Sjoerd W. H. de Haan, "Ride-through of Wind Turbines with Doubly-Fed Induction Generator During a Voltage Dip" IEEE Transactions on Energy Conversion, Vol. 37, No. 2, June 235.

[26] I. Erlich, , and F. Shewarega " Modeling of Wind Turbines Equipped with Doubly-Fed Induction Machines for Power System Stability Studies" Power and Energy Society General Meeting - Conversion and Delivery of Electrical Energy in the 21st Century, 2008 IEEE, Page(s):1 – 8.

[27] C. Ghiţă1, D. I. Deaconu1, A. I. Chirilă1, V. Năvrăpescu1 and D. Ilina1 "Lab Model for a Low Power Wind Turbine System" International Conference on Renewable Energies and Power Quality (ICREPQ'09) Valencia (Spain), 15th to 17th April, 2009.

[28] J. G. Slootweg, S. W. H. de Haan, H. Polinder, and W. L. Kling, "General Model for Representing Variable Speed-Wind Turbines in Power System Dynamics Simulations" IEEE Transactions on Power Systems, Vol. 18, No. 1, February 2003.

[29] Yazhou Lei, Alan Mullane, Gordon Lightbody, and Robert Yacamini "Modeling of the Wind Turbine with a Doubly Fed Induction Generator for Grid Integration Studies" IEEE Transaction on Energy Conversion, Vol. 28, No. 1, March 2006.

[30] Miguel Garcıa-Gracia, M. Paz Comech, Jesus Sallan, Andres Liombart "Modelling wind farms for grid disturbance studies" Elsevier journal, 2008.

[31] Bingchang Ni, Constantinos Sourkounis, "Power Output Characteristics Analysis of Wind Energy Converter Control

Methods" 13th International Power Electronics and Motion Control Conference (EPE-PEMC 2008).

[32] Fei Ye, Xueliang Huang, Chaoming Wang, Gan Zhou, Ping Luo " The impact and simulation on large wind farm connected to power system" Electric Utility Deregulation and Restructuring and Power Technologies. DRPT 2008. Nanjuing, China.

[33] Wei Qiao, Member, IEEE "Dynamic Modeling and Control of Doubly Fed Induction Generators Driven by Wind Turbines" IEEE Power & Energy Society, March 15-18, 2009.

[34] Pavlos S. Georgilakis "Technical challenges associated with the integration of wind power into power systems"Elsevier journal, 2008.

[35] I. Erlich, , J. Kretschmann, S. Mueller-Engelhardt, F. Koch, J. Fortmann, "Modeling of Wind Turbines based on Doubly-Fed Induction Generators for Power System Stability Studies" Power and Energy Society General Meeting - Conversion and Delivery of Electrical Energy in the 28st Century, IEEE 2008.

[36] Marcus V. A. Nunes, J. A. Peças Lopes, Hans Helmut Zürn, Ubiratan H. Bezerra, and Rogério G. Almeida "Influence of the Variable-Speed Wind Generators in Transient Stability Margin of the Conventional Generators Integrated in Electrical Grids" IEEE Transactions on Energy Conversion, Vol. 36, No. 4, December 2004.

[37] J.G. Slootweg, H. Polinder W.L. Kling, "Dynamic Modeling of a Wind Turbine with Doubly Fed Induction Generator" Power Engineering Society Summer Meeting, 2001, IEEE.

[38] Janaka B. Ekanayake, Lee Holdsworth, XueGuang Wu, and Nicholas Jenkins, " Dynamic Modeling of Doubly Fed Induction Generator Wind Turbines" IEEE Transactions On Power Systems, Vol. 18, No. 2, May 2003.

[39] Lucian Mihet-Popa, Frede Blaabjerg, and Ion Boldea," Wind Turbine Generator Modeling and Simulation Where Rotational Speed is the Controlled Variable" IEEE Transactions On Industry Applications, Vol. 58, No. 1, January/February 2004.

[40] Iov, F.; Blaabjergg, F.; Hansen, A.D.; Chen, Z. "Comparative study of different implementations for induction machine model in Matlab/Simulink for wind turbine simulations" Computers in Power Electronics, 2102. Proceedings.2102 IEEE Workshop on Volume, Issue, 3-4 June 2002 Page(s): 67 – 66.

[41] L.M. Fernández, C.A. García, J.R. Saenz, F. Jurado "Equivalent models of wind farms by using aggregated wind turbines and equivalent winds" Elsevier journal, 2008.

[42] Hamed H. H. Aly, and M. E. El-Hawary "State of the Art for Tidal Currents Electrical Energy Resources", 24th Annual Canadian IEEE Conference on Electrical and Computer Engineering, Niagara Falls, Ontario, Canada, 2011.

[43] F. Wu, X.-P. Zhang, and P. Ju "Small signal stability analysis and control of the wind turbine with the direct-drive permanent magnet generator integrated to the grid" Journal of Electric Power and Engineering Research, 2009.

[44] Kala Meah; Yi Zhang; Sadrul Ula "Wind Energy Resources in Wyoming and Simulation for Existing Grid Connection" Power Systems Conference and Exposition. PSCE apos; 06. IEEE PES Volume, Issue, Oct. 2006-Nov. 2006

Page(s):3879 – 3883.

[45] Vilchez, E.; Stenzel, J.;" Wind energy integration into 110 kV system Impact on power quality of MV and LV networks" Transmission and Distribution Conference and Exposition: Latin America, IEEE/PES 13-15 Aug. 2008 Page(s):1 – 6. Summary: The amount of electrical energy produced by wind farms is constantly increasing. Nowadays detailed analyses considering the impact of wind energy integration on the transmission system are required. Therefore several wind impact studies have been car.Mark L. Ahlstrom, and Robert M. Zavadil, "The Role of Wind Forecasting in Grid Operations & Reliability" IEEE/PES Transmission and Distribution Conference & Exhibition 2005: Asia and Pacific Dalian, China.

[47] Fox, B.; Flynn, D.;" Wind Intermittency - Mitigation Measures and Load Management" Power Tech, IEEE Russia 52-55 June 2005 Page(s):1 – 3.

[48] http://www.ewec3820proceedings.info/allfiles2/552_Ewec3 820fullpaper.pdf.

[49] Zhenyu Fan; Enslin, J.H.R.;" Wind Power Interconnection Issues in the North America" Transmission and Distribution Conference and Exhibition, IEEE PES May 2006 Page(s):561 – 568.

[50] Pavlos S. Georgilakis "Technical challenges associated with the integration of wind power into power systems"Elsevier journal, 2008.

[51] Matevosyan, J.; "Wind power integration in power systems with transmission bottlenecks" Power Engineering Society General Meeting. IEEE June-2007-Page(s): 1–7. Summary: The best conditions for the development of wind farms are in remote, open areas with low population density. The transmission system in such areas might not be dimensioned to accommodate additional large-scale power infeed. Furthermore a part of the .

[52] Karki, R.; Billinton, R.;" Cost-effective wind energy utilization for reliable power supply" Energy Conversion, IEEE Transaction on Volume 36, Issue 2, June 2004 Page(s):625-622.

[53] K. A. Folly and S. P. N. Sheetekela, "Impact of Fixed and Variable Speed Wind Generators on the Transient Stability of a Power System Network" Power Systems Conference and Exposition, IEEE/PES, 15-18 March 2009.

[54] Shuhui Li; Haskew, T.; Challoo, R. "Characteristic study for integration of fixed and variable speed wind turbines into transmission grid" Transmission and Distribution Conference and Exposition, 2008. T&D. IEEE/PES Volume, Issue, 21-24 April 2008 Page(s):1 – 9.

[55] Eping, C.; Voelskow, M.;" Enhancement of the Probability of Occurrence for Off-Shore Wind Farm Power Forecast " Power Tech, IEEE Lausanne 1-5 July 2007 Page(s):642 – 646.

[56] Bingchang Ni; Sourkounis, C.; " Influence of Wind Energy Converter Control Methods on the Output Frequency Components" Industry Applications Society Annual Meeting. IAS '08. IEEE 5-9 Oct. 2008 Page(s):1 – 7.

[57] Abbey, C.; Joos, G.;" Coordination of Distributed Storage with Wind Energy in a Rural Distribution System" Industry Applications Conference. Conference Record of the 2007 IEEE 42-52 Sept. 2007 Page(s):1087 – 1092.

[58] Voller, S.; Al-Awaad, A.-R.; Verstege, J.F.;" Benefits of energy storages for wind power trading" Sustainable Energy Technologies. ICSET. IEEE International Conference on 43-52 Nov. 2008 Page(s):702 – 706.

[59] Hong Huang; Liuchen Chang "A New DC Link Voltage Boost Scheme of IGBT Inverters for Wind Energy Extraction" Electrical and Computer Engineering, Canadian Conference, Volume 1, 7-10 March 230 Page(s):679 - 683 vol.1.

[60] Muljadi, E. Mills, Z. Foster, R. Conto, J. Ellis, A. "Fault analysis at a wind power plant for one year of observation" Power and Energy Society General Meeting - Conversion and Delivery of Electrical Energy in the 21st Century, 2008 IEEE July 2008, page(s): 1-7.

[61] Salman K. Salman, Senior Member, IEEE, and Anita L. J. Teo "Windmill Modeling Consideration and Factors Influencing the Stability of a Grid-Connected Wind Power-Based Embedded Generator" IEEE Transactions On Power Systems, Vol. 18, No. 2, MAY 2003.

[62] Muljadi, E., Butterfield, C.P., Parsons, B., Ellis, A. "Characteristics of Variable Speed Wind Turbines Under Normal and Fault Conditions" Power Engineering Society General Meeting, 2007. IEEE Volume, Issue, 25-29 June 2007 Page(s):1 - 7.

[63] Erlich, I.; Wrede, H.; Feltes, C. "Dynamic Behavior of DFIG-Based Wind Turbines during Grid Faults" Power Conversion Conference - Nagoya, 2007. PCC '072-5 April 2007 Page(s):1195 – 1200.

[64] L.M. Fernandez, C.A. Garcia, F. Jurado "Comparative study on the performance of control systems for doubly fed induction generator (DFIG) wind turbines operating with power regulation" Elsevier journal 2008.

[65] F. Michael Hughes, Olimpo Anaya-Lara, Nicholas Jenkins, and Goran Strbac "A Power System Stabilizer for DFIG-Based Wind Generation" IEEE TRANSACTIONS ON POWER SYSTEMS, VOL. 21, NO. 2, MAY 2006.

[66] Aditya P. Jayam, Badrul H. Chowdhury "Improving the Dynamic Performance of Wind Farms With STATCOM" Power Systems Conference and Exposition, 2009. PES '09. IEEE/PES, 17 March 2009.

[67] Aditya P. Jayam, Nikhil K. Ardeshna, Badrul H. Chowdhury "Application of STATCOM for improved reliability of power grid containing a wind turbine" Power and Energy Society General Meeting - Conversion and Delivery of Electrical Energy in the 21st Century, 2008 IEEE 20-25 July 2008 Page(s):1 – 7.

[68] Eping, C.; Stenzel, J.;" Control of offshore wind farms for a reliable power system management" Power Tech, IEEE Russia 52-55 June 2005 Page(s):1 – 4.

[69] http://www.icrepq.com/full-paper-icrep/539-Eping.pdf.Ch. Eping, J. Stenzel "Energy Management System for Offshore Wind Farms".

[70] Luis M. Fernandez, Francisco Jurado, Jose Ramon Saenz "Aggregated dynamic model for wind farms with doubly fed induction generator wind turbines" Elsevier Journal 2007.

[71] J.G. Slootweg, W.L. Kling, "Aggregated Modelling of Wind Parks in Power System Dynamics Simulations" IEEE Bologna Power Tech. Conference, June 23-26, 2003, Bologna, Italy.

[72] Sheth, S., Shahidehpour, M.;" Tidal energy in electric power systems" Power Engineering Society General Meeting. IEEE 12-16 June 2005 Page(s):682 - 687 Vol. 1.

[73] Bryans, A.G.; Fox, B.; Crossley, P.A.; O'Malley, M.; "Impact of tidal generation on power system operation in Ireland" Power Systems, IEEE Transactions on Volume 37, Issue 4, Nov. 2005 Page(s):3863 – 3862.

[74] Muljadi, E. Mills, Z. Foster, R. Conto, J. Ellis, A. "Fault analysis at a wind power plant for one year of observation" Power and Energy Society General Meeting - Conversion and Delivery of Electrical Energy in the 21st Century, 2008 IEEE July 2008, page(s): 1-7.

[75] Hammons, T.J.; "Tidal Power in the United Kingdom" Universities Power Engineering Conference, 61rd International 1-4 Sept. 2008 Page(s):1 - 8.

[76] Jones, A.T., Westwood, A. " Recent progress in offshore renewable energy technology development" Power Engineering Society General Meeting, IEEE 12-16 June 235, Vol. 2.

[77] S.E. Ben Elghali, M.E.H. Benbouzid, and J.F. Charpentier, "Marine Tidal Current Electric Power Generation Technology: State of the Art and Current Status" Electric Machines & Drives Conference, IEMDC '07. IEEE International Volume 2, 3-5 May 2007 Page(s):1593 – 1595.

[78] Eleanor Denny, "The economics of tidal energy" Elsevier journal, Volume 62, Issue 5, May 2009, Pages 3714-3744.Hamed H. Aly "Forecasting, Modeling, and Control of Tidal currents Electrical Energy Systems"PhD thesis, Halifax, Canada, 2012.

[80] Smit, J.J.;" Trends in emerging technologies in power systems" Future Power Systems, International Conference on Nov. 2005 Page(s):7.

[81] http://www.pstidalenergy.org/Tidal_Energy_Projects/Misc/E PRI_Reports_and_Presentations/EPRI-TP 001_Guidlines_Est_Power_Production_14Jun06.pdf. Summary: Tidal stream generation is a form of renewable energy that is predictable but variable in nature. The paper initially identifies the tidal resource around Ireland, utilizing the most appropriate and developed tidal energy technology, thus providing a.

[82] Salman, S.K.; Gibb, J.; Macdonald, I.;" Integration of tidal power based electrical plant into a grid" Universities Power Engineering Conference, 61rd International 1-4 Sept. 238 Page(s):1-4.

[83] Khan, J.; Bhuyan, G.; Moshref, A.; Morison, K.; Pease, J.H.; Gurney, J.; "Ocean wave and tidal current conversion technologies and their interaction with electrical networks" Power and Energy Society General Meeting - Conversion and Delivery of Electrical Energy in the 40st Century, IEEE 37-43 July 238 Page(s):1 – 8. Summary: The environmental impacts of tidal stream energy extraction are not yet understood. What is known is that the ecological effects of tidal mixing are both direct and indirect. The direct effects of changes in mixing affect the location and timing of f.Hongda Liu, Dian-pu Li,Yao-hua Luo, Zhong-li Ma "The Grid-connection Control System of the Tidal Current Power Station", The 35rd Annual Conference of the IEEE Industrial Electronics Society (IECON), Nov. 5-8, 2007, Taipei, Taiwan. Summary: This paper presents a brief review of the state of ocean energy technologies, with special attention to

ocean wave and tidal current systems. Schematic outline of a set of selected converter technologies is presented with a view to inferring their op.

[85] http://www.tidalgeneration.co.uk/background.html.

[86] http://www.johnarmstrong1.pwp.blueyonder.co.uk/CostEffe ctiveness.htm.

[87] http://oceanenergy.epri.com/attachments/streamenergy/repor ts/008_Summary_Tidal_Report_06-10-06.pdf.

[88] http://peswiki.com/index.php/Directory:Marine_Current_Tu rbines_Ltd#How_it_Works.

[89] http://www.tidalstream.co.uk/html/background.html.

[90] http://www.openhydro.com/techOCT.html.

[91] M.J. Khan, G. Bhuyan, M.T. Iqbal, J.E. Quaicoe "Hydrokinetic energy conversion systems and assessment of horizontal and vertical axis turbines for river and tidal Elsevier journal applications: A technology status review" Applied Energy, Volume 86, Issue 10, October 2009, Pages 1824-1837.

[92] George Lemonis "Wave and Tidal Energy Conversion Encyclopedia of Energy", Elsevier journal 2004, Pages 405-416.

7

Dynamic active power control in Mosul city ring system using ESTATCOM

Dhiya A. Al-Nimma, Majid S. M. Al-Hafidh, Saad Enad Mohamed

Electrical Engineering Department, University of Mosul, Mosul- IRAQ

Email address:

dalnimma@ieee.org(Dhiya A. Al-Nimma), el_noor2000@yahoo.co(Majid S. M. Al-Hafidh),
saadenadmohamed@yahoo.com(S. E. Mohamed)

Abstract: Static synchronous compensation (STATCOM) is an application that utilizes a voltage source converter (VSC) to provide instantaneous reactive power support to the connected power system. Conventionally, STATCOMs are employed for reactive power support only. However, with the integration of energy storage (ES) into a STATCOM, it can provide active power support in addition to the reactive power support. The control method of a STATCOM with an energy storage device is discussed in this paper. To determine the switching level control, SPWM (Sinusoidal Pulse Width Modulation) methods is used. This paper will introduce an integrated STATCOM/BESS for the improvement of dynamic and transient stability and transmission capability. This work suggests a simple and easy to implement PI controller to control the operation of ESTATCOM placed at the weak point of Mosul city 132 kV ring system. The whole system including the ESTATCOM and its controller has been simulated after some kinds of disturbances and the results show improvements in the dynamic active and reactive power capabilities of the system.

Keywords: ESTATCOM, Active Power Control, Dynamic Active and Reactive Power Capabilities, Battery Energy Storage

1. Introduction

The Iraqi national grid is divided into sub grids on geographical bases. The Iraqi Northern Region National Grid (INRNG) is one of these sub grids. Mosul city is part of this (INRNG) and it is supplied by one 400 kV substation at Bus 6 and seven 132 kV substations. These substations form a ring consists of seven 132 kV buses. The lines connecting the buses are mainly of two circuits. There are three single circuit lines. This part of INRNG suffers from problems such as loading on these three lines can exceed the accepted loading percent during peak load periods. This causes outage due to over load.

STATCOM as reactive power compensation has been used in a pervious study to improve the voltage profile in Mosul Ring System, which can also be used to determine the optimal value of the reactive power needed at different load conditions [1].

Later study used STATCOM with Energy storage device as active power compensation [2].

In this work a PI controllers has been suggested and added to the same system to control its operation during some kinds of disturbances. The transient behavior of the whole system has been studied via modeling and simulation. The results so obtained were compared with those obtained without the controller and the results are discussed.

2. Modeling of Mosul City Ring System

The largest part of Iraqi Northern Region National Grid (INRNG) is Mosul ring and this part of the grid is modeled by using MATLAB SIMULINK program. The MUSOL ring includes two generator units one of them is hydraulic and the other is Gas Station and the Grid connected with power lines and this grid is shown in figure (1). In previous study we found that the weak point in MOSUL grid is located at BUS 37 and depending on this study we suggested to put the STATCOM with energy storage system at that location [3]. The SIMULINK model of the grid with STATCOM is shown in appendix (2).

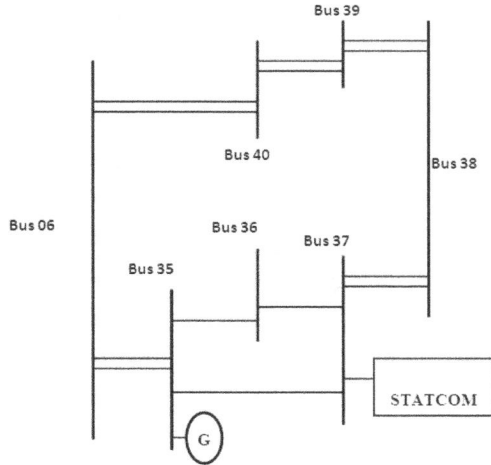

Figure (1) the largest part of northern Iraqi Grid (Mosul ring system)

3. ESTATCOM Control Algorithm

The schematic of STATCOM with a battery (ESTATCOM) is shown in Figure (2). The ESTATCOM is connected to the power system via a transformer. v_a, v_b, v_c represent the three phase line to neutral system voltages at the connection point. e_a, e_b, and e_c represent the fundamental components of the three phase line to neutral output voltage of the ESTATCOM 's inverter. L and R represent the resistance and reactance of the transformer. The battery is represented by an ideal DC voltage source Vs and a resistor R_s. R_s can also accounts for any losses in inverter. i_a, i_b, i_c represent line currents [4]. A simple model to represent the battery is used because the STATCOM is used in a transmission system for improving system's transient stability. In the short of the system transient, there should be no significant variation to the potential of the battery. The main objective of the controller is to meet the desired performance of the system requirement.

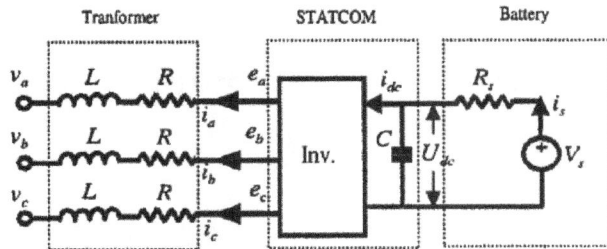

Figure (2) Equivalent Circuit of ESTATCOM

If, we assume the ESTATCOM is working in a balance condition, then, we can define a reference frame transformation and make the attained dynamic model of the STATCOM and battery simple. The reference frame coordinate is defined where the d-axis is always coincident with the instantaneous system voltage vector and the q-axis is in quadrature with it. The transformation of variables is defined in equation (1).

$$[c] = \frac{2}{3} \begin{bmatrix} \cos\theta & \cos(\theta - \frac{2\pi}{3}) & \cos(\theta + \frac{2\pi}{3}) \\ -\sin\theta & -\sin(\theta - \frac{2\pi}{3}) & -\sin(\theta + \frac{2\pi}{3}) \\ \frac{1}{\sqrt{2}} & \frac{1}{\sqrt{2}} & \frac{1}{\sqrt{2}} \end{bmatrix} \quad (1)$$

$$[c]^{-1} = \frac{3}{2}[c]^T, \begin{bmatrix} i_d \\ i_q \\ 0 \end{bmatrix} = [c]\begin{bmatrix} i_a \\ i_b \\ i_c \end{bmatrix}, \begin{bmatrix} e_d \\ e_q \\ 0 \end{bmatrix} = [c]\begin{bmatrix} e_a \\ e_b \\ e_c \end{bmatrix}, \begin{bmatrix} |v| \\ 0 \\ 0 \end{bmatrix} = \begin{bmatrix} v_a \\ v_b \\ v_c \end{bmatrix}$$

Where $\square$ is the angle between instantaneous system voltage vector and the a-phase axis of the abc coordinate.

In the reference frame coordinate, the equations of the AC side circuit in Figure (2) can be written as:

$$p\begin{bmatrix} i_d \\ i_q \end{bmatrix} = \begin{bmatrix} -\frac{R}{L} & \omega_o \\ \omega_o & -\frac{R}{L} \end{bmatrix}\begin{bmatrix} i_d \\ i_q \end{bmatrix} + \begin{bmatrix} \frac{(e_d - |v|)}{L} \\ \frac{e_q}{L} \end{bmatrix} \quad (2)$$

Where:

$$p = d/dt, \omega_o = 2\pi f_o, f_o = 50\ Hz$$

The DC side circuit equation can be written as:

$$pV_{dc} = \frac{1}{C}(i_s - i_{dc})$$
$$= \frac{1}{R.C}(V_s - U_{dc}) - \frac{1}{C}i_{dc} \quad (3)$$

The instantaneous active power on the ac side of the inverter is calculated by:

$$P_{ac} = e_a i_a + e_b i_b + e_c i_c$$

The relationship between the direct abc-axis and the transformed dq-axis is shown in figure (3). From this figure and after transform the abc frame to dq frame we obtained:

$$P_{ac} = \frac{3}{2}(e_d i_d + e_q i_q) \quad (4)$$

And the active power on the dc side of the inverter can express by:

$$P_{dc} = U_{dc} i_{dc} \quad (5)$$

Considering that the instantaneous active power exchanged between the ac and the dc side of the inverter should be the same, equation (6) must hold:

$$P_{ac} = P_{dc}, \quad U_{dc} i_{dc} = \frac{3}{2}(e_d i_d + e_q i_q) \quad (6)$$

So,

$$i_{dc} = \frac{3}{2}\left(\frac{e_d i_d + e_q i_q}{U_{dc}}\right) \qquad (7)$$

When an inverter of ESTATCOM operates in SPWM mode, its output voltage must satisfy the following equations:

$$e_d = \frac{1}{2}U_{dc}M\cos\alpha$$

$$e_q = \frac{1}{2}U_{dc}M\sin\alpha \qquad (8)$$

Where M is the modulation index and α is the firing angle of the sinusoidal reference wave referring to the system voltage vector.

Combining equation (3) with equation (2) and substituting equation (8) and equation (7) into them, we can set up a dynamic model for a ESTATCOM:

$$p\begin{bmatrix} i_d \\ i_q \\ U_{dc} \end{bmatrix} = [A]\begin{bmatrix} i_d \\ i_q \\ U_{dc} \end{bmatrix} + \begin{bmatrix} \dfrac{U_{dc}}{2L}M\cos\alpha \\ \dfrac{U_{dc}}{2L}M\sin\alpha \\ -\dfrac{3i_d}{4C}M\cos\alpha - \dfrac{3i_q}{4C}M\sin\alpha \end{bmatrix} + \begin{bmatrix} -\dfrac{|v|}{L} \\ 0 \\ \dfrac{V_s}{R_sC} \end{bmatrix} \qquad (9)$$

Where

$$[A] = \begin{bmatrix} -\dfrac{R}{L} & \omega_o & 0 \\ -\omega_o & -\dfrac{R}{L} & 0 \\ 0 & 0 & -\dfrac{1}{R_sC} \end{bmatrix}$$

After linearization in the neighborhood of equilibrium point, the control system in equation (9) can be transformed to a linear system as shown in equation (10):

$$p\begin{bmatrix} \Delta i_d \\ \Delta i_q \\ \Delta U_{dc} \end{bmatrix} = [A_o]\begin{bmatrix} \Delta i_d \\ \Delta i_q \\ \Delta U_{dc} \end{bmatrix} + [B_o]\begin{bmatrix} \Delta M \\ \Delta\alpha \end{bmatrix}$$

$$\begin{bmatrix} \Delta i_d \\ \Delta i_q \\ \Delta U_{dc} \end{bmatrix} = \begin{bmatrix} i_d - i_{do} \\ i_q - i_{qo} \\ U_{dc} - U_{dco} \end{bmatrix}, \begin{bmatrix} \Delta M \\ \Delta\alpha \end{bmatrix} = \begin{bmatrix} M - M_o \\ \alpha - \alpha_o \end{bmatrix} \qquad (10)$$

Where

$$[A_o] = \begin{bmatrix} -\dfrac{R}{L} & \omega_o & \dfrac{M_o\cos\alpha_o}{2L} \\ -\omega_o & -\dfrac{R}{L} & \dfrac{M_o\sin\alpha_o}{2L} \\ -\dfrac{3M_o\cos\alpha_o}{4C} & -\dfrac{3M_o\sin\alpha_o}{4C} & -\dfrac{1}{R_sC} \end{bmatrix}$$

$$[B_o] = \begin{bmatrix} \dfrac{U_{dco}\cos\alpha_o}{2L} & -\dfrac{U_{dco}M_o\sin\alpha_o}{2L} \\ \dfrac{U_{dco}\sin\alpha_o}{2L} & \dfrac{U_{dco}M_o\cos\alpha_o}{2L} \\ -\dfrac{3(i_{do}\cos\alpha_o + i_{qo}\sin\alpha_o)}{4C} & \dfrac{3M_o(i_{do}\sin\alpha_o - i_{qo}\cos\alpha_o)}{4C} \end{bmatrix}$$

Where $[\Delta i_d \ \Delta i_q \ \Delta U_{dc}]^T$ is the state variable vector, and $[\Delta M \ \Delta\alpha]T$ are the control variables. All the symbols with a subscription 0 in equation (10) represent the values at the equilibrium point.

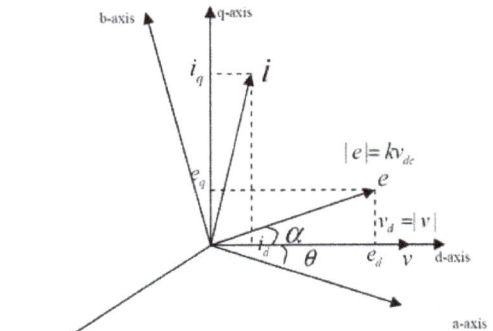

Figure (3) *Relationship between the abc-axis and the dq-axis*

The block diagram of ESTATCOM controller is shown in figure (4), while Figure (5) shows the block diagram of decoupled controller [5]. the decoupled proportional–integral(PI) type of controller is considered K_{P1} and K_{I1} are the gains of the reactive power regulator, where K_{P2} and K_{I2} are the gains of the Active power regulator and K_{P3}, K_{I3} are the gains of the AC current decoupled regulator. The strategy of ESTATCOM controller is to keep the active power constant while controlling the angle (α) by which the VSC voltage leads (or lags) the line voltage vector to compensate the active power during the disturbance interval [6].

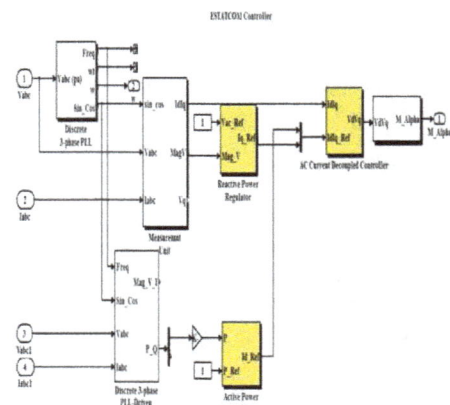

Figure (4) *Simulink Model of the ESTATCOM*

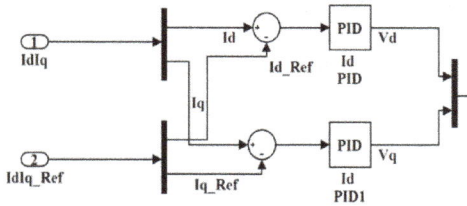

Figure (5) AC Current Decoupled Controller

4. Simulation Results

The effectiveness of ESTATCOM controller has been verified on bus 37 in the system discussed in section II (Mosul grid). A model of the system with ESTATCOM with the controllers has been developed in MATLAB / SIMULINK. Sudden changes in Load (figure 6) were studied here to show the controller response of ESTATCOM.

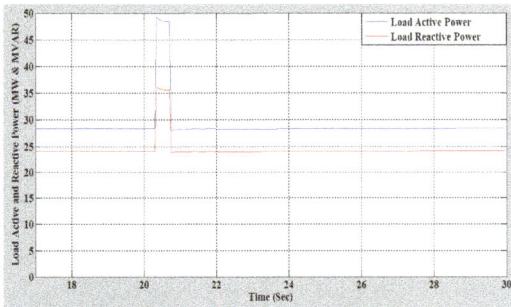

Figure (6) Load Active and Reactive Power

Figure (7) shows the active power drowns from the source1 (Mosul Dam Synchronous Generator) at load disturbance (load increased) during the interval 20.3 to 20.7 sec. This type of disturbance occurs with and without ESTATCOM. From this figure we observed that when ESTATCOM not installed the active power drawn from the source increased during that interval and there is an oscillation in active power after the load returned to its normal condition. Where installing the ESTATCOM cause the active power drown from source approximately not changing during the interval of disturbance and there is no oscillation after the load returned to its normal condition.

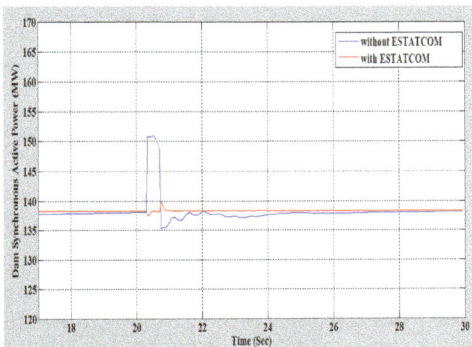

Figure (7) Mosul Dam Station Active Power

Figure (7) shows the active power drowns from the source1 (Mosul Dam Synchronous Generator) at load disturbance (load increased) during the interval 20.3 to 20.7 sec. This type of disturbance occurs with and without ESTATCOM. From this figure we observed that when ESTATCOM not installed the active power drawn from the source increased during that interval and there is an oscillation in active power after the load returned to its normal condition. Where installing the ESTATCOM cause the active power drown from source approximately not changing during the interval of disturbance and there is no oscillation after the load returned to its normal condition.

Where figure (8) shows the active power drown from the source2 (Mosul Gas Synchronous Generator) at load disturbance (load increased) during the interval 20.3 to 20.7 sec. This type of disturbance occurs with and without ESTATCOM.

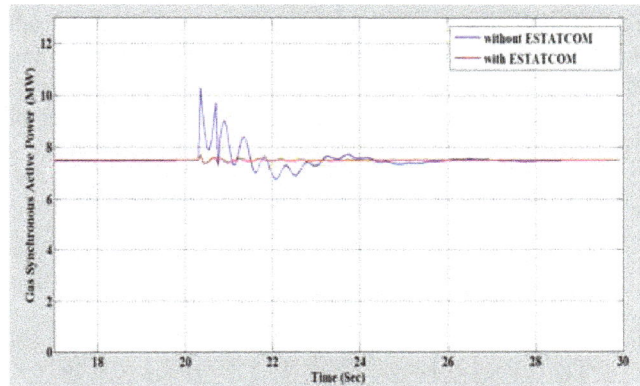

Figure (8) Mosul Gas Station Active Power

From this figure, we observed that when ESTATCOM not installed the active power drawn from the source increased during that interval and there is an oscillation in active power after the load returned to its normal condition. Where installing the ESTATCOM cause the active power drown from source approximately not changing during the interval of disturbance and there is no oscillation after the load returned to its normal condition.

Figure (9) shows the reactive power drowns from the source1 (Mosul Dam Synchronous Generator) at load disturbance (load increased) during the interval 20.3 to 20.7 sec. This type of disturbance occurs with and without ESTATCOM. From this figure we observed that when ESTATCOM not installed the reactive power drawn from the source increased during that interval, where installing the ESTATCOM cause the reactive power drown from source approximately not changing during the interval of disturbance.

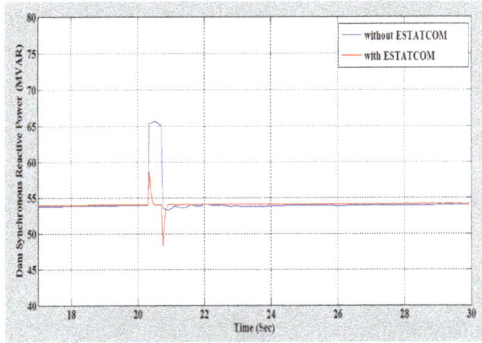

Figure (9) Mosul Dam Station Reactive Power

Where figure (10) shows the reactive power drown from the source2 (Mosul Gas Synchronous Generator) at load disturbance (load increased) during the interval 20.3 to 20.7 sec. This type of disturbance occurs with and without ESTATCOM. From this figure, we observed that when ESTATCOM not installed the reactive power drawn from the source increased during that interval. Where installing the ESTATCOM cause the reactive power drown from source approximately not changing during the interval of disturbance.

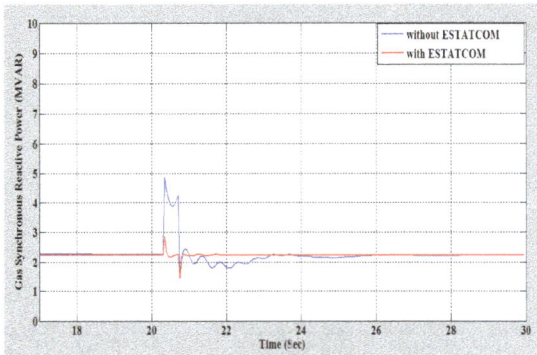

Figure (10) Mosul Gas Station Reactive Power

Figure (11) shows the ESTATCOM active and reactive power when the load increased between the interval 20.3 to 20.7 sec. From this figure, we observed that the ESTATCOM compensate the active and reactive power drown from the load during the interval of disturbance.

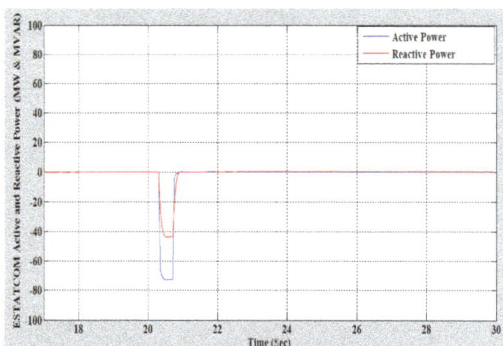

Figure (11) ESTATCOM Active and Reactive Power.

Where figure (12) and figure (13) shows the firing angle and the modulation index of the inverter used in ESTATCOM at the same condition above.

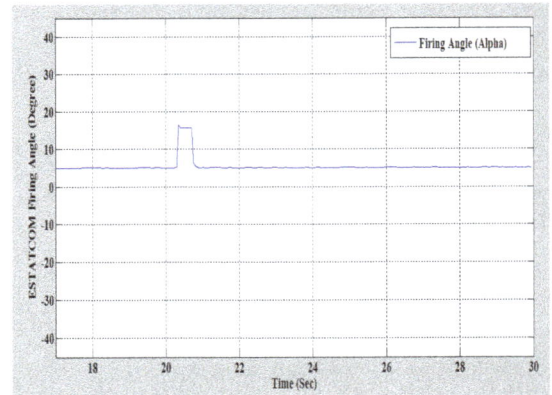

Figure (12) ESTATCOM firing angle

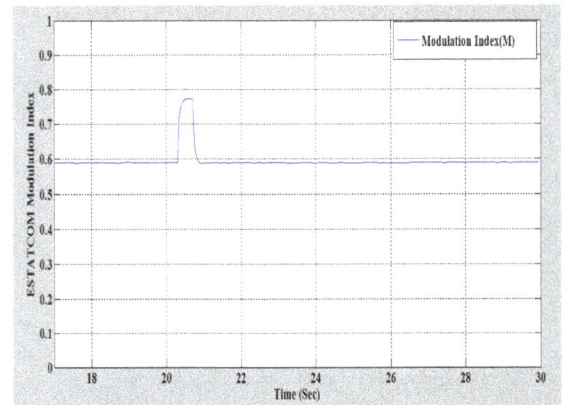

Figure (13) ESTATCOM Modulation Index

Figure (14) shows the current of energy storage element at load disturbance.

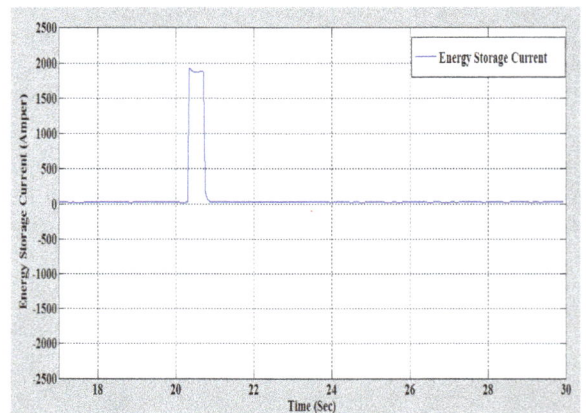

Figure (14) Energy storage current

From the figure above we observed that the ESTATCOM inject current to the system to compensate the active power and keep it constant during the interval of disturbance and prevent the oscillation occurs after the disturbance clearing.

5. Conclusions

A three phase two level PWM ESTATCOM with its controller has been inserted in the Mosul ring system and some disturbances have been studied via modeling and simulation. A conventional PI controller strategy which is simple and easy to implement has been adopted.

The simulation results have shown that the ESTATCOM with the proposed controller can decrease the active power oscillation of the power system during and after the load disturbances and keep the reactive power constant during the disturbance interval. ESTATCOM model has been developed with all the necessary components and controllers in order to demonstrate its effectiveness in maintaining a fast active and reactive power regulation at any bus bar. The simulation results were compared with that of the system without compensation. Simulation results have proved the ability of the ESTATCOM to respond to the system active and reactive power requirements with satisfactory performance. The Addition of the ESTATCOM at bus 37 caused the THD of the voltage at that bus to be 1.47% which is within IEEE-519 standards [7].

The performance of ESTATCOM with its controller was very close (within $\pm$ 98%) of the nominal value of voltage. The response time took about two cycles.

Appendix (1)

Parameters (Base- 100MVA)

IGBT-VSCs based two-level 12-Pulse, $\pm$ 100MVAR ESTATCOM.

A. Converter Parameters

IGBT-VSC Converters; No. of pulses-12; Nominal AC voltage-132kV; DC link voltage-12kV.

B. Transformer Units

Each 3-winding Transformer Rating: (100/3) MVA, 50Hz, 132kV/12kV, 12% (X).

C. PI-controller Gain

Reactive Power Regulator: $K_p = 0$, $K_I = 100$; AC Current Decoupled Regulator: D-axis controller: $K_{pd} = 3$, $K_{Id} = 1$; Q-axis controller: $K_{pq} = 3$, $K_{Iq} = 1$; Active Power Regulator: $K_p = 1$, $K_i = 70$.

D. Thevenin's Equivalent Voltage Source

Nominal Voltage: 132kV; Source impedance:1% System frequency: 50Hz; Short circuit level: 1000MVA; X/R ratio-7.

Appendix (2)

The model of Mosul ring with ESTATCOM using MATLAB SIMULINK

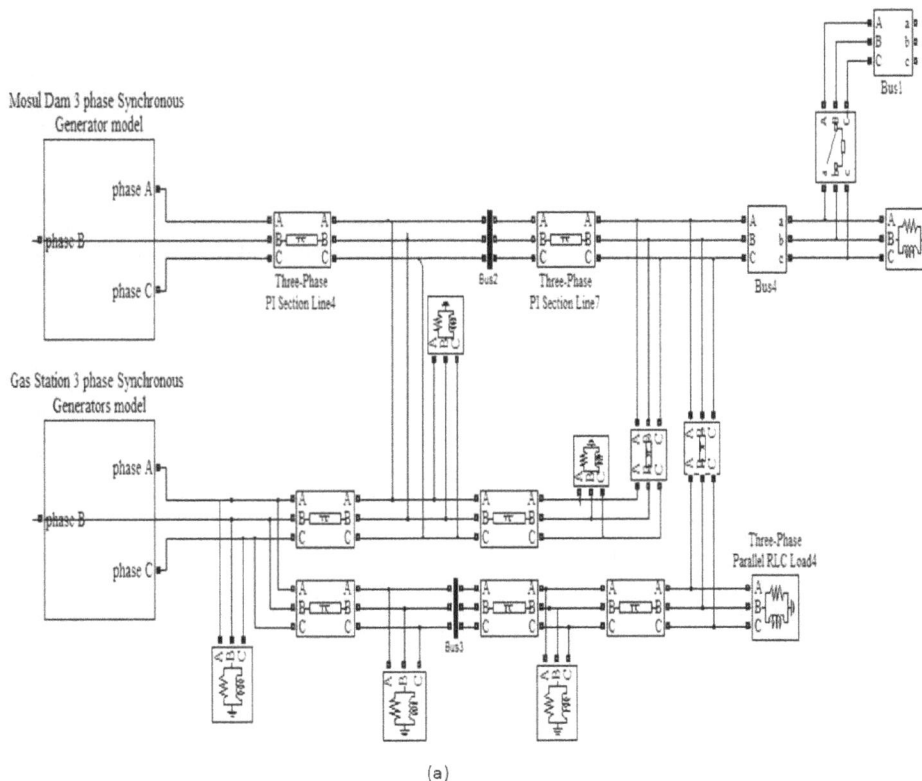

(a)

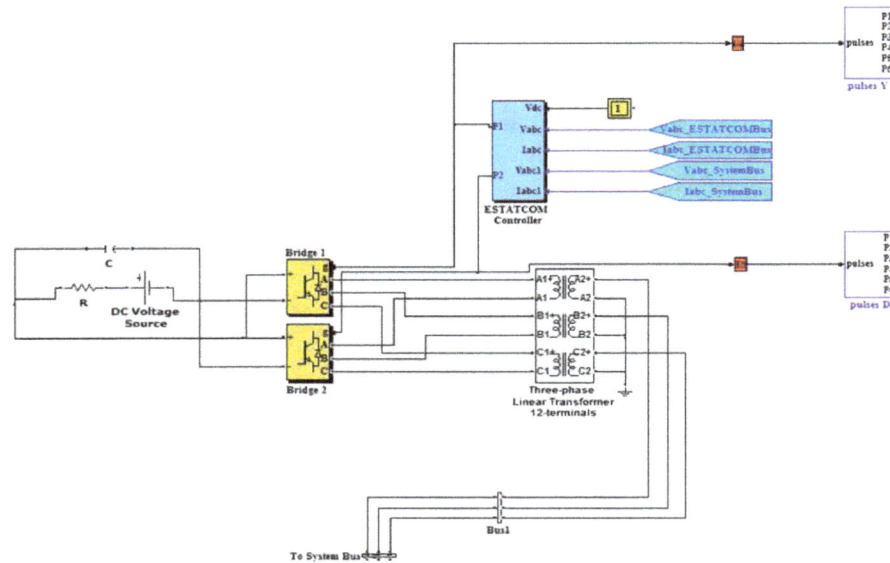

(b)

(a) The model of Mosul ring using MATLAB/SIMULINK
(b) ESTATCOM model using MATLAB/SIMULINK

References

[1]   Dr. Dhiya A. Al-Nimma, Dr. Majed S. M. Al-Hafid and Saad
       Enad Mohamed "Voltage Profile Improvements of Mosul
       City Ring System by STATCOM Reactive Power Control",
       Acemp – Electromotion 2011, 8 – 10 September 2011
       Istanbul – Turkey.

[2]   Dhiya A. Al-Nimma, Majed S. M. Al-Hafid and Saad Enad
       Mohamed "Simultaneous Active and Reactive Power
       Control of ESTATCOM for Improving Mosul City Ring
       System" The First National Conference for Engineering
       cience, Baghdad-Iraq, 7-8 Nov. 2012.

[3]   Majed S. M. Al-Hafid "Simulation of a Static Synchronous
       Compensator in the 132 KV Mosul Ring System" a PHD
       Thesis , University of Mosul - Iraq, 2006.

[4]   R. Kuiava, R. A. Ramos and N. G. Bretas, "Control Design
       of a STATCOM with Energy Storage System for Stability
       and Power Quality Improvements," IEEE International
       Conference on Industrial Object Identifier, 2009.

[5]   Vasudeo B. Virulkar and Mohan V. Aware, "Modeling and
       Control of DSTATCOM with BESS for Mitigation of
       Flicker, " Asian Power Electronics Journal, Vol. 4 No.1
       April 2010.

[6]   B. Singh and R. Saha, "A New 24-Pulse STATCOM for
       Voltage Regulation," International Conference on Power
       Electron. Drives and Energy Systems, 2006. PEDES '06,
       12-15 Dec. 06, pp. 1-5.

[7]   Alper Çetin and Muammer Ermi¸s, Member, IEEE,
       "VSC-Based D-STATCOM With Selective Harmonic
       Elimination," IEEE TRANSACTIONS ON INDUSTRY
       APPLICATIONS, VOL. 45, NO. 3, MAY/JUNE 2009.

8

Simulation of non linear adaptive observer for sensorless induction motor control

Benheniche Abdelhak[1,2], Bensaker Bachir[2]

[1]Electrotechnical Department, Mouloud Mammeri Tizi-Ouzou University
[2]Electronic Department, Badji Mokhtar Annaba University

Email address:

benheniche2006@yahoo.fr(B. Abdelhak), bensaker_bachir@yahoo.fr(B. Bachir)

Abstract: This paper presents a model reference adaptive system based sensorless induction motor drive. In this scheme, an adaptive full order flux observer is used. The simulation result show that with the large PI gain for the adaptive scheme, the convergence for the speed estimation is fast and very well, however higher harmonics and noises are included in the estimated speed. Usually noises caused by inverter. Simulation results show that proposed scheme can estimate the motor speed under various adaptive PI gains and estimated speed can replace to measured speed in sensorless induction motor.

Keywords: Induction Motor, Sensorless Control, Pole Placement, MRAS Speed Estimation, Lyapunov Function

1. Introduction

Three phase induction motor is widely used in many industries, mainly due to its rigidness, maintenance free operation, and relatively low cost. In contrast to the commutation dc motor, it can be used in aggressive or volatile environments since there are no risks of corrosion or sparks. However, induction motor constitute a theoretically challenging control problem since the dynamical system is non linear, the electrical rotor variable are not measurable, and the physical parameters are most often imprecisely known[1-2-3-6].

Machine model based methods of speed estimation have found a great interest among different speed estimation methods for their simplicity. They include different methods such as Luenberger observer (LO)[10-13], Model Reference Adaptive System (MRAS)[1-2-4-10-15]; Adaptive Flux Observer (AFO)[2-12-14-16]; Sliding Mode Observer (SMO)[17-18] Artificial Intelligence Techniques (AI)[19-20-21]; and Kalman Filer (KF)[22-23].Machine model-based methods are characterized by their simplicity and good performance at high and medium speeds. However at low speeds, they are problematic. The main limitations arise from instability problems associated with most speed estimation schemes at low speeds due to the change of machine parameters.

Adaptive flux observer is one of the machine model based methods of speed estimation of sensorless induction

motor drive. Parameters variations, low speed operation and the difficulty encountered in the design in the feedback gain and the adaptation mechanism are the most crucial aspects affecting the accuracy and stability of this method. This unstable region of AFO can be reduced by proper design of both the observer feedback gain and adaptive law using several techniques. . Instability problems of low speed regenerative mode of reduced-order observers and their remedies have been proposed in[24].

Many researches have been devoted to yielding better speed estimation of sensorless induction motor drives using AFO. However, there is a well known unstable region encountered at low speeds. One of the techniques to study the stability analysis of the speed estimation and simplify the structure of the sensorless control system by means of the using Routh–Hurwitz criterion[12].Or using Lyapunov theory[13]. Stability analysis of both rotor speed and stator resistance estimators for stable AFO[8] and in the regenerative mode at low speeds has been presented in[9-11].and parallel speed and stator resistance estimation algorithm based on a sliding mode current observer which combines variable structure control and Popov's hyper stability theories[7].

It is well known from control theory that a state estimator, called also state observer, is a dynamic system that is driven by the input-output of the considered system, estimate asymptotically its un-measurable state variable. It uses an adaptive mechanism involving as input, the error between the measured and estimated output value of the system. It is a "software sensor» that plays an important role in the estimation of the un-measurable state variables that are essential not only in the sensorless control techniques.

As au result, the drive has a wider adjustable speed range and can be operated at very low speed. This paper presents the simulation results of the proposed scheme which has been implanted on a 0.75KW induction motor driver. The systematic of this paper consist of six sections. Section 1 describes the overview of the observation and control system structure of induction motor under studies. Section 2 discusses with the model of adaptive flux observer; the model reference adaptive system is presented in section 3 while section 4 describes the adaptive scheme for speed estimation. Section 5 depicts the simulation results of the designed induction motor control. Finally, section 6 gives the conclusion of this paper.

2. Mathematical Models for Induction Motor and Adaptive Flux Observer

2.1. Dynamic Model of Induction Motor

For an induction motor, if the stator current i_s and rotor flux $\emptyset_r$ are selected as the state variables equations can be expressed as (1) in the stationary reference frame[2]

$$\frac{d}{dt}\begin{bmatrix} i_s \\ \emptyset_r \end{bmatrix} = \begin{bmatrix} A_{11} & A_{12} \\ A_{21} & A_{22} \end{bmatrix} \begin{bmatrix} i_s \\ \emptyset_r \end{bmatrix} + \begin{bmatrix} B_1 \\ 0 \end{bmatrix} v_s \qquad (1)$$

$$i_s = Cx \qquad (2)$$

Where

$i_s = [i_{ds} \quad i_{qs}]^T$ is stator current

$\emptyset_r = [\emptyset_{dr} \quad \emptyset_{qr}]^T$ is rotor flux

$v_s = [v_{ds} \quad v_{qs}]^T$ is stator voltage

$$x = [i_s \quad \emptyset_r]^T$$

$$A_{11} = -\left\{ \frac{R_s}{\sigma L_s} + \frac{1-\sigma}{\sigma \tau_r} \right\} I = a_{r11} I$$

$$A_{12} = \frac{L_m}{\sigma L_s L_r} \left\{ \frac{1}{\tau_r} I - \omega_r J \right\} = a_{r12} I + a_{i12} J$$

$$A_{21} = \left(\frac{L_m}{\tau_r} \right) I = a_{r21} I$$

$$A_{22} = \left(\frac{1}{\tau_r} \right) I + \omega_r J$$

$$B_1 = \frac{1}{\sigma L_s} I, \qquad C = [I \quad 0]$$

$\sigma = 1 - \frac{L_m^2}{L_s L_r}$ Is the inductance leakage coefficient

$$I = \begin{bmatrix} 1 & 0 \\ 0 & 1 \end{bmatrix} \quad and \quad J = \begin{bmatrix} 0 & -1 \\ 1 & 0 \end{bmatrix}$$

Where R_s, R_r and L_s, L_r are stator and rotor resistances and self inductances, respectively, L_m is mutual inductance, τ_r is the rotor time constant $\frac{L_r}{R_r}$ and ω_r is electrical motor angular speed.

The electromechanical equation of induction motor is given by

$$T_e = \frac{3}{2} \frac{P}{2} \frac{L_m}{L_r} \left(\emptyset_{dr} i_{qs} - \emptyset_{qr} i_{ds} \right) \qquad (3)$$

3. Model Reference Adaptive System

The model reference adaptive system (MRAS) is one of the major approaches for adaptive control[4][6]. The model reference adaptive system (MRAS) is one of many promising techniques employed in adaptive control. Among various types of adaptive system configuration, MRAS is important since it leads to relatively easy-to-implement systems with high speed of adaptation for a wide range of applications. Theoretically MRAS computes a desired state (called as the functional candidate) using two different models (i.e. reference and adjustable models). The error between the two models is used to estimate an unknown parameter (here speed is the unknown parameter). A condition to form the MRAS is that the adjustable model should only depend on the unknown parameter. Here, the reference model is independent of rotor speed, whereas the adjustable model is dependent on the same. The error signal is fed to the adaptation mechanism. The output of the adaptation mechanism is the estimated quantity ($\hat{\omega}_{rest}$), which is used for the tuning in adjustable model and also for feedback. The stability of such closed loop estimator is achieved through Popov's Hyper stability criterion[1-2].

Several other approaches such as variable structure-based technique, passivity based technique, etc. are also reported to estimate the speed of a PMSM drive. The more recent approach based on Artificial Intelligence (AI) are the Artificial Neural Networks (ANN)[21] and Fuzzy Logic[22] for speed estimation. But, the AI-based methods require huge memory and involve computational complexity.

Out of all the techniques discussed so far, MRAS is widely accepted for speed estimation due to its simplicity and good stability. Also the method does not require any extra hardwire or signal injection or huge memory like EKF or ELO.

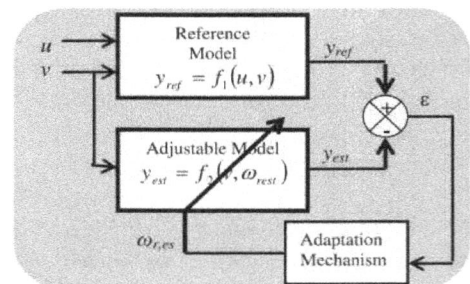

Figure 1. *Basic configuration of a parallel adaptive system.*

2.2. Adaptive Flux Observer

The APFO flux observer can be written as fellows[1]:

$$\frac{d}{dt}\hat{i}_s = \hat{A}_{11}\hat{i}_s + \hat{A}_{12}\hat{\emptyset}_r + Bv_s + G(\hat{i}_s - i_s) \quad (4)$$

$$\frac{d}{dt}\hat{\emptyset}_r = \hat{A}_{21}\hat{i}_s + A_{22}\hat{\emptyset}_r \quad (5)$$

Where i_s and v_s are measured values of stator current vector and stator voltage vector, respectively, G is the full order observer gain matrix which is also determined to make (3) stable and "^"denote the estimated values. The observer is the closed loop system, which is obtained by driving estimated model of the induction motor by the residual of the current measurement, e_{is}.

$$e_{is} = i_s - \hat{i}_s \quad (6)$$

The estimation of stator currents is conducted by a closed loop observer with[4X2] feedback gain matrix G as in (3), whereas the estimation of the rotor fluxes is carried out by an open loop observer of (4) without the flux error. Therefore, the real and estimated rotor fluxes are assumed the same.

$$\emptyset_r = \hat{\emptyset}_r \quad (7)$$

The observer gain matrix is chosen as:

$$G = \begin{bmatrix} g_1 & g_2 & g_3 & g_4 \\ -g_2 & g_1 & -g_4 & g_3 \end{bmatrix}^T \quad (8)$$

Where the observer gain matrix G is calculated based on the pole placement technique. The selection of the observer poles is a compromise between the rapidity of error responses and the sensitivity to disturbances and measurement noises. In practice, the eigen-values of the observer are selected to be negative, so that the state of the observer will converge to the state of the observed system, and they are chosen to be somewhat more negative than the eigen-values of the observed system so that convergence is faster than other system effects. Based on the above mentioned criteria the author chose[1],

$$g_1 = (k-1)(a_{r11} + a_{r22}) \quad (9)$$

$$g_2 = (k-1)a_{i11} \quad (10)$$

$$g_3 = (k^2-1)(ca_{r11} + a_{r21}) - c(k-1)(a_{r11} + a_{r22}) \quad (11)$$

$$g_4 = c(k-1)a_{i22} \quad (12)$$

$$c = \sigma L_s L_r / L_m \quad (13)$$

4. Adaptive Scheme for Speed Estimation

Consider the Lyapunov function candidate[9]:

$$V = V_1 + V_2 \quad (14)$$

$$V_1 = e^T e \quad V_2 = \frac{e_\omega^2}{\lambda} \quad (15)$$

With $(\lambda > 0)$, is the positive constant ensuring the positive definiteness of V_2 and which will be tuned in (19) to improve observer dynamics. $e_\omega = \omega_r - \hat{\omega}_r$ and

$$e^T = \begin{bmatrix} i_{sd} - \hat{i}_{sd} & i_{sq} - \hat{i}_{sq} & 0 & 0 \end{bmatrix}$$ Because we supposed that $\emptyset_r = \hat{\emptyset}_r$

The derivatives of this lyapunov candidate function in thus:

$$\frac{dV}{dt} = e^T[(A-GC)^T + (A-GC)]e$$
$$-2(\omega_r - \hat{\omega}_r)\left[K(e_{isd}\hat{\emptyset}_{rq} - e_{isq}\hat{\emptyset}_{rd}) - \frac{1}{\lambda}\frac{d}{dt}\hat{\omega}_r\right] \quad (16)$$

$$e^T[(A-GC)^T + (A-GC)]e < -Q \quad (17)$$

With $Q = \varepsilon I_n$ and $\varepsilon > 0$

The stability of adaptive observer has proved if we respect two conditions as follows:

The eigen-value of the observer are selected to have negative real parts so that the states of the observer will converge to the desired states of the observed system. The term in factor of $(\omega_r - \hat{\omega}_r)$ in the equation (16) must be zero. The expression of the derivative of estimated speed becomes then:

$$K(e_{isd}\hat{\emptyset}_{rq} - e_{isq}\hat{\emptyset}_{rd}) - \frac{1}{\lambda}\frac{d}{dt}\hat{\omega}_r = 0 \quad (18)$$

$$\frac{d}{dt}\hat{\omega}_r = \lambda K(e_{isd}\hat{\emptyset}_{rq} - e_{isq}\hat{\emptyset}_{rd}) \quad (19)$$

However this adaptive law of the speed

$$\hat{\omega}_r = K_i \int_0^t (e_{isd}\hat{\emptyset}_{rq} - e_{isq}\hat{\emptyset}_{rd})dt \quad (20)$$

Has obtained for the satatorique frame his dynamic has adjusted by K_i (finite positive constant). For augmented the dynamic of this observer during the transitory phase of rotor speed, we estim the speed by large PI regulator; we added a supplementary term proportional of error. Then

$$\hat{\omega}_r = K_p(e_{isd}\hat{\emptyset}_{rq} - e_{isq}\hat{\emptyset}_{rd}) + K_i \int_0^t (e_{isd}\hat{\emptyset}_{rq} - e_{isq}\hat{\emptyset}_{rd})dt \quad (21)$$

Where K_p and K_i are adaptive gains for speed estimator. An identification system for speed is shown in Fig.2, which is constructed from a linear time-invariant forward block and a nonlinear time-varying feedback block.

5. Simulation Results

The basic configuration of speed estimation of sensorless induction motor drive is shown in figure (2).this configuration will be used for both simulations. All reference or command preset values are subscripted with a " * " in the diagram.IM speed will be estimated by (21) and will be compared with the reference speed in order to create the error speed. The proposed full order flux observer for induction motor states estimations has been developed and applied in the direct field oriented control of induction motor.

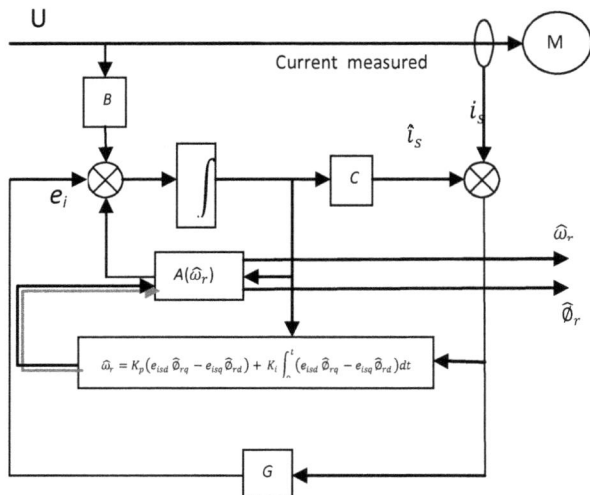

Figure 2. *Speed adaptive observer*

Table 1. *Induction motor parameters*

Symbol	Quantity	N. Values
P_a	Power	O.75KW
F	Supply frequency	50HZ
P	Number of pair poles	2
V	Supply voltage	220V
R_s	Stator resistance	10Ω
R_r	Rotor resistance	6.3Ω
L_s	Stator inductance	0.4642H
L_r	Rotor inductance	0.4612H
L_m	Mutual inductance	0.4212H
ω_r	Rotor angular velocity	157rd/s
J	Inertia coefficient	0.02Kg²/s
f	Friction coefficient	0N.s/rd

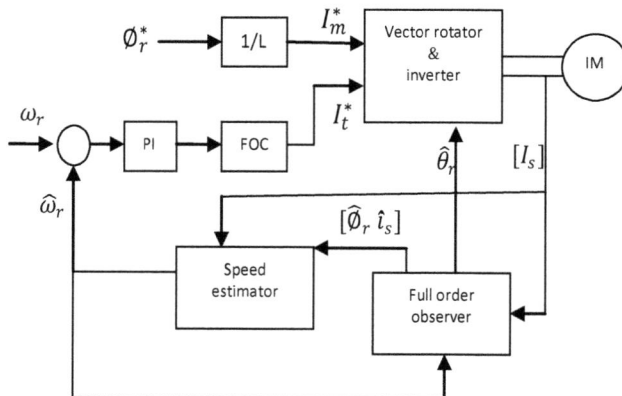

Figure 3. *Block diagram of sensorless IM drive*

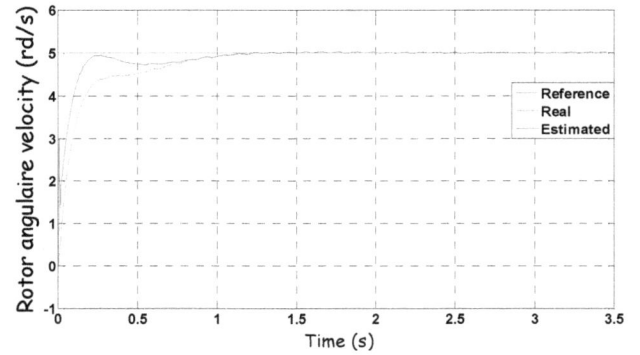

Figure 4. *Reference, measured and observed rotor angular velocity (very low speed)*

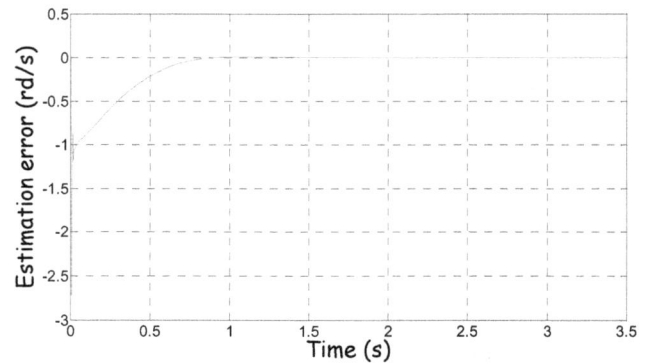

Figure 5. *Speed estimation error*

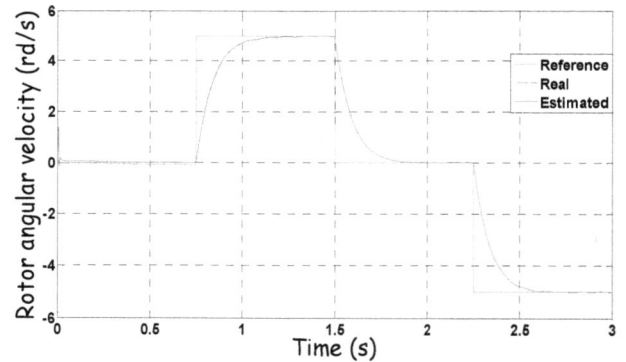

Figure 6. *Reference measured and observed rotor angular velocity during step change of speed reference from[0:5:0:-5] (very low and zero speed)*

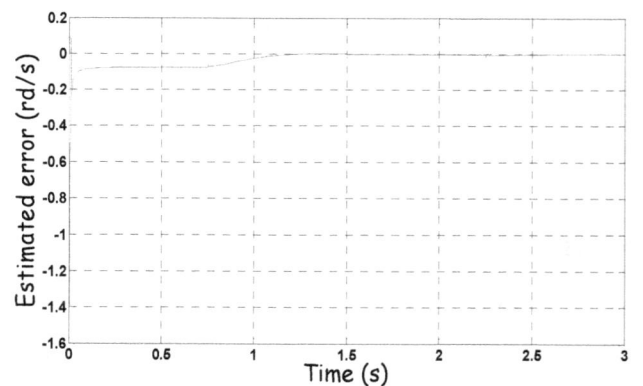

Figure 7. *Speed estimation error*

In fig.4-5 show the behavior of IM speed estimation and estimation error where the induction motor rotates at a constant very low speed (5 rd/s), the observed speed also converge to real speed at the time (0.8s for 5 rd/s) .these figures show that with large PI gain for adaptive scheme, the convergence for speed estimation is fast.

In fig.6-7 show the behavior of IM speed estimation and estimation error under various command speed , where the command speed is first set at a zero speed (0 rd/s), at (0.75s) the reference speed is changed to (5 rd/s) (very low speed),.at (1s) the reference speed is changed to (0rd/s) (zero speed),finally at (2.25s) the command speed is changed to (-5rd/s) .the proposed observer confront no problem in the low speed region. Furthermore the algorithm of speed identification scheme is characterized by their simplicity and small computation time.

6. Conclusion

This paper presents a MRAS-based adaptive full-order observer (AFFO) sensorless induction motor drive. This method has been applied to a direct field-oriented induction motor control without speed sensor. The simulation results demonstrated that with PI gains for the adaptive regulators, the convergence for the speed estimation is fast and well. The proposed observer can work accurately at very low speed region.

References

[1]   H. M. Kojabadi, L. Chang, "Comparative study of pole placement method in adaptive flux observers" control engineering practice 13 (2005) [749-757]

[2]   H. M. Kojabadi, "Simulation and experimental studies of model reference adaptive system for sensorless induction motor drive", Simulation Modeling Practice and Theory 13 (2005) [451-464]

[3]   H. Kubota, K. Matsuse, T. Nakano, " DSP-based speed adaptive flux observer of induction motor", IEEE Trans. Ind. Appl. 29 (1993) [344-348]

[4]   D. P. Marcetic, S. N. Vukasavic, " Speed-Sensorless AC Drive With the Rotor Time Constant Parameter Update", IEEE Trans on industrial electronics, VOL. 54,NO. 5, October 2007.

[5]   J. Boker, "State of the Art of Induction Motor Control", IEEE, Shashidhar Mathapati University Peterborn,Warburger Str. 100 D-33098 Pederborn, Germany (2007).M. Ale

[6]   E. LeKsono and PratiKto, "Adaptive Speed Control of Induction Motor With DSP Implementation", The 30th Annual Conference of the IEEE industrial electronics society,November 2-6, 2004,Busan, Koria?

[7]   M. S. Zaki, M.Khater, H. Yasin, S.S. Shokralla, "Very low speed and zero speed estimations of sensorless induction motor drives", Electric Power Systems Research 80 (2010) 143-151.Contents lists available at ScienceDirect.

[8]   M. S. Zaki, "A stable adaptive flux observer for a very low speed-sensorless induction motor drives insensitive to stator resistance variations", Ain Shams Engineering Journal (2011) 2, 11-20 Production and hosting by Elsevier.

[9]   Zaky MS. Stability analysis of simultaneous estimation of speed and stator resistance for speed-sensorless induction motor drives. In: Proceedings of 13th International Middle-East Power Systems Conference (MEPCON'2010), Cairo University, Egypt, 2010. p. 364–70.

[10] A. Paladugu, B. H. Chowdhury,"Sensorless control of inverter-fed induction motor drives", Electric Power Systems Research 77 (2007) 619-629.

[11] A. Bouhenna, C. Chaigne, N. Bensiali, E. Etien, "Design of speed adaptation law in sensorless control of induction motor in regenerating mode", Simulation Modiling Practice and Theory 15 (2007) 847-863.

[12] Kubota H, Sato I, Tamura Y, Matsuse K, Ohta H, Hori Y. Regenerating-mode low-speed operation of sensorless induction motor drive with adaptive observer. IEEE Trans Ind Appl 2002;38(4):1081–6.

[13] M.Juili, K.Jarray, Y.Koubaa, M.Boussak, "Lenberger state observer for speed sensorless ISFOC induction motor drives", Electric Power Systems Research 89 (2012) 139-147.

[14] Suwankawin S, Sangwongwanich S. Design strategy of an adaptive full-order observer for speed-sensorless induction-motor drives-tracking performance and stabilization. IEEE Trans Ind Electron 2006;53(1):96–119

[15] Kowalska TO, Dybkowski M. Stator-current-based MRAS estimator for a wide range speed-sensorless induction-motor drive. IEEE Trans Ind Electron 2010;57(4):1296–308

[16] Etien E, Chaigne C, Bensiali N. On the stability of full adaptive observer for induction motor in regenerating mode. IEEE Trans Ind Electron 10;57(5):1599–608

[17] S.M.Kim, W.Y.Han, S.J.Kim,"Design of a new adaptive sliding mode observer for sensorless induction motor drive", Electric Power Systems Research 70 (2004) 16-22.

[18] N.Inanc."A robust sliding mode flux and speed observer for speed sensorless control of an idirect field oriented induction motor drives" ", Electric Power Systems Research 77 (2007) 1681-1688.

[19] Solly aryza;Ahmed N Abdellah, ZulkefleeKhalidin, Zulkarnian Loubis, "Adaptive speed estimation of induction motor based on neural network inverse control"advanced in control engineering and information science,procedia engineering 15(2011)4188-4193

[20] Hasan A. Yousef*, Manal A. Wahba "Adaptive fuzzy mimo control of induction motors" Expert Systems with Applications 36 (2009) 4171–4175

[21] Bharat Bhushan, Madhusudan Singh, Prem Prakash "Performance Analysis of Field Oriented Induction Motor using Fuzzy PI and Fuzzy Logic based Model Reference Adaptive Control", International Journal of Computer Applications (0975 – 8887) Volume 17– No.4, March 2011

[22] Salomón Chávez Velázquez, Rubén Alejos alomares, Alfredo Nava Segura "Speed Estimation for an Induction Motor Using the Extended Kalman Filter", Proceedings of

the 14th International Conference on Electronics, Communications and Computers (CONIELECOMP'04) 0-7695-2074-X/04 $ 20.00 © 2004 IEEE

[23] Americo Vicente Leite, Rui Esteves Araujo, , and Diamantino Freitas "A New Approach for Speed Estimation in Induction Motor Drives Based on a Reduced-Order Extended Kalman Filter" 0-7803-8304-4/04/$20.00 C02004

IEEE

[24] Harnefors L. "Instability phenomena and remedies in sensorless indirect field oriented control". IEEE Trans Power Electron 2000;15(4):733–43.

9

Critical clearing time evaluation of Nigerian 330kV transmission system

Adepoju Gafari Abiola[1], Tijani Muhammed Adekilekun[2]

[1]Electronic and Electrical Engineering Department, LAUTECH, Ogbomoso, Nigeria
[2]Electrical and Electronics Engineering Department, Federal Polytechnic, Ede, Nigeria

Email address:
agafar@justice.com (A. Gafari), muhammedtijani@gmail.com (T. Muhammed)

Abstract: Critical Clearing Time (CCT) is the largest possible time for which a power system is allowed to remain in fault condition without losing stability. Appropriate CCTs settings of protective equipments on power system greatly determine the reliability of power supply. This paper determines the CCTs for all the transmission lines in the Nigerian 24-bus, 39-lines 330kV transmission system. The Transient Stability Analysis (TSA) program adopted used the method of partitioned approach with explicit integration method. The result of TSA was considered satisfactory since about 87% of the values obtained fall within acceptable international range. It was concluded that the determination of appropriate CCTs for the Nigerian power system will enhance the operation of the power system by limiting effects of faults on the power system.

Keywords: Power System, Transient Stability Analysis, Critical Clearing Time

1. Introduction

The Nigerian 330kV transmission network links the generating stations and distribution system. Interruptions in this network hinder the flow of power to the load. The cost of losing synchronism through transient instability is extremely high [1]. The quality of electricity supply is measured, amongst other factors, by the ability of the power system to clear faults before they cause damage to the power system equipments. The time at which fault is cleared before it causes damage on the power system is known as Critical Clearing Time (CCT). CCT is the largest possible time for which a power system can remain in fault condition without losing stability once the fault is cleared [2, 3, 4]. Constant faults on Nigerian power system have had adverse effects on the Nigerian economy and its citizenry. The yearnings and aspirations of the Nigerians for constant supply of electricity can partly be met if Nigerian 330kV National Grid is operated to clear faults before damages are caused by the faults given adequate generation, transmission and distribution facilities. Hence, the need to determine appropriate CCT for the circuit breakers on the power system. CCT is determined through the performance of Transient Stability Analysis (TSA) of the power system.

TSA is the evaluation of the stability of a power system when there is large and sudden disturbance on the power system. This disturbance can be a fault which includes; transmission line short-circuit, loss of generator, load or a part of the transmission network and gain of load [5]. The generators in the power system respond to the occurrence of these disturbances with large swinging of their rotor angles. Initial operating conditions and the severity of the disturbance determine the stability of a power system [6].

Transient stability problems are concerned mainly with the behaviors of synchronous machines in power system after they have been perturbed. When a fault occurs on the power system, an imbalance is created between the generator output and the load. The rotor angle of the machine accelerates beyond the synchronous speed which is the reference speed, for a time greater than zero. When this happens, the machine is said to be "swinging" and two possibilities are identified when the rotor angle is plotted as a function of time [7].

i. The rotor angle swings and eventually settles at a new angle. The system is said to be stable i.e. in synchronism.
ii. The rotor angle swings and the relative rotor angle diverge as time increases. This condition is considered unstable i.e. losing synchronism.

The single line diagram of the Nigerian 24-bus, 330kV transmission system considered is shown in Figure 1 [8].

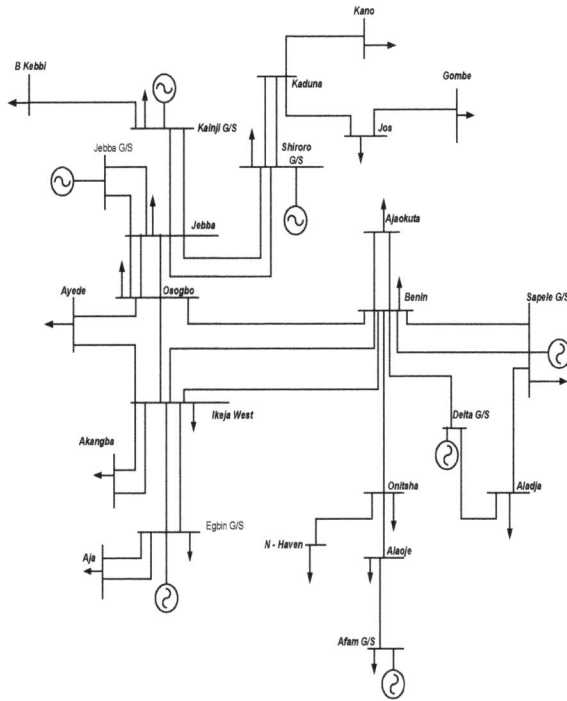

Figure 1. Nigerian 24 Bus 330kV National Grid [8].

2. Methodology

The initiation of fault and its removal by circuit breakers in a power system is considered that the system is going through a change in system configuration in three stages: pre-fault, fault and post-fault stages. The dynamics of the power system during fault and post-fault periods are nonlinear and the exact solution is too complex. In order to reduce the complexity of the transient stability analysis, the following simplified assumptions are made [7];

i. Every synchronous machine in the power system is modeled by a constant voltage source at the back of direct axis transient reactance.

ii. The governor's actions for the automatic control generation are neglected and the input powers are assumed to remain constant.

iii. Using the pre-fault bus voltages, all loads are converted to equivalent admittances to ground and are assumed to remain constant.

iv. Damping or asynchronous powers are ignored.

v. The mechanical rotor angle of each machine coincides with the angle of the voltage behind the machine reactance.

vi. Machines belonging to the same station swing together and are said to be coherent. A group of coherent machines is represented by one equivalent machine.

In transient stability studies, particularly those involving short periods of analysis in the order of a second or less, a synchronous machine can be represented by a voltage source behind transient reactance that is constant in magnitude but changes its angular position [9]. This representation neglects the effects of saliency and assumes a constant flux linkages

and a small change in speed. The voltage behind the transient reactance is determined from the following equation (1) [7, 9, 10].

$$E_i' = V_i + jX_d'I_i \qquad (1)$$

Where
E_i' = Voltage behind transient reactance.
$V_{i,}$ = Machine terminal voltage
X_d' = direct axis transient reactance.
I_i = Machine terminal current

The rotor mechanical dynamics are represented by the following equations [3, 11]..

$$2H\frac{d\omega}{dt} = T_m - T_e - D\omega \qquad (2)$$

$$\frac{d\delta}{dt} = \omega \qquad (3)$$

Where H = per unit Inertial Constant
D = Damping coefficient
ω = rotor angle of the generator
δ = angular speed of the generator
T_m = Mechanical Torque Input
T_e = Electrical Torque output

Numerical integration techniques are used to solve the swing equation for multimachine stability problems. The Modified Euler's method is used to compute machine power angles and speeds in this research work. The real electrical power output of each machine is computed by the following equations.

$$P_e = Real[E_nI_n^*], \quad n = 1, 2, \dots, m \qquad (4)$$

$$P_e = \sum_{j=1}^n |E_i'||E_j'||Y_{ij}|\cos(\theta_{ij} - \delta_i - \delta_j) \qquad (5)$$

The above equations (1) to (5) are very crucial to transient stability studies because they are used to calculate the output power of each machine in the power system.

The individual models of the generators and the system load given by the differential and algebraic equations have been stated. Together, these equations form a complete mathematical model of the system, which are solved numerically to simulate system behaviors. To develop a power system dynamic simulation, the equations used to model the different elements are collected together to form: [6, 10, 12].

(i)   A set of differential equations

$$\dot{x} = F(x, y) \qquad (6)$$

That describes the system dynamics, primarily contributed by the generating units and the dynamic loads.

(ii)   A set of algebraic equations

$$0 = g(x, y) \qquad (7)$$

That describes the network, static loads and the algebraic

equations of generators.

The solutions of these two sets of equations define the electromechanical state of the power system at any instant in time. Equations (6) and (7) can be solved using either a Partition Solution method or a Simultaneous Solution method. In the partitioned solution method, the differential equations are solved using a standard explicit numerical integration method with algebraic equation (7) being solved separately at each time step. The simultaneous solution uses implicit integration methods to convert the differential equations of (6) into a set of algebraic equations which are combined with the algebraic network equations of (7) to be solved as one set of simultaneous algebraic equations [6, 11, 12].

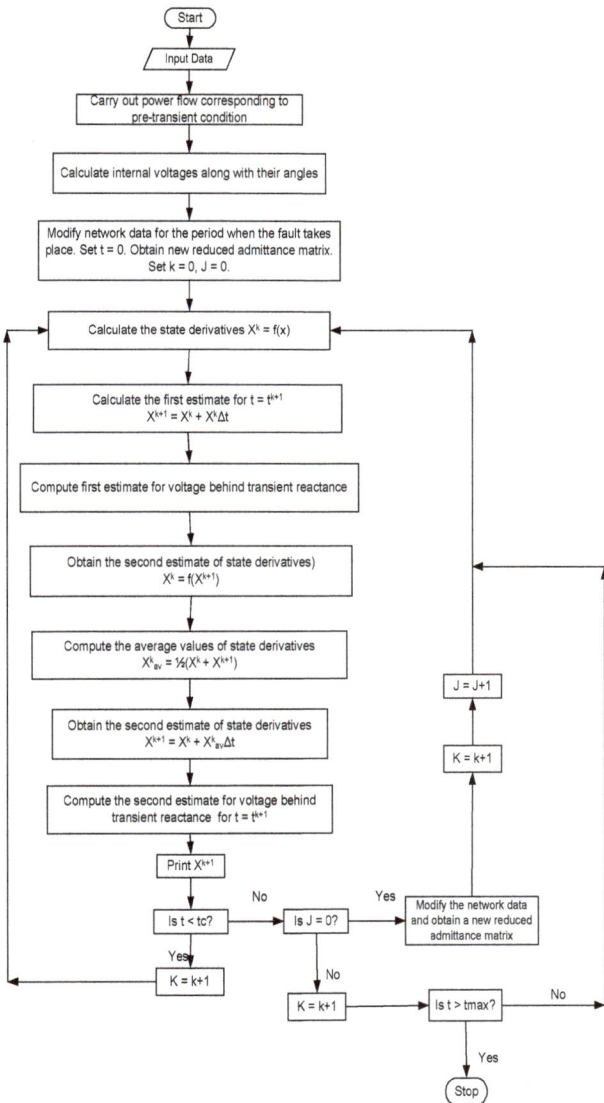

Figure 2. Modified Euler's Method Applied to Transient Stability sstability Problems [13]

The partitioned approach with explicit integration method is the traditional approach used widely in production grade stability program [6]. The transient stability program adopted in this research work uses the method of partitioned approach with explicit integration method. The advantages of the method include programming flexibility and simplicity, reliability and robustness. Figure 2 shows the flowchart for the transient stability solution using the Modified Euler's method [13].

3. Results and Discussion

The Nigerian power system considered is a 24 bus system which has seven (7) generators and thirty-nine (39) transmission lines. Three phase fault was simulated on each bus and different lines removed to determine the stability or otherwise of the power system. The critical clearing time, which is a measure of the stability, was determined by varying the fault clearing times. At any time greater than the critical clearing time, the system becomes unstable. The stability and instability of the power system at a given fault is determined by the behavior of the generators. If the rotor

Angles of the generators diverge, the system is unstable and if otherwise, the system is stable.

Figures 3, and 4 show the behaviors of the generators, with generator 7 (Egbin) as reference, on the power system when three-phase faults occur on Buses 3 (AJA) and 9 (AYEDE) and lines L1 (AJA to EGBIN) and L20 (AYEDE to OSOGBO) are removed respectively for faults cleared at the critical clearing times. The generators swing together to show stable equilibrium. Figures 5, and 6 show the behaviors of the generators, with generator 7 as reference, on the power system when faults occur on Buses 3 (AJA) and

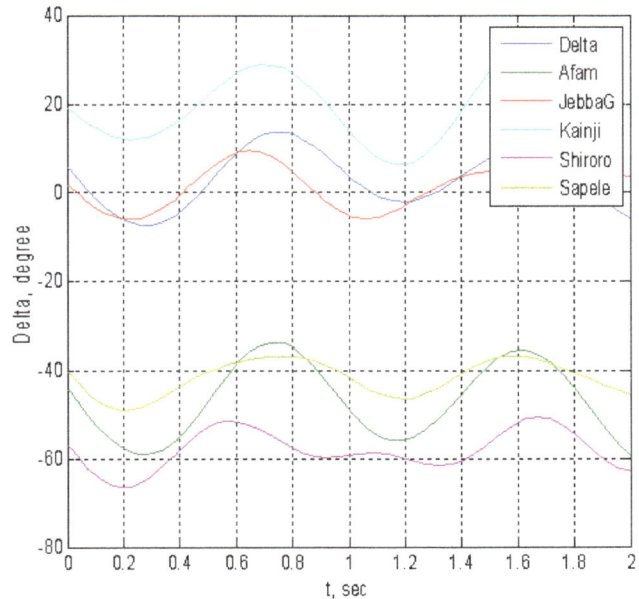

Figure 3. Generator rotor angle behaviour for fault at bus3 and line L1 removed at 0.01s critical clearing time (stable condition)

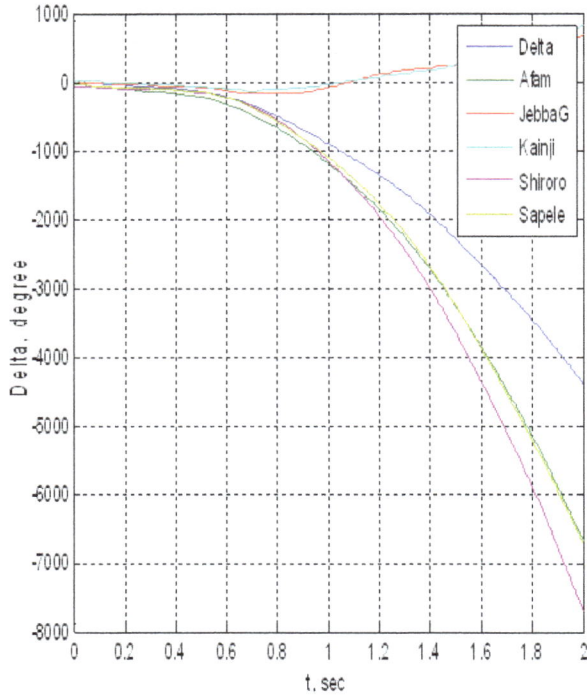

Figure 4. *Generator rotor angle behaviour for fault at bus 3 and line L1 removed at 0.05s critical clearing time (unstable condition)*

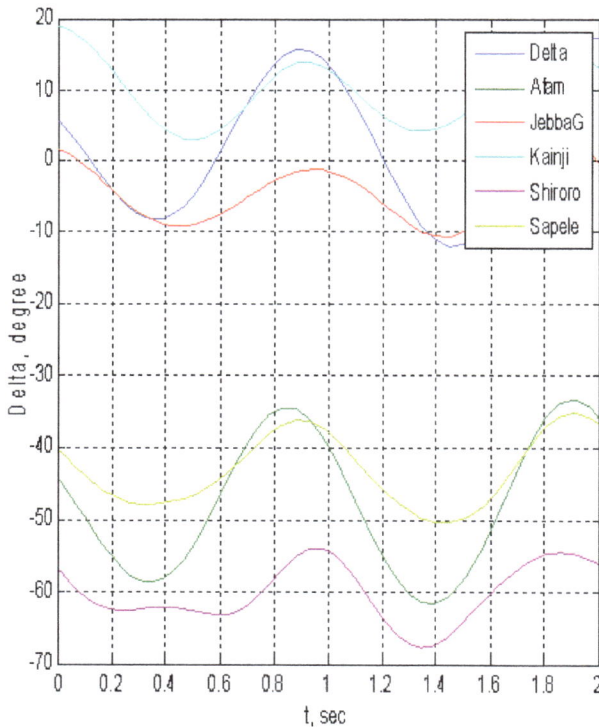

Figure 6. *Generator rotor angle behaviour for fault at bus 3 and line L1 removed at 0.08s critical clearing time (unstable condition)*

9 (AYEDE) and lines L1 (AJA to EGBIN) and L20 (AYEDE to OSOGBO) are removed respectively for faults cleared at times greater than the critical clearing times. It was observed that the rotor angle diverges and hence the system is unstable.

Figures 3 and 4 show that the generators on the Nigerian 330kV Transmission Grid System are grouped into two. Buses 2 (Delta), 18 (JebbaG) and 21 (Kainji) formed the first group while buses 11 (Afam), 23 (Shiroro) and 24 (Sapele) are in the second group. Machines in each group swing together with respect to the reference machine (Bus 1). The system is stable from these figures. Figures 5 and 6 show the angular positions of machines for a clearing time greater than the critical clearing time. In these cases, the clearing times increase beyond the critical clearing times and the machines go out of steps as seen in the figures. It is found that the critical clearing time of the fault on Bus 18 (JebbaG) is lower than those of buses 3 (Aja) and 9 (Ayede). This can be interpreted in such a way that faults closer to generating stations must be cleared rapidly than faults on the lines far from the generating stations.

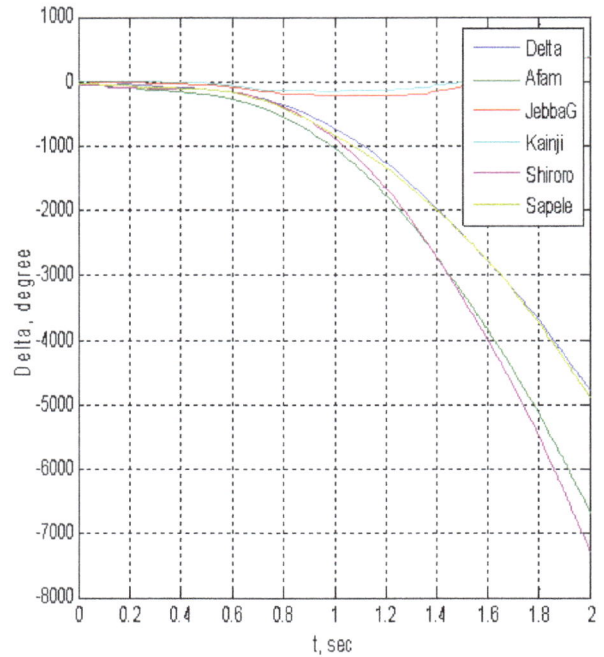

Figure 5. *Generator rotor angle behaviour for fault at bus 3 and line L1 removed at 0.01s critical clearing time (stable condition)*

Table 1 shows the critical clearing times in milliseconds determined for all the thirty-nine lines on the Nigerian 330kV National Grid. Starting with a clearing time of 0.01seconds, transient stability analysis program is run and if the conditions proved stable, another run is made with a higher value; if the second run showed the system to be unstable, then the clearing time of the first run gives the desired result. If the second run is still stable then more runs are made until the system becomes unstable. In some cases, a lower value of time is tested if 0.01 proved to be unstable. Comparing the results in Table 1 with Table 2 [14] which is a standard Range of typical EHV Relay-Breaker Clearing

Time, it was observed that all the relay breaker clearing times fall below the normal Critical Clearing Times but 87% fall below the total back-up clearing time as shown in Table 1.

4. Conclusion

The transient stability analysis was carried out to determine the critical clearing time for the circuit breakers on the power system. The partitioned solution method with explicit Modified Euler's integration method was used in

Table 1. Critical clearing time for faults on Nigerian 330kV transmission system

Faulted bus	Removed lines	Critical clearing time, t_{cr} (milliseconds)
3	L1	20
	L2	20
4	L3	20
	L4	20
1	L5	20
	L6	20
5	L7	20
	L8	20
9	L9	60
10	L10	60
6	L11	40
	L12	40
2	L13	40
7	L14	20
24	L15	100
14	L16	60
8	L17	40
8	L18	20
	L19	20
9	L20	60
15	L21	20
17	L22	60
	L23	60
	L24	60
11	L25	60
	L26	60
12	L27	60
13	L28	100
16	L29	100
18	L30	20
	L31	20
23	L32	20
	L33	20
21	L34	20
	L35	20
20	L36	100
22	L37	100
20	L38	20
	L39	20

Table 2. Range of typical EHV relay-breaker clearing time [14]

Function	Time in cycles	Time (milliseconds)
Primary relay	0.85 – 1.25	17 – 25
Breaker clearing	2.5	50
Total normal clearing time	**3.35 – 3.75**	**67 – 75**
Breaker-failure detection	0.425 – 1.25	8.5 25
Coordination time	2.5 – 4.25	50 – 85
Auxiliary relay	0.425 – 0.85	8.5 – 17
Backup breaker clearing	1.75	35
Total backup clearing	**8.45 – 11.85**	**169 – 237**

solving the transient stability equations. The results of the transient analysis are considered satisfactory since the entire relay breaker clearing times fall below the normal Critical Clearing Times. The system is well protected in the first instant but the protection has to be improved upon so that 100% total back-up clearing will be achieved. It has been demonstrated in this research work that determination of appropriate CCTs for the Nigerian power system will enhance the operation of the system by limiting the effects of faults on the power system. The CCTs for three-phase faults of circuit breakers installed on the Nigerian 330kV Transmission Grid were established which serves as reference data for the use of power system experts and researchers.

References

[1] O. Anthony, and O. Ameze. "Transient stability assessment of the Nigerian 330kV network,". International Journal of Engineering, Vol. 4, Issue 5, pp 357 – 367, 2011.

[2] M.H. Haque, and A.H. Rahim. "Determination of first swing stability limit of multimachine power system through taylor series expansion," IEEE Proceedings on Power Application, Vol. 136, No. 6, November 1989, pp. 374 – 379.

[3] B. Boussahoua, and M. Boudour. "Critical clearing time evaluation of power system with UPFC by energetic method," Journal of Electrical Systems, Special Issue, No. 1, pp 85 – 88, 2009.

[4] Z. Eleschova, M. Smitkova, and A. Belan. "Evaluation of power system transient stability and definition of basic criterion," International Journal of Energy, Issue 1, Vol. 4, pp 9 – 16, 2019.

[5] P. Iyambo, and R. Tzenova. "Transient stability analysis of the IEEE 14 test bus electric power system," IEEE Transaction on Power Systems, Vol. 25, No. 3, pp 1 – 7, 2007.

[6] S. Ravi, and S. Siva. "Transient stability improvement using UPFC and SVC," ARPN Journal of Engineering and Applied Sciences, Vol. 2, No. 3, pp 28 – 45, 2007.

[7] H. Saadat. Power System Analysis. McGraw-Hill, New York, USA, 1999.

[8]   G. Adepoju, O. Komolafe and A. Aborishade. "Power flow analysis of the Nigerian transmission system incorporating FACTS controller," International Journal of Applied Science and Technology, Vol. 1, No. 5, pp 186 – 200, September 2011.

[9]   G.W. Stagg, and H. El-Abiad. Computer Methods in Power System Analysis. McGraw-Hill, Kogakusha, Tokyo, 1968.

[10] X. Wang, S. Yonghua, and I. Malcolm. Modern Power System Analysis. Springer , New York, USA, 2008.

[11] K. Prabha. Power System Stability and Control. McGraw-Hill, New York, USA, 1994.

[12] J. Machowsky, J. Bialek, S. Robak, and J. Bumby. Power System Dynamics Stability and Control. John Willey & Sons Ltd, Wiltshire, Great Britain, 2008.

[13] C.L. Wadhwa. Electrical Power System. John Wiley and Sons, New Delhi, India, 1991.

[14] G. Adepoju. Transient stability: a case study of NEPA 330kV transmission grid system. M.Sc, University of Lagos, Akoka, Nigeria, 2000.

10

A test for solid phase extracted polychlorinated biphenyls (PCBs) levels in transformer oil

E. A. Kamba, A. U. Itodo, E. Ogah

Department of Chemical Sciences, Federal University Wukari, Nigeria.

Email address:

eacambah@yahoo.com (E. A. Kamba), itodoson2002@gmail.com (A. U. Itodo)

Abstract: Polychlorinated biphenyls (PCBs), the synthetic electrical insulation fluid in transformers and capacitors, known to reduce the risk of fire hazards due to their high chemical stability and low flammability turned out to be environmentally hazardous. In this research, different techniques used to analyses Polychlorinated biphenyls (PCBs) in transformer oil including their qualification and identification processes have been outlined. 12 samples of transformer oils collected from various transformers from Kebbi state, Nigeria were investigated. Solid Phase Extraction (SPE) method was used to treat the oil samples and extracts were analyzed on Gas Chromatography with Electron Capture Detection (GC/ECD). No detectable PCBs were observed. To check the efficiency the SPE tubes, oil sample A10 was spiked with Mix 525 of PCB standard and treated. A full recovery of all the PCBs of Mix 525 was made even at a level as low as 50ng/ul with GC/MS. The results obtained in this study confirm that transformer oil from Kebbi state is well within the safe level of PCBs in accordance with EPA PCB Regulatory limited.

Keywords: Kebbi State, Transformer Oil, Polychlorinated Biphenyls, GC/MS, GC/ECD, Solid Phase Extraction

1. Introduction

Polychlorinated biphenyls (PCBs) are a group of organic compounds in which about 1-10 hydrogen atoms of biphenyl are replaced by chlorine atoms [1]. There are 209 distinct PCB compounds (known as congeners) each of which consists of a biphenyl molecule with a specific number of chlorine atoms attached[1]. PCBs have a general chemical formula of $C_{12}H_{10-x}CI_x$ (the x represents the 1-10 hydrogen atoms that can be replaced with chlorine atoms) [2][3][4]. PCBs were widely used as a dielectric fluid in electrical transformers and capacitors [4][5]. It was discovered in the 1930s, that transformers and capacitors needed a synthetic electrical insulation fluid to reduce the risk of fire hazards. This led to the production of PCBs which proved to have some performance and electrical strength to resist fire, and as expressed by electrical engineers "it enabled the transformers to be positioned anywhere" [6].

Due to their high chemical stability and low flammability and some other desirable physical properties, their commercial utility increased [2]. Unfortunately, PCBs turned out to be environmentally hazardous which led to the banning of their production [7].

PCBs are classified as Persistent Organic Pollutants (POP), which when in contact with the environment are not easily degradable [8][5]. PCBs find their way into the environment mainly through industrial processes, urban incineration, improper disposal; practices, re-use of incompletely reconditioned oil and hazardous waste accidents; a large amount is also believed to have leaked out of transformers into the environment unintentionally, [9][10][5][3]. Previously PCBs were referred to as just 'Phenyls' or by different trade names such as Aroclor, Kennechlor, Pyrenol etc.

1.1. Chemistry of PCBs

Benzene is a very stable compound, but heating it to a very high temperature can cause disruption of the Carbon-hydrogen bonds[7]. In the presence of a catalyst biphenyl is obtained (figure 1), which itself is a molecule with two benzene rings linked by a single bond between the carbon atoms which have lost an atom of hydrogen each. [11]

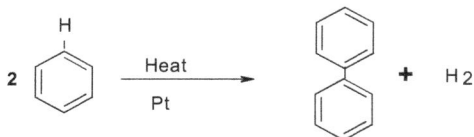

Figure 1: Reaction between two benzene rings to give biphenyl

If biphenyl is combined with chlorine in the presence of a ferric chloride as catalyst, like benzene, depending on the number of chlorine atoms introduced, hydrogen atoms are displace. The products obtained are referred to as polychlorinated biphenyls (PCBs)[12]. The number and position of the chlorine atoms on the biphenyl molecule determine the congener's nomenclature as well as dictate its environmental fate and toxicity. The more chlorinated congeners are generally the most persistent and toxic [1].

When numbering an individual PCB congener, certain rule is followed. It starts with the carbon of one ring that is joined to a carbon of the other ring, given the first number 1, and other the carbons of the ring are given the other numbers (2-6).

However, the number in the second ring is distinguished by adding a prime to the numbers (figure 2). The second position in the second ring lies on the same side of the carbon-carbon bond that joins the rings as the second position in the first ring. The same applies to the other positions [12].

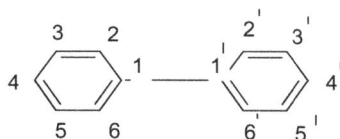

Figure 2: Positions of Carbon atoms of biphenyl

The two rings Chlorinated biphenyl molecule are used not equivalent as the patterns of substitution differ. The ring with unprimed numbers is chosen to give a substituent with the lowest carbon number (Figure 3). [12]

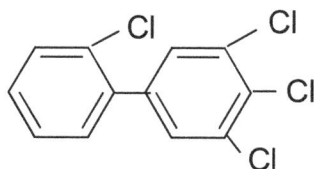

Figure 3: 2, 3', 4', and 5'-tetrachlorobiphenyl

There tends to be a quick rotation around carbon-carbon single bonds in most organic molecules. This includes the carbon-carbon link between the two rings of biphenyls, and indeed to most PCBs [12]. It is usually not possible to isolate compounds that correspond to different orientations of the two rings in PCB relative to each other. For instance, 3, 3'-and 3,5'-dichlorobiphenyl are not individual isolatable compounds; one form is constantly being converted into the other and back again by rapid rotation about the carbon-carbon bond joining the rings figure 4)[12]

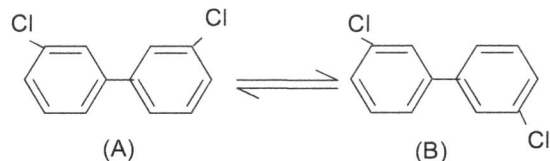

Figure 4: (A)3,3'-dichlorobiphenyl.(B)3,5'-dichlorobiphenyl

Several attempts to identify particular plants as a source of PCBs contamination have proved abortive; however, the only reliable means of achieving this has been through chemical analysis of oils used in plants [13]. There are several ways of determining the presences of PCBs in electrical insulating oils. Two of such ways are:

The used of enzyme immunoassays[13] and iconography-high resolution mass spectrometry[14]. Impurities in transformer oil need to be removed before effective determination of PCBs can be carried out[15]

PCB analysis is divided into two categories; specific and non-specific methods[16]. The specific methods particular test for PCB molecules while the non-specified methods identify the classes of compounds (e.g. chlorinated hydrocarbons) to which PCBs belong[16]. Specific method includes gas chromatography (GC) and mass spectrometry (MS). None-specific methods include Choir-N-Oil PCB screening kits, x-ray fluorescence spectrometry, microcoulemetric titration, and total organic halogen (TOX) test[16].

Generally, PCB analysis is performed with such analytical equipment as packed-column Gas Chromatography coupled with Electron Capture Detector (GC-ECD), capillary–column GC–ECD, quadrupled low-resolution Gas chromatography/mass spectrometry (GC/MS)[18][15] and double-focus high-resolution GC/MS[18][17]. An electron capture detector can detect any chlorinated compound including PCBs, when a GC is used to separate the components of the mixture (oil sample).

Takada, et al, in 2001 observed that efficient and cost-effectiveness of PCB analysis can be improved if the *dimethylsulfoxide (DMSO) portion*solid phase extraction (SPE) were used in the clean-up of the PCB contaminated transformer oils. They used quadrupole GC/MS to determine PCB level in transformer oil and the analytical data were compared with those of the standard method using statistics.

A suitable solvent was added to the sample containing PCB to be analyzed to dilute it, which was then treated to remove substances (impurities) that may otherwise alter the determination process. The treated sample solution was then analyzed using gas chromatography procedure with an electron capture detector. The result obtained was then evaluated by comparing it with the standard for the PCB sources known[15]. The data collected can be used to identify the type of PCB or their mixtures and their quantity in the sample. The result obtained are usually reported in parts per million (ppm), of the PCB present[19]

Table 1: Classification of electrical equipment and insulating fluids according to the level of PCB contamination

PCB Concentration (ppm)	Classification
Less than 50	Uncontaminated equipment
Greater than 50, less than 500	PCB contaminated equipment/fluid
Greater than 500	PCB equipment/fluid

Source: (NTT, 1999)

The overwhelming release of PCBs into environment posed a serious danger owing to their persistency and toxicity. The hydrophobicity of PCBs causes them to bio accumulate in the fatty issues of organisms, hence, their toxicity[20][8][3] the amount and concentration of PCBs in an organism determines the extent of damage caused to the organism exposed to them[8]. Although the actual extent of damage in human is still not certain, it appeared the PCBs do not immediately exhibit significant hazards upon exposure to them[8][21]. However, they cause effects on the neurological and immunological systems, liver and thyroid gland[8][2].Other effects caused by PCBs include Chloracne, which is another form of acne and is common to people who are exposed to organochlorine compounds for a long period of time[12]. [22] reported that "taken as a whole, the prepondence of evidence appears to support the argument that PCBs constitute a minimum threat to human health. Action phasing out their use is justified based on their capacity to accumulate in the environment"

Table 2: Average concentrations of organochlorine

Matrix Dimension	PCB	HCH	HCB	DDT
Air ng/M^3	0.014	0.58	0.19	<0.001
Snow ng/L	0.086	1.72	<0.002	<0.01
Seawater (surf.) ng/L	0.007	4.3	0.028	<0.001
Seawater (deep) ng/l	<0.014	0.51	0.01	<0.002
Zooplankton ug/g lipid	0.11	0.08	0.02	0.06
Amphipods ug/g Lipid	<0.44	0.5	0.17	<0.35
Cod ug/g lipid	0.23	0.58	0.2	0.26
Belugaug/g lipid	3.79	025	0.5	2.82
Ringed seal ug/g lipid	055	0.23	0.03	0.5
Polar bear ug/g lipid	5.4	0.51	0.27	0.4
Human milk ug/g lipid	1.05	NA	0.14	1.21

Sourve: [9]

1.2. Distribution of PCBs

It is a common belief these days that many chlorinated organic compounds including PCBs are distributed globally through the atmosphere[9][10][23]. The distribution occurs mainly by a phenomenon in which substances evaporate and spread to the atmosphere at latitudes with warmer climates and then condense and fall- out closer to the poles. As a result, regions near to the Northern and Southern Poles receive a disproportionate share of this fall-out[9]. PCBs and some other polychlorinated compounds were detected in marine organisms, lakes and sea sediments in remote areas such as the Arctic and the Antarctic [9].

Table 2 below shows average concentrations of polychlorinated biphenyls (PCBs) hexachlorohexanes (HCHs), hexachlorobenzene (HCB), dichlorodiphenylchoroethene (DDT) and chlordane[9].These result were obtained from experiments carried out in Arctic environment which show variation in concentration against experimental results obtained in the Antarctica [9].

About 94,800 kg of PCBs was estimated discharged into Hudson River by general electric company from two capacitor manufacturing plants located in Hudson falls, New York [2]. Anextensive toxic waste problem has been created ever since owing to the PCBs spread throughout the river and its food chain. This led to banning of fishing in the river by the New York state[2]. In Africa a worse scenario is expected as the wide distribution of PCB containing equipment, the poor storage of PCB-contaminated oils and the high cost of incineration make it likely that there will be further releases of PCBs to the (global) environment [24][25]. In the UK the sale and new usage of PCBs was prohibited in 1986, although a part of it is still encountered in some part of the country due to the fact that not every electrical unit have been refilled with PCB-free insulating oils [13]. There has been a significant reduction of the use of PCBs in the UK between 1990 and 1998 (table 3) [26].

Table 3: PCB emissions in the UK for 1990 and 1998

Sources of emissions	1990 emissions (kg)	1998 emissions (kg)
Electrical equipments	6,228	2,193
Application of sewage sludge	70	33
Power stations	89	44
Industrial/domestic combustion	32	22
Iron and steel industry	529	441
Other sources	28	14
Total	6,976	2,747

Source:UK national atmospheric emissions inventory (in weight, 2003)

This work is not aimed at comparing methods but to understand the various techniques use in determining the presence, quantifying the amount and identifying the type of PCBs in transformer oils and to Identify transformers capable of releasing harmful PCB containing fluids into the environment.

2. Materials and Methods

2.1. Sample Collection /Inventory of Transformers

Twenty two samples of transformer oils from various

areas were collected randomly by power Holding Company of Nigeria (PHCN) in Kebbi State. The collection or sampling activities were conducted in accordance with procedure presented in the sampling method by Shin, et. al; 2004 (that is the Ministry of environment in Korea, Korean Waste Official method, 2004). Prior to sampling the transformer oils, the power was disconnected by electricians at the working region. An inventory of the transformers from which the sample was collected was taken. The transformer valves were opened and 60ml oil samples collected in polypropylene bottles. The containers were filled to the top and then the transformer valves were closed carefully and the container caps tightly fitted unto the containers. Finally the containers were labeled and samples were transported to the laboratory at the University of Greenwich in the UK and stored at room temperature

2.2. Oil Pre-treatment:Solid Phase Extraction (SPE) Method.

In this study Solid Phase Extraction Method (SPE) for polychlorinated biphenyls (PCBs) cleanup was used Supelco's Application Note 67 Solid" Phase Extraction of PCBs from Transformer Oil and Waste Oil, and Analysis by Capillary GC" (Supelco's Bellefonte, USA). This procedure was chosen because of its significant advantages which include; avoiding the use of concentrated sulphuric acid, less need for glass wares, reducing solvent costs and disposal problems, it is faster than the acid extraction method, and many samples can be processed simultaneously on the SPE vacuum manifold

A 500µL aliquot of isooctane (2, 2, 4-Trimethlpentane) was added to a Supelelclean^{TM} LC-Florisil$^®$ solid phase extraction tube (Supelco). 0.2g of transformer oil sample was added onto the frit. PCB was flushed through the tube by adding five 2ml aliquots of isooctane. The eluant was collected in a glass vial. This was repeated for all the 12 transformer oil samples using a new SPE tube for each sample

2.3. Chromatographic Operating Conditions

The collected eluant was analyzed using Agilent 6890 GC coupled with an Electron Capture Detector (ECD) and connected to a PC with EZChrom Elite software. A DB5 GC Column; 30 x 0.25µp internal diameter (J & W Scientific) was used with helium carrier gas for the analysis of the sample eluants. The oven initial temperature was set at $50^{o}C$ for 2 minutes, and then programmed at $8^{o}C$ min $^{-1}$to $320^{o}C$. The GC was equipped with a 7683 auto sampler in which all the samples were put and allowed to run automatically. The injector point and the ECD temperature were set at $300^{o}C$.

2.4. Verification

Sample A01 was spiked a solution of PCB standard Mix 525 (Catalog No. 48246). Then the same SPE was carried out as was done to the other oil samples. Eluant was collected and run in GC/ECD. In the standard, each PCB is

500µg/ml. 100µl of the standard was added to 10ml of oil to make the concentration of each PCB in the oil 5ng/µl that is 1/10 of the safe level.

Mix 525 was diluted to 50ng/µl. SPE was then carried out and eluant collected and run in GC/MS.

3. Results and Discussions

3.1. Transformer Survey Result

The transformer inventory results which include information on the types of transformer, serial number, manufacture year, manufacturer and type of commercial dielectric fluids used are presented in Table 4.

Table 4: Transformer inventory/Sampling information

Sample ID	Serial No	Man. year	Commercial Dielectric Fluid
A1	TP 7716. 100/48113	1979	Minel
A2	121612	1991	Hawker Siddeley
A3	E81-349	1999	NI
A4	WB 90540	1990	Seoul Korea
A5	A 08233197	1981	Mitsubishi
A6	02/ 82/2370	1983	ONAN
A7	10104	2001	ELCO
A8	A 08252556	1981	Mitsubishi
A9	HD 118707/4	1977	Woden
A10	D 1109498	1952	GCE
A11	37888	1991	NI
A12	26894	1991	NI

Different transformers use different dielectric fluids (as recommended by the manufacturer). Only 2 transformers were made in Japan, 2 in Yugoslavia and 3 in the UK. The number of transformers sample was 10.6% of the total number of transformers around the Kebbi state capital. The survey results show that both the transformers and the dielectric fluids used in them were imported. Therefore, since transformers oils are the major source of PCB contamination, it could be said that greater percentage of the PCBs presence in Kebbi state were imported to be used as commercial products from Japan, UK and Yugoslavia.

3.2. Analytical Results

In the analytical process, PCBs were fist separated from the samples by solid Phase Extraction procedure. This was adopted because analytical techniques which are highly sensitive (such as GC/MS) require pre-treatment of samples of transformer oil to be analyzed in order to separate the PCBs from oil[17].

All the tables showing retention times for the major peaks of all the 12 transformers oil samples obtained from the ECD are presented below. To verify the SPE procedure, sample

A01 was spiked with 200μl of Mix 525 and treated the same way. When the extract was injected into GC/ECD all the PCB peaks contained in Mix 525 did appeared. So it shows that the PCBs (if any) certainly were separated from the oil and did come through SPE tubes.

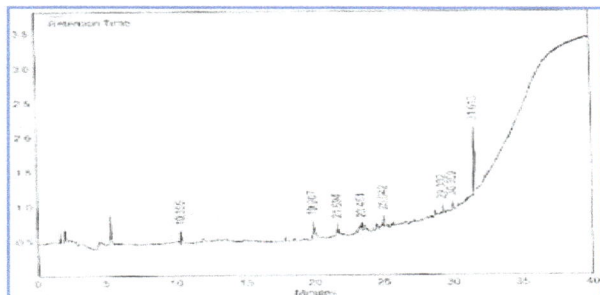

Figure 5: Chromatogram for sample A01

Table 5: Sample ID A01 spiked EDC Results

Peak No.	Retention Time	Area	Area %
1	19.91	2141691	25.86
2	21.69	1061280	12.81
3	25.04	1042343	12.59
4	31.24	56249	0.68
5	31.61	3980245	48.06

Table 5 shows retention times, names, areas and percentages of PCBs of spiked sample A01. Note that all the PCBs in the Mix 525 have been recovered. This indication confirms that every step of the procedure have worked through efficiently. Although minor peaks were observed in the chromatograms of the oil samples (which could be PCBs). If these peaks were PCBs, they are probably only about 1/100 of the peaks in the standard and the latter are at 1/10 of the permitted level

Table 6: Summary of peaks information

Peak	R/Time	Name	PCB No	Area	Area %
1	18.38	a	1	1943059	0.67
2	21.58	b	5	26370050	9.10
3	23.65	c	29	32422069	11.19
4	25.15	d	47	34155209	11.79
5	26.59	e	98	4408093	11.87
6	28.13	f	154	38552820	13.31
7	31.22	g	171	60187586	20.77
8	31.36	h	199	59787791	20.63

a= 2-chlorobiphenl , b=2,3-dichlorobiphenl, c=2,4,5-Trichlorobiphenl, d=-2,'3,'4,'4-Tetrachlorobiphenl;e=2,2,4,'6-Pentachlorobiphenl;f=2,2'4,'4,'5,6'-Hexachlorobiphenl;g=2,2,'3,3,4,4,6'-Heptachlorobiphenl, ,2,3,3,'4, 5,6,6'-octacchlorobiphenl.

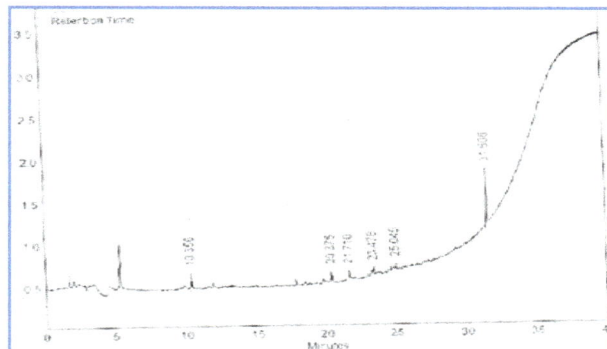

Figure 6: Chromatogram for sample A02

Table 7: Sample ID A02 spiked EDC Results

Peak No.	Retention Time	Area	Area %
1	31.61	2751701	100.0

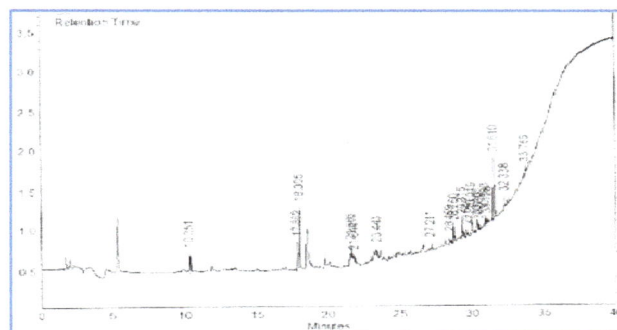

Figure 7: Chromatogram for sample A03

Table 8: Sample ID A03 ECD Results

Peak No.	Retention Time	Area	Area %
1	5.32	3019611	15.96
2	17.86	1465805	7.75
3	18.01	2792218	14.76
4	18.52	5445988	28.79
5	21.67	1513306	8.00
6	30.02	1070652	5.66
7	31.13	264177	1.40
8	31.24	176876	0.94
9	31.45	115198	0.61
10	31.61	3053175	16.14

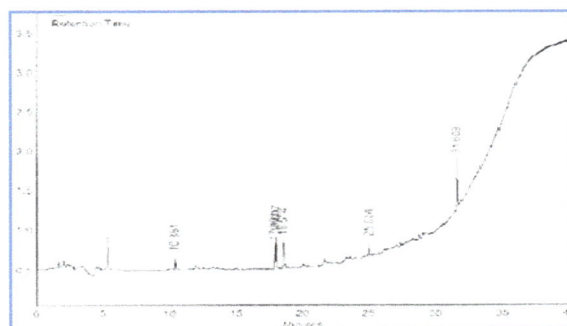

Figure 8: Chromatogram for sample A04

Table 9: *Sample ID A04 spiked EDC Results*

Peak No.	Retention Time	Area	Area %
1	17.86	1210893	14.89
2	18.01	1563269	19.23
3	18.51	2664922	32.78
4	31.61	2690762	33.10

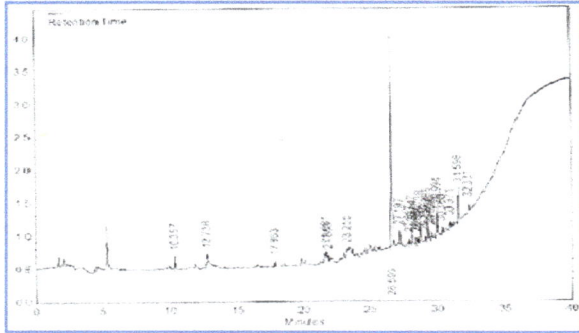

Figure 9: *Chromatogram for sample A05*

Table 10: *Sample ID A05 ECD Results*

Peak No.	Retention Time	Area	Area %
1	12.74	1349960	4.32
2	21.68	1329313	4.26
3	23.67	1242685	3.98
4	26.59	17618101	56.42
5	27.20	1322943	4.24
6	28.15	1116933	3.58
7	28.74	1546409	4.95
8	29.29	1151251	3.69
9	29.99	1790312	5.73
10	31.11	213389	0.68
11	31.23	245924	0.79
12	31.60	2296880	7.36

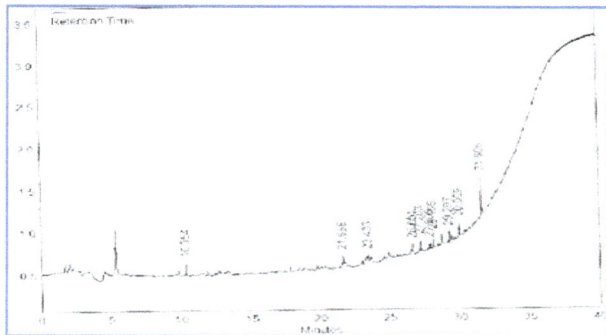

Figure 10: *Chromatogram for sample A06*

Table 11: *Sample ID A06 ECD Results*

Peak No.	Retention Time	Area	Area %
1	5.34	2677290	57.86
2	31.61	1950076	42.14

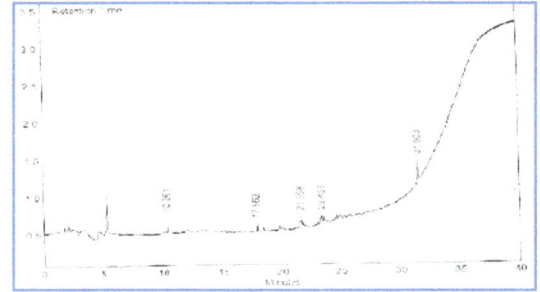

Figure 11: *Chromatogram for sample A07*

Table 12: *Sample ID A07 ECD Results*

Peak No.	Retention Time	Area	Area %
1	21.70	1157232	47.91
2	31.61	1258125	52.09

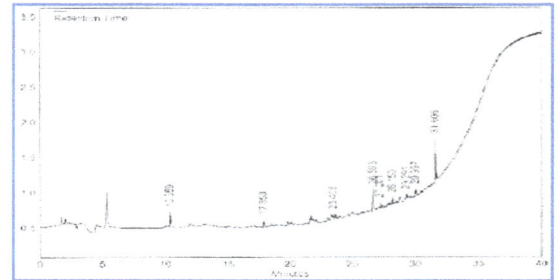

Figure 12: *Chromatogram for sample A08*

Table13: *Sample ID A08 ECD Results*

Peak No.	Retention Time	Area	Area %
1	26.59	1096457	23.73
2	26.60	1120362	24.25
3	31.61	2403758	52.02

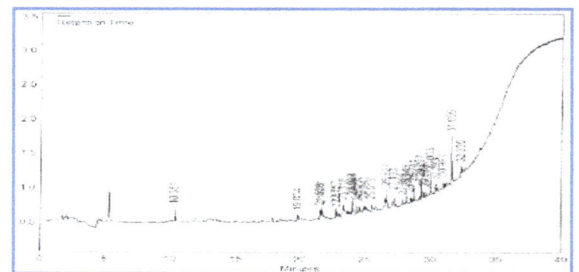

Figure 13: *Chromatogram for sample A09*

Table 14: *Sample ID A09 ECD Results*

Peak No.	Retention Time	Area	Area %
1	5.32	2725811	23.88
2	23.32	1309205	11.47
3	26.63	1698453	14.88
4	28.74	1233489	10.81
5	30.00	1281414	11.23
6	31.12	193556	1.70
7	31.23	215540	1.89
8	31.61	2757781	24.16

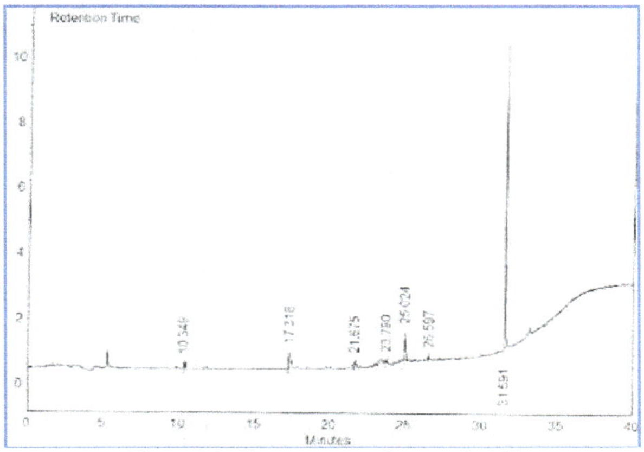

Figure 14: Chromatogram for sample A10

Table 15: Sample ID A10 ECD Results

Peak No.	Retention Time	Area	Area %
1	17.32	5234234	10.66
2	21.68	1328705	2.71
3	23.45	1250498	2.55
4	24.65	1253454	2.55
5	25.02	4297787	8.75
6	31.13	158364	0.32
7	31.59	35572045	72.46

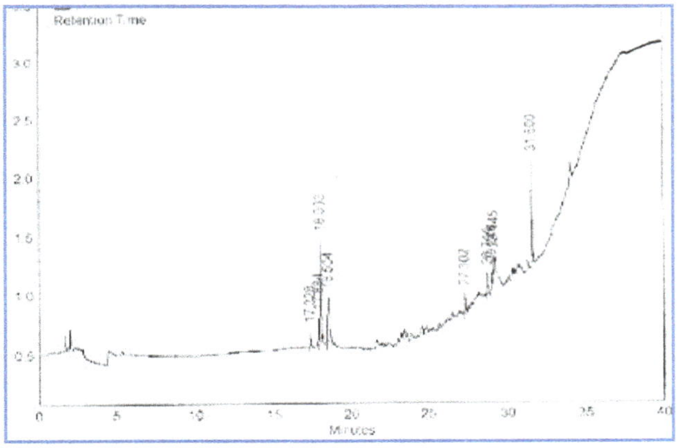

Figure 15: Chromatogram for sample A11

Table 16: Sample ID A11 ECD Results

Peak No.	Retention Time	Area	Area %
1	17.88	1353579	6.49
2	18.00	4426194	21.23
3	18.50	5049856	24.22
4	27.30	1180478	5.66
5	28.74	1311892	6.29
6	29.05	1227439	5.89
7	29.15	2056201	9.86
8	31.31	418184	2.01
9	31.35	93304	0.45
10	31.60	3733953	17.91

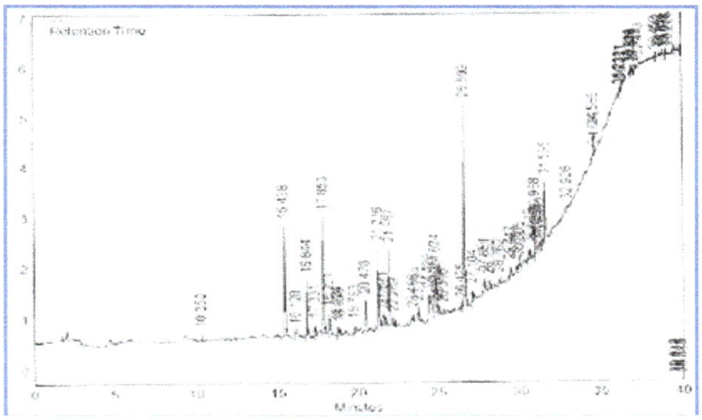

Figure 16: Chromatogram for sample A12

Table 17: Sample ID A12 ECD Results

Peak No.	Retention Time	Area	Area %
1	15.44	10564042	9.22
2	16.84	5958207	5.20
3	17.33	1085908	0.95
4	17.63	1122901	0.98
5	17.85	11679157	10.20
6	18.20	2853566	2.49
7	20.48	3337403	2.91
8	21.24	9379817	8.19
9	21.57	2875707	2.51
10	21.71	1366775	1.19
11	21.99	8297043	7.24
12	22.27	1032256	0.90
13	23.47	1025086	0.89
14	23.71	1377927	1.20
15	23.80	1270827	1.11
16	24.44	5247210	4.58
17	24.82	5489181	4.79
18	25.03	1126382	0.98
19	26.60	21548352	18.81
20	27.10	1414019	1.23
21	27.85	1589961	1.39
22	28.15	1590149	1.39
23	30.62	1163440	1.02
24	30.97	3353034	2.93
25	31.06	271510	0.24
26	31.10	92479	0.80
27	31.16	86802	0.80
28	31.22	352128	0.31
29	31.23	202755	0.18

Considering mass spectroscopy as an analytical technique that specifically identifies the chemical composition of a compound or sample base on the ratio of mass to charge of charged particles. It involves chemical fragmentation of a sample into ions (or charged particles) while measuring two main properties, mass and charge, of the end particles. When the particles are passed through electric and magnetic fields in a mass spectrometer, the mass-charge ratio can be obtained. A mass spectrometer has been designed with three essential modules, each performing a specific function. They are:

1) An ion source: This performs the function of transforming the molecules in a sample into ionized fragments.

2) A mass analyzer: which performs the function of arranging the ions according to their masses using electric and magnetic fields and;

3) A detector: this measures the value of some indicator quantity thereby providing data essential for calculating the abundances of each ion fragment present in a compound or sample. Mass spectroscopy is particularly important as it can be sued for both qualitative and quantitative measurements including identification of unknown compounds.

Upon dilution of the Mix 525 to 50ng/ul (of each component), the peaks were still detectable and identifiable at that level under mass spectrometer (Figure 17).

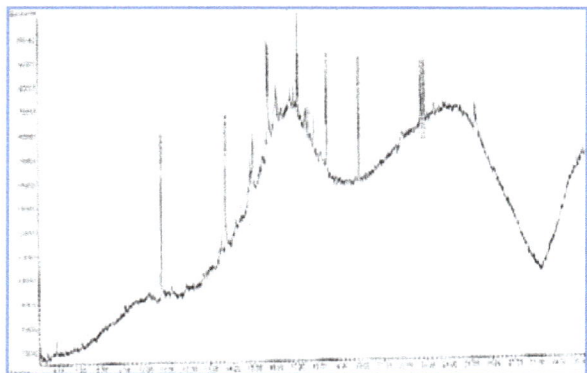

Figure 17: Chromatogram for Mix525 sample

4. Conclusion

PCBs are compounds primarily found in transformer oils and capacitors. Recently, these oils are being replaced with PCB-free transformer oils. The SPE procedure as well as the GC/ECD used in this study was efficient enough to produce required result which conforms to the EPA PCB Regulatory Limits. This can be seen in the result obtained by spiking sample A01, which was used as a reference to the other samples. This study did not find any detectable levels of PCBs in the 12 transformer oils sampled from Kebbi state. These results were validated after analysis when it was subsequently discovered that the transformers had recently been refilled with PCB-free oils after clean up.

References

[1] Bernhard, T. and Petron, S. (2001). Analysis of PCB Congeners vs. Arcolors in Ecological Risk Assessment http://Web.ead.anl.gov/ecorisk/issue/pdf/PCB%20IssuePaperNavy.pdf Accessed on 12th February 2008.

[2] ANZECC, (1997). Identifying PCB-Containing Capacitors.Australian and New Zealand Environment and Conservation Council. Free encyclopedia Wikipedia http://en.wikipedia.org/wiki/polychlorinatedbishyenylAccessed on 22nd December, 2007

[3] Shin,S.K. and Kim., T.S.,Kim,J.K., Chung,Y.H.AndChung,R.(2004). Analytical method of polychlorinated biphenyls (PCBs) in transformer oil.Organohalagen Compounds 66.

[4] UNEP,(1999). Guidelines for the identification of PCBs and

materials containing PCBs. UNEP Chemicals, 11-13 chemin des Anemones CH-1219 Chatelaine (Geneva), Switzerland.Http://www.chem .unep.ch/pops/pdf/PCBident/ pcbid 1.pdf Accessed on 22ndFebuary 2008.

[5] Kukharchyk, T. I. and Kakareka, S. V.(2008).Polychlorinated biphenyls inventory in Belarus. Journal of environmental Management 88(4).1657-1662

[6] Emmerson, A. And BVWS.(2002).Technical-Polychlorinatedbiphenyl. 405-Alive .http ://www.bvws.org.uk/405alive/tech/pcbs.html. Accessed on 22nd November, 2007.

[7] Barbalace. R.C (1995).The chemistry of polychlorinated biphenyls. http://environmentalchemistry.com/yogi/chemistry /pcb.html.Accessed on 28th November, 2007.

[8] Hill, M.K.(2004) Understanding Environmental Pollution. Cambridge University Press. UK. 2nd edition pg 339-349.

[9] Fiedler, H. (1997). Polychlorinated Biphenyl (PCBS): Uses and Environmental Releases http:www.cheunep.ch/pops/POPs inc/proceedings/cartagena/FIEDLER1 html Accessed on 2nd February 2008.

[10] Ferrario, J., Byrne, C. and Dupuy, A.E(1997). Background contamination by coplanar polychlorinated biphenyl (PCBs) in trace level high resolution gas chromatograph/high resolution mass spectrometry (HRGC/HRMS) analytical methods.Chemosphere 34(11). 2451-2465.

[11] Bunce, N.J. (1994). Environmental Chemistry.Wuerz Publishing Ltd. Canada.2nd edition. Pg 300-305.

[12] Baird, C. (1995). Environmental Chemistry.W.H. Freeman and company. Pg 254-256.

[13] Kim, S., Setford, S.J. and Sain, S (2000). Determination of polychlorinated biphenyl compounds in electrical insulating oils by enzyme immunoassay. Analytical ChimicaActa 422(2).167-77.

[14] Takasuga, T.,Senthilkumar,K., Matsumura,T., Shiozaki, K. and Sakai, S. (2006). Isotope dilution analysis of polychlorinated biphenyls (PCBs) in transformer oil and global commercial PCB formulation by hiresolution gas chromatography-high resolution mass spectrometry.Chemosphere 62(3).469-484.

[15] Takada ,M., Toda,H.andUchida,R .(2001).A new repid method for qualification of PCBs in transformer oil. Chemosphere.43(4-7).455-459.

[16] Finch, S. (1990) Alternative methods of PCB analysis. DEXSIL CORPORATION. www.dexsil.com/uploads/docs/dtr10 01.pdf Accessed on 31st March 2008.

[17] Shin and Kim. (2006) Levels of Polychlorinated Biphenyls(PCBs) in transformer from Korea. Journal of Hazardous Materials 137(3).1514-1522.

[18] Buthe A. and Denker E (1995) .Qualitattive and quantitative determination of PCB congeners by a HT-5 GC columin and an efficient quadrupole Ms. Chemosphere ,30(4). 753-771.

[19] Hoof, P.V. and Hsieh, J. (1996). Analysis of Polychlorinated Biphenyls and Chlorinated Pesticides by Gas Chromatograph

with Electron Capture Detection. http://www.epa.gov/grtlakes/Immb/methods/sop-501.pdt Accessed on 23rd November, 2007.

[20] Arisawa, K., Takeda, H., and Mikasa, H.(2005). Background exposure to PCDDs/PCDFs/PCBs and its potential health effects: a review of epidemiologic studies. The journal of medical investigation 52.10-21.

[21] Bailey, R.A., Clark, H.M., Ferris, J.P., Krause, S., and Strong, R.L. (1978). Chemistry of the environment.Academic press Inc. London Ltd. Pg 157-158.

[22] NTT, (1999).NTT-Technical Bulleting:PCB Testing and Analysis. http://www.nttworldwide.com/pcb.htm. Accessed on 22nd November, 2007.

[23] Huang, P., Gong, S.L., Zhao, T. L., and Barrie, L. A. (2007). GEM/Pops: a global 3-D dynamic model for semi- volatile persistent organic pollutants-part 2: Global transports and budgets of \PCBS. Atmos. Chem. Phys., 7. 4015-4025.

[24] O'Neil, p. (1998).Environmental Chemistry.3rd edition, Thomson Blakie Science Academy and professional.pg 239-240.

[25] Osibanjo,O. (ed.Fielder, H) (2002). Organochlorine in Nigeria and Africa.The hand book of Environmental Chemistry.Persistent organic pollutants.Springer Verlag Berlin Heidelberg Volume 3.321-354.

[26] Wright, j. (2003). Environmental Chemistry.Routledge London. Pg 121-122.

11

Investigation of nigerian 330 kv electrical network with distributed generation penetration: deterministic and probabilistic analyses

Funso K. Ariyo

Department of Electronic and Electrical Engineering, Ile-Ife, Nigeria
Obafemi Awolowo University, Ile-Ife, Nigeria

Email address:
funsoariyo@yahoo.com (F. K. Ariyo)

Abstract: The concluding part of this work (Part III) presents the non-probabilistic (deterministic) assessment of failure effects under given contingencies and reliability analysis is an automation and probabilistic extension of contingency evaluation. Also, PowerFactory generation adequacy tool is design specifically for testing of system adequacy using Monte-Carlo method. Running adequacy analysis produces convergence plots, distribution plots and Monte-Carlo draw plots. PowerFactory's contingency analysis module offers two distinct contingency analysis methods: single time phase and multiple time phase contingency analysis, while an analytical assessment of the network reliability indices is initiated by the following actions (failure modeling, load modeling, system state production, failure effect analysis (FEA), statistical analysis and reporting) within PowerFactory. Lastly, voltage sag analysis is a calculation that assesses the expected frequency of voltage sags within a network.

Keywords: Deterministic, Assessment, Probabilistic, Contingencies, Generation Adequacy, Monte-Carlo Method, Reliability, Failure, Failure Effect Analysis, Statistical Analysis, Voltage Sag, Powerfactory

1. Introduction

This paper uses probabilistic and non-probabilistic assessment to solve contingency cases using PowerFactory by DIgSILENT. The assessment of reliability indices for a power system network or of parts thereof, is the assessment of the ability of that network to provide the connected customers with electric energy of sufficient availability, as one aspect of power quality. The reliability assessment module of PowerFactory offers two distinct calculation functions for the analysis of network reliability under probabilistic scenarios [1]:

• network reliability assessment: The probabilistic assessment of interruptions during an operating period of the power system;

• voltage sag assessment: the probabilistic assessment of the frequency and severity of voltage sags during an operation period.

Contingency analysis is performed to ascertain the risks that contingencies pose to an electrical power system.

PowerFactory's contingency analysis module offers two distinct contingency analysis methods: single time phase and multiple time phase contingency analysis, while generation adequacy is the ability of the power system to be able to supply system load under all possible load conditions is known as 'System Adequacy'. This relates to the ability of the generation to meet the system demand.

Contingency analysis is critical in many routine power system and market analyses to show potential problems with the system. However, contingency analysis is computationally very expensive as many different combinations of power system component failures must be analyzed. Analyzing several million such possible combinations can take inordinately long time and it is not be possible for conventional systems to predict blackouts in time to take necessary corrective actions. To address this issue, PowerFactory software provides a probabilistic contingency analysis scheme that processes severe and most probable contingencies.

The liberalization of electricity markets in countries all over the world has lead to tremendous changes for electric utilities. This evolution calls for enhanced power system planning tools. The software used in this paper can provide reliability indices at any system bus, while voltage sags caused by the short-circuit faults in transmission and distribution lines have become one of the most important power quality problems facing industrial customers and utilities. Voltage sags are normally described by characteristics of both magnitude and duration. A simple and practical method is proposed in this paper which is discussed in Section 6.

2. Methodology

The proposed Nigerian 330 kV electrical network (37-bus system shown in Figure 1) was built in PowerFactory 14.1 software and the following analyses were carried out using PowerFactory tools. PowerFactory works with three different classes of graphics: single line diagrams, block diagrams, and virtual instruments. They constitute the main tools used to design new power systems, controller block diagrams and displays of results. In order to meet today's power system analysis requirements, the DIgSILENT power system calculation package was designed as an integrated engineering tool which provides a complete 'walk-around' technique through all available functions, rather than a collection of different software modules [1].

Data used are stated in Table 1 (Appendix), while solar farm (minimum value of 50 MW per state) was proposed for every state, having potentials to produce energy from the sun because of high solar radiation. Offshore wind power was proposed for states along the coast which include: Lagos, Ondo, Delta, Bayelsa, and Akwa-Ibom (minimum value of 50 MW per state) and some Northern states.

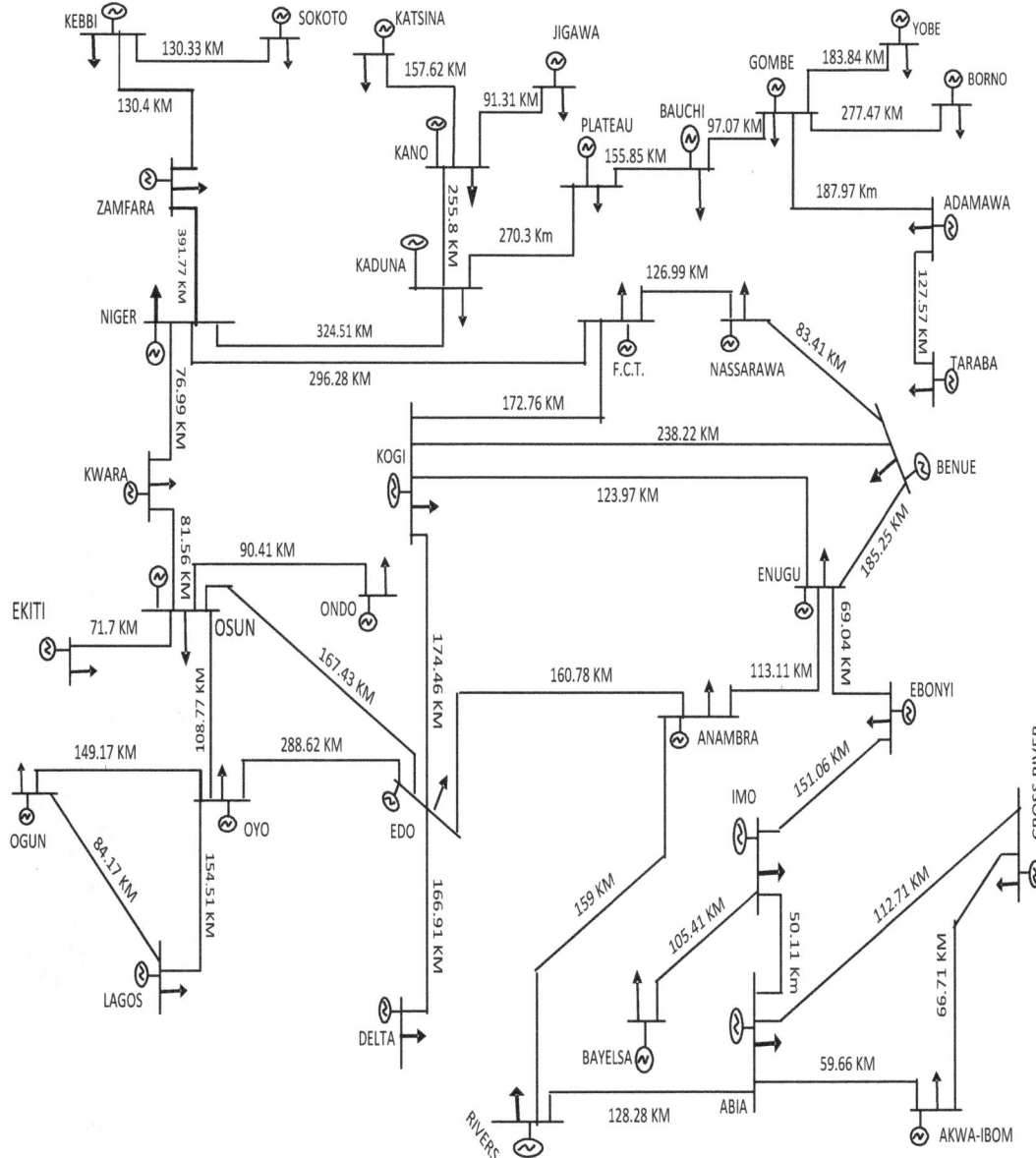

Figure 1. Proposed Nigerian 330 kV electrical network (37-bus system).

Table 1. Proposed Power generation and Allocation per State.

S/N	State	Total Capacity Per State (Mw)	Population Size	Real Power Allocation (P)	Reactive Power Allocatn (Q)
1	F.C.T.	535	1,405,201	906.79	363
2	Abia	2,404	2,833,999	820.42	328
3	Adamawa	100	3,168,101	917.14	367
4	Akwa-Ibom	1,790	3,920,208	1,134.87	454
5	Anambra	1,705	4,182,032	1,210.67	484
6	Bauchi	742.6	4,676,465	1,353.80	542
7	Bayelsa	350	1,703,358	493.11	197
8	Benue	2,130	4,219,244	1,221.44	489
9	Borno	120.8	4,151,193	1,201.74	481
10	Cross River	705	2,888,966	836.33	335
11	Delta	5,900	4,098,391	1,186.45	475
12	Ebonyi	230	2,173,501	629.21	252
13	Edo	1,000	3,218,332	931.68	373
14	Ekiti	70	2,384,212	690.21	276
15	Enugu	1,050	3,257,298	942.96	377
16	Gombe	400	2,353,879	681.43	273
17	Imo	425	3,934,899	1,139.12	456
18	Jigawa	146.2	4,348,649	1,258.90	504
19	Kaduna	379.2	6,066,562	1,756.22	702
20	Kano	246	9,383,682	3,216.50	1,287
21	Katsina	111	5,792,578	1,676.91	671
22	Kebbi	240	3,238,628	937.56	375
23	Kogi	1,804	3,278,487	949.10	380
24	Kwara	90	2,371,089	686.41	275
25	Lagos	1,616	9,013,534	3,609.35	1,444
26	Nassarawa	196	1,863,275	539.40	216
27	Niger	2,710	3,950,249	1,143.57	457
28	Ogun	2,125	3,728,098	1,079.26	432
29	Ondo	920	3,441,024	996.15	398
30	Osun	65	3,423,535	991.09	396
31	Oyo	3,800	5,591,589	1,618.72	647
32	Plateau	245.4	3,178,712	920.21	368
33	Rivers	3,924	5,185,400	1,501.13	600
34	Sokoto	133.6	3,696,999	1,070.25	428
35	Taraba	3,735	2,300,736	666.05	266
36	Yobe	140	2,321,591	672.08	269
37	Zamfara	246	3,259,846	943.70	377
	Total	**42,529.95**	**140,003,542**	**42,529.95**	**17,012**

3. Generation Adequacy

The ability of the power system to be able to supply system load under all possible load conditions is known as 'System Adequacy'. This relates to the ability of the generation to meet the system demand, while also considering typical system constraints such as:

• generation unavailability due to fault or maintenance requirements;

• variation in system load on a monthly, hourly and minute by minute basis;

• variations in renewable output (notably wind generation output), which in turn affects the available generation capacity.

PowerFactory generation adequacy tool is design specifically for testing of system adequacy. This tool is used to determine the contribution of wind and solar generations to overall system capacity and to determine the probability of 'Loss of Load' (LOLP) and the 'Expected Demand Not Supplied' (EDNS) [1, 2-4].

The analytical assessment of generation adequacy requires that each generator in the system is assigned a number of 'probabilistic states' which determine the likelihood of a generator operating at various output levels.

Likewise, each of the system loads is assigned a time-based characteristic that determine the actual system load level for any point of time. However, as the number of generators, generator states, loads and load states increase, the degrees of freedom for the analysis rapidly expands so that it becomes impossible to solve in a reasonable amount of time. Such a problem is ideally suited to Monte Carlo simulation [5]. Monte Carlo methods are a class of computational algorithms that rely on repeated random sampling to compute their results. Monte Carlo methods are often used in computer simulations of physical and mathematical systems.

These methods are most suited to calculation by a computer and tend to be used when it is infeasible to compute an exact result with a deterministic algorithm. In the Monte Carlo method, a sampling simulation is performed. Using uniform random number sequences, a random system sate is generated. This system state consists random generating operating states and of random time points. The generating operating states will have a corresponding generation power output, whereas the time points will have a corresponding generation power output, whereas the time the time points will have a corresponding power demand [5].

The value of demand not supplied (DNS) is then calculated for such state for such state. This process is done for a specific number of draws (iterations). At the end of the simulation, the values of the loss of load probability (LOLP), loss of load expectancy (LOLE), expected demand not supplied (EDNS), and loss of energy expectancy (LOEE) indices are calculated as average values from all the iterations

performed.

There are several database objects in PowerFactory specifically related to the generation adequacy analysis such as:

• stochastic model for generation object (StoGen);

• power curve type (TypPowercurve); and

• Meteorological station.

Stochastic model for generation object was used for this work. Generation object (StoGen) was used for defining the availability states of a generator. An unlimited number of states is possible with each state divided into:

• availability of generation (in %);

• probability of occurrence (in %).

This means that for each state, the total available generation capacity in % of maximum output must be specified along with the probability that this probability that this availability occurs. The probability column is automatically constrained, so that the sum of the probability of all states must equal 100%.

The generator maximum output is calculated as:

$$S_{nom} \quad \cos \theta .$$

where S_{nom} is the nominal apparent power and $\cos \theta$ is the nominal power factor [2].

4. Contingencies Analysis

Contingency analysis is performed to ascertain the risks that contingencies pose to an electrical power system. PowerFactory's contingency analysis module offers two distinct contingency analysis methods:

• single time phase contingency analysis: the non-probabilistic (deterministic) assessment of failure effects under given contingencies, within a single time period.

• multiple time phase contingency analysis: the non-probabilistic (deterministic) assessment of failure effects under given contingencies, performed over different time periods, each of which defines a time elapsed after the contingency occurred. It allows the definition of user defined post-fault actions.

Contingency analyses can be used to determine power transfer margins or for detecting the risk inherent in changed loading conditions [1, 6-8].

5. Reliability Assessment

Reliability analysis is an automation and probabilistic extension of contingency evaluation. The planner is not required to pre-define outage events, but can optionally select that all possible outages to be considered for analysis. The relevance of each outage is considered using statistical data about the expected frequency and duration of outages according to component type. The effect of each outage is analyzed in an automated way, meaning that the software simulates the protection system and the network operator's actions to re-supply interrupted customers. As statistical data regarding the frequency of each event is available, the results can be formulated in probabilistic terms. An analytical assessment of the network reliability indices (transmission, sub-transmission or distribution level) is initiated by the following actions within PowerFactory:

- failure modeling;
- load modeling;
- system state production;
- failure effect analysis (FEA);
- statistical analysis; and
- reporting.

The system state production module uses the failure models and load models to build a list of relevant system states. Each of these system states may have one or more faults. It is the task of the FEA module to analyze the faulted system states by imitating the system reactions to these faults, given the current load demands. The FEA will normally take the power system through a number of operational states which may include:

- fault clearance by tripping protection breakers;
- fault separation by opening separating switches;
- power restoration by closing normally open switches;
- overload alleviation by load transfer and load shedding.

The basic task of the FEA functions is to find out whether system faults will lead to load interruptions and if so, which loads will be interrupted and for how long. The results of the FEA are combined with the data that is provided by the system state production module to update the statistics. The system state data describes the expected frequency of occurrence of the system state and its expected duration. The duration of these system states should not be confused with the interruption duration. A system state with a single line on outage (that is, due to a short- circuit on that line), will normally have a duration equal to the time needed to repair that line. In the case of a double feeder, however, no loads may suffer any interruption. In the case that loads are interrupted by the outage, the power may be restored by network reconfiguration (that is, by fault separation and closing a back-stop switch). The interruption duration will then equal the restoration time, and not the repair duration (=system state duration).

A stochastic model describes how and how often a certain object changes. A line, for example, may suffer an outage due to a short-circuit. After this kind of outage, repair will begin and the line will be put into service again following successful repair. If two states for line 'A' are defined (that is, "in service" and "in repair"). The repair durations are also called the "Time To Repair" (TTR). The service durations are called the "life-time" or "Time To Failure" (TTF). Both the TTR and the TTF are stochastic quantities. By gathering failure data about a large group of similar components in the power system, statistical information about the TTR and TTF, such as the mean value and the standard deviation, can be calculated. The statistical information is then used to define a stochastic model. There are many ways in which to define a stochastic model. The so-called "homogenous

Markov-model" is a highly simplified but generally used model. A homogenous Markov model with two states is defined by:

a constant failure rate lambda (λ), and

a constant repair rate mu (μ).

These two parameters can be used to calculate the following quantities:

- mean time to failure, $TTF = \frac{1}{\lambda}$;

- mean time to repair, $TTR = \frac{1}{\mu}$; (1.0)

- availability, $P = \frac{TTF}{(TTF + TTR)}$;

- unavailability $Q = 1 - P = TTR/(TTF + TTR)$.

The availability gives the fraction of time during which the component is in service; the unavailability gives the fraction of time during which it is in repair; and $P + Q = 1.0$.

These equations also introduce some of the units used in the reliability assessment:

frequencies are normally expressed in [1/a] = "per annum";

- lifetimes are normally expressed in [a] = "annum";
- repair times are normally expressed in [h] = "hours";
- probabilities or expectancies are expressed as a fraction or as time per year ([h/a],[min/a]) [1, 9-12].

6. Voltage Sag Analysis

Voltage sag analysis is a calculation that assesses the expected frequency of voltage sags within a network. The PowerFactory voltage sag tool calculates a short-circuit at the selected load points within the system and uses the failure data of the system components to determine the voltage sag probabilities. Voltage sag analysis has a lot in common with probabilistic reliability analysis. Both use fault statistics to describe the frequency of faults and then use these statistics to weight the results of each event and to calculate the overall effects of failures.

Reliability analysis looks for sustained interruptions as one aspect of quality of supply, whereas voltage sag analysis calculates the voltage drop during the fault until the protection system has disconnected the defective component. The voltage sag analysis simulates various faults at all relevant busbars. It starts with the selected load points, and proceeds to neighboring busbars until the remaining voltage at all load points does not drop below the defined Exposed area limit. The remaining voltages and the short-circuit impedances for all load points are written to the result file specified by the Results parameter. After all relevant busbars have been analyzed, the sag table assessment continues by analyzing short-circuits at the midpoint of all lines and cables that are connected between the relevant busbars. Again, the remaining voltages and short-circuit impedances for all load points are written to the result file.

After the complete exposed area has been analyzed in this way, the result file contains the values for the two ends of all relevant lines and cables and at their midpoints. The written impedances are interpolated between the ends of a line and the middle with a two-order polynomial. From them, and from the written remaining voltages, the various source impedances are estimated. These estimated impedances are also interpolated between the ends and the midpoint. The interpolated impedances are then used to estimate the remaining voltages between the ends and the midpoints of the lines or cables. This quadratic interpolation gives very good results also for longer lines, and also for long parallel or even triple parallel lines. The main advantage is a substantial reduction in computation and an increase in the overall calculation speed [1, 13-17].

7. Results and Discussion

For voltage sag, busbars at Ekiti, Kano, Cross River,

Enugu, Jigawa and Delta were first defined in the network and then the voltage sag table assessment tool was used to carry out the analysis. Single-phase to ground fault (phase-b) was considered using complete short circuit method. The break time is 0.1 seconds and the fault clearing time of 0.4 seconds. The results are shown Figures 2-5. The voltage sag plot shows the annual frequency of occurrence on the y-axis.

Figure 2 shows minimum line-to-ground voltage with x-variable which is short-circuit type. The burbars could be seen to suffer deep sag, while Figure 3 displays minimum line-line voltages with x-variable of fault clearing time. Plots in Figures 4 and 5 show the voltage sag of minimum line-line and line-ground voltages; and positive-sequence voltage.

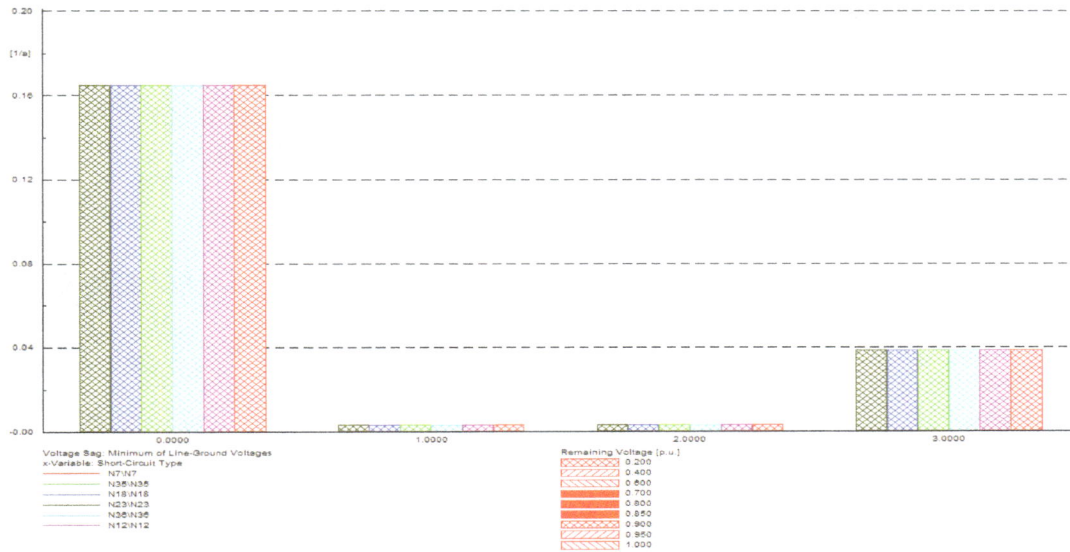

Figure 2. Voltage sag of minimum line-ground voltages.

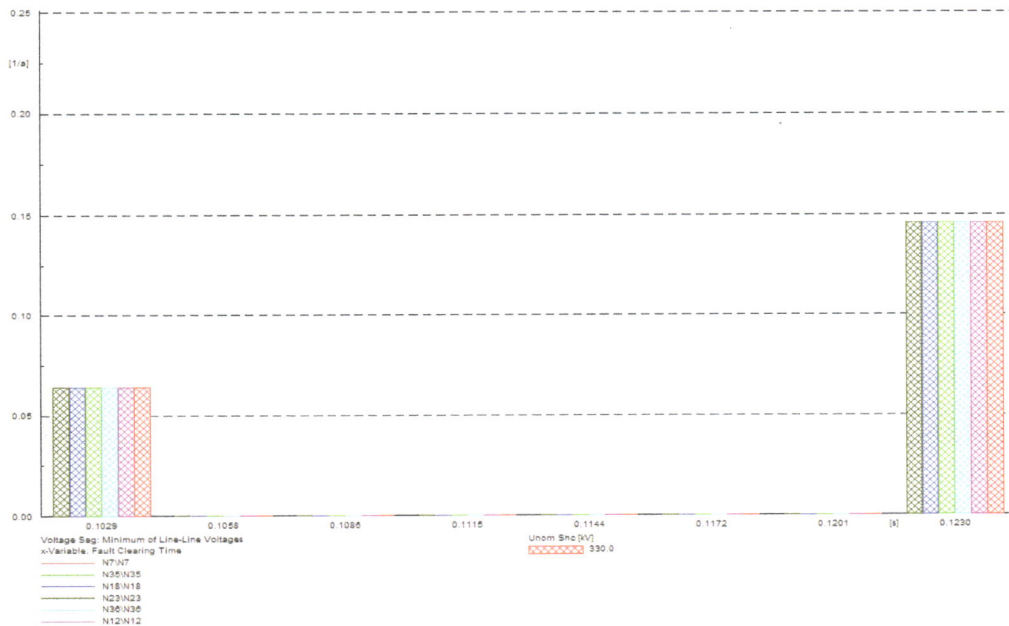

Figure 3. Voltage sag of minimum line-line voltage.

Figure 4. *Voltage sag of minimum line-line and line-ground voltages.*

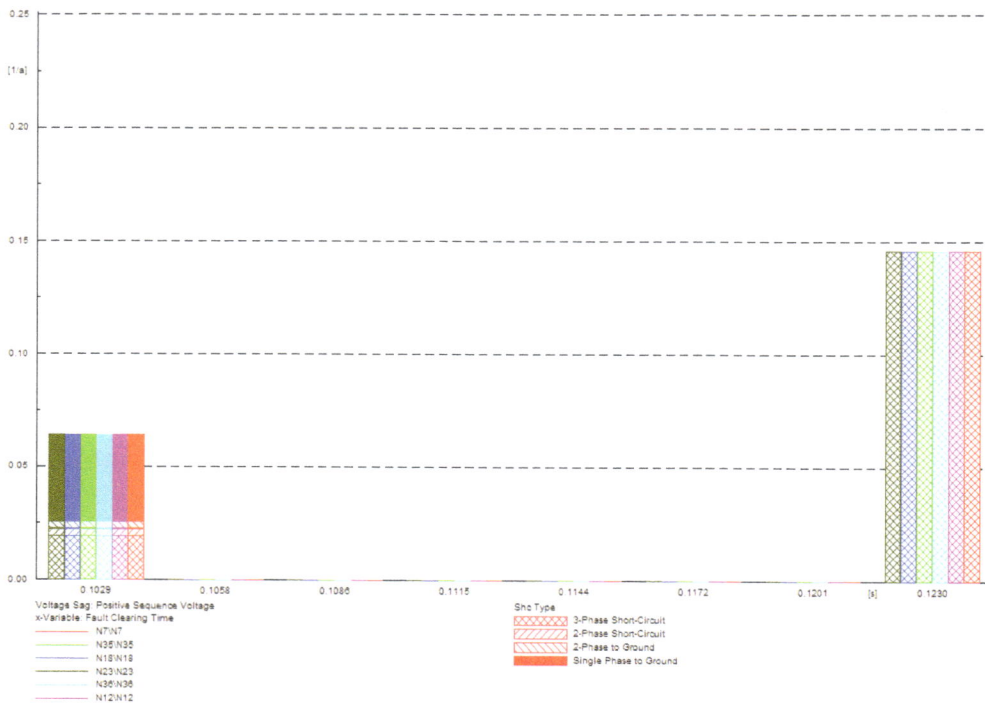

Figure 5. *Voltage sag of positive-sequence voltage.*

Furthermore, the simulation was first initialized before a generation adequacy analysis. The load flow command used was set to AC load-flow balanced, positive sequence. For fixed demand level selection (where all load characteristics were ignored and the total demand was calculated at the initial iteration and used for subsequent iterations), maintenance plans were considered on line 28 which connected Niger to F.C.T. The system losses were set to 3%. The period considered for generation adequacy was 2010. This variable does not influence the wind speed or wind power data of the wind model for the generator references time series data.

Running adequacy analysis produces convergence plots, distribution plots and Monte-Carlo draw plots as shown in Figures 6 – 15.

Convergence plots (Fig. 6 and 7) show loss of load probability and expected demand not supplied. These two plots converge towards the final value as the number of iterations increases. The distribution plots are (Fig. 8-11) essentially the data from 'Draws' plot sorted in descending order, the data then becomes a cumulative probability distribution. The loss of load probability index was obtained by inspection directly from the plots.

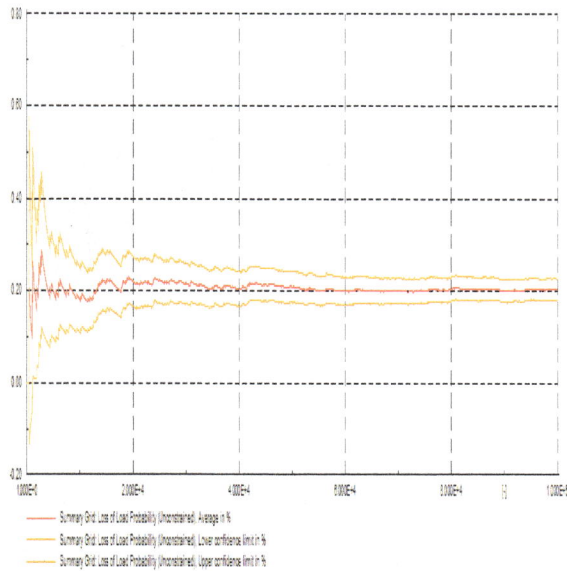

Figure 6. *Convergence plot of fixed demand for loss of probability.*

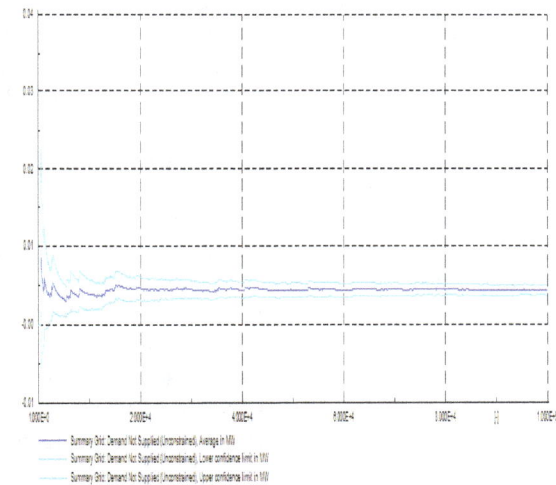

Figure 7. *Convergence plot of fixed demand for demand not supplied.*

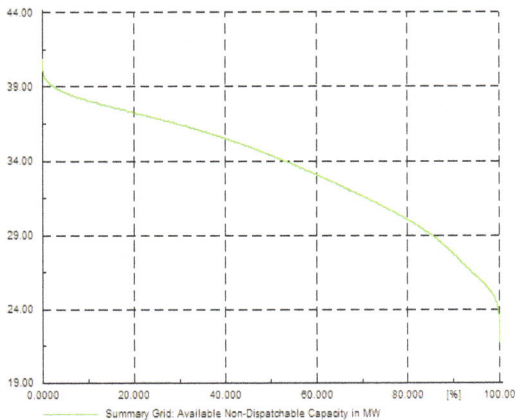

Figure 8. *Distribution plot of fixed demand for available non-dispatchable capacity (MW).*

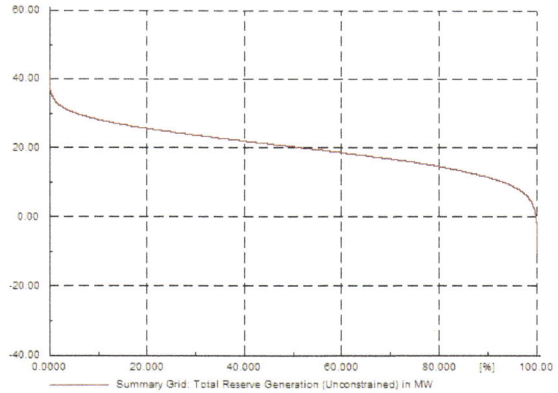

Figure 9. *Distribution plot of fixed demand for total reserve generation (MW).*

For time characteristics, any time characteristics assigned to loads was automatically considered in the calculation, therefore, the total demand varied at each iteration. In Fig. 10, LOLP index can be obtained by inspection – read from the intersection of total demand and available disputable capacity, while in Fig. 11, the intersection of residual demand with x-axis gives the LOLP index.

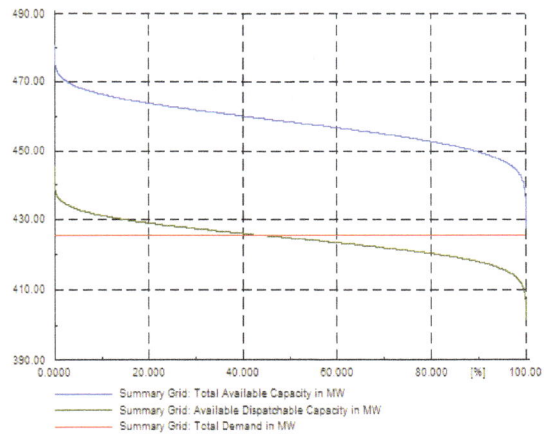

Figure 10. *Distribution plot of fixed demand for available dispatchable capacity/total demand (MW).*

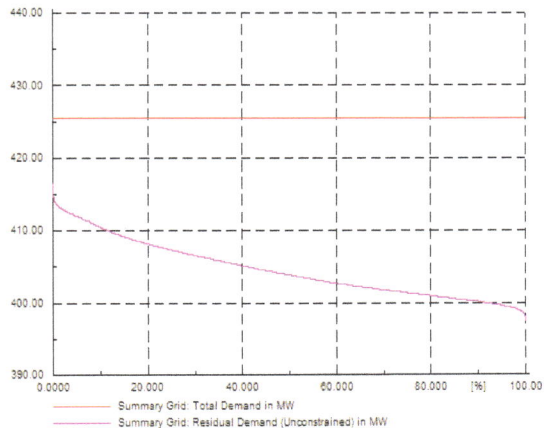

Figure 11. *Distribution plot of fixed demand for total demand/residual demand (MW).*

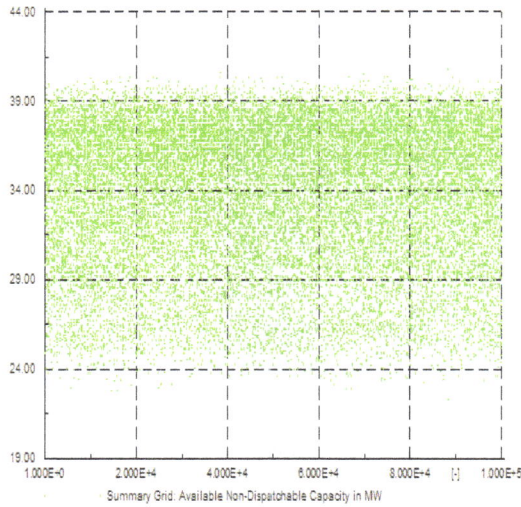

Figure 12. *Monte-Carlo draw plots of fixed demand for: available dispatchable capacity (MW).*

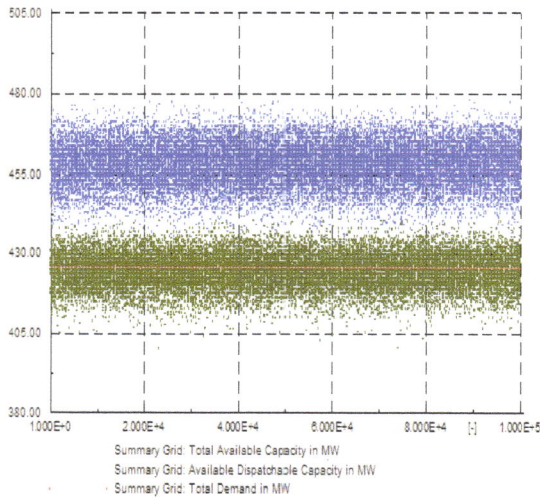

Figure 13. *Monte-Carlo draw plots of fixed demand for total demand/ available dispatchable capacity (MW)/total available capacity.*

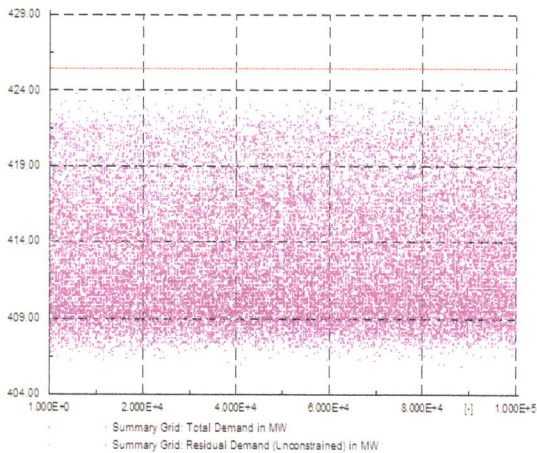

Figure 14. *Monte-Carlo draw plots total demand/residual demand (MW).*

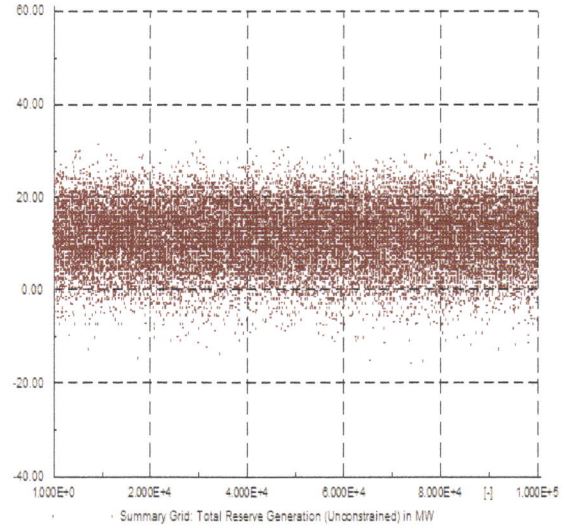

Figure 15. *Monte-Carlo draw plots total reserved generation (MW).*

For reliability assessment, failure models were made for the following PowerFactory objects:
• busbars and
• lines – the objects defined for stochastic failure and repairs model are: line 28   line 42   line 43 and double earth faults defined on: lines 28 and 29. Also, outages and maintenance period each assigned to the following lines: line 1, 2, 28 and 35 as shown in figure 16.0;
• synchronous generators – receive active and reactive powers limits;
• loads – the following loads are defined for shedding: while numbers of Customers were entered into all loads and creating load states for each load.

All failure models define how often a component will suffer an active failure. All active failures must be cleared by protection. When a failure cannot be separated from all generators or external networks by protection, a warning message will be issued. Repair of the faulted component is assumed to start directly after the fault has been cleared. The repair duration (which is also defined in the failure models) is the time needed to restore the functionality of the component. The time needed to begin the repair (that is, if spare parts need to be ordered first) and all other delays are therefore to be included in the total repair time.

There are two methods used for this analysis: connectivity analysis (without considering constraints) and load flow analysis (considers constraints by completing load-flows for each contingency). The calculation period for the year 2010 (specified). The results are shown in Tables 2-5.

Lastly contingency analysis was carried out which include single- and multiple-time phases. In the former, A.C. load flow calculation was used to calculate the power flow and voltages per contingencies cases. Bauchi wind farm (BauchiW), Benue conventional station (Benue), Edo solar farm (EdoS), transmission line linking Niger and F.C.T. generating stations (line 28) were defined as contingency cases for the single-time phase over a time sweep as shown in Figure 17. The time sweep must be enabled to define a post contingency time. This value defines the time phase under consideration for the update of contingencies. This means that all switch-open events with an event time less than or equal to this are considered in the update. The contingency load flow is calculated at the post contingency time. The buses reports are the same for the two methods.

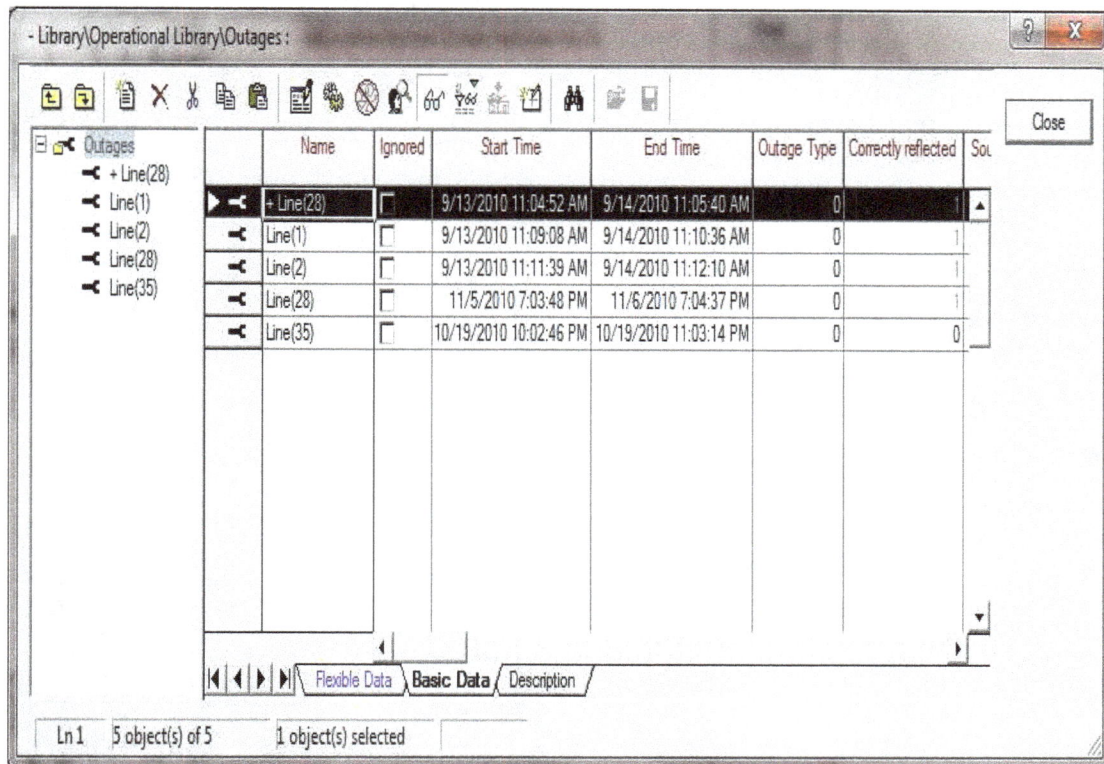

Figure 16. *Command dialogue for outages.*

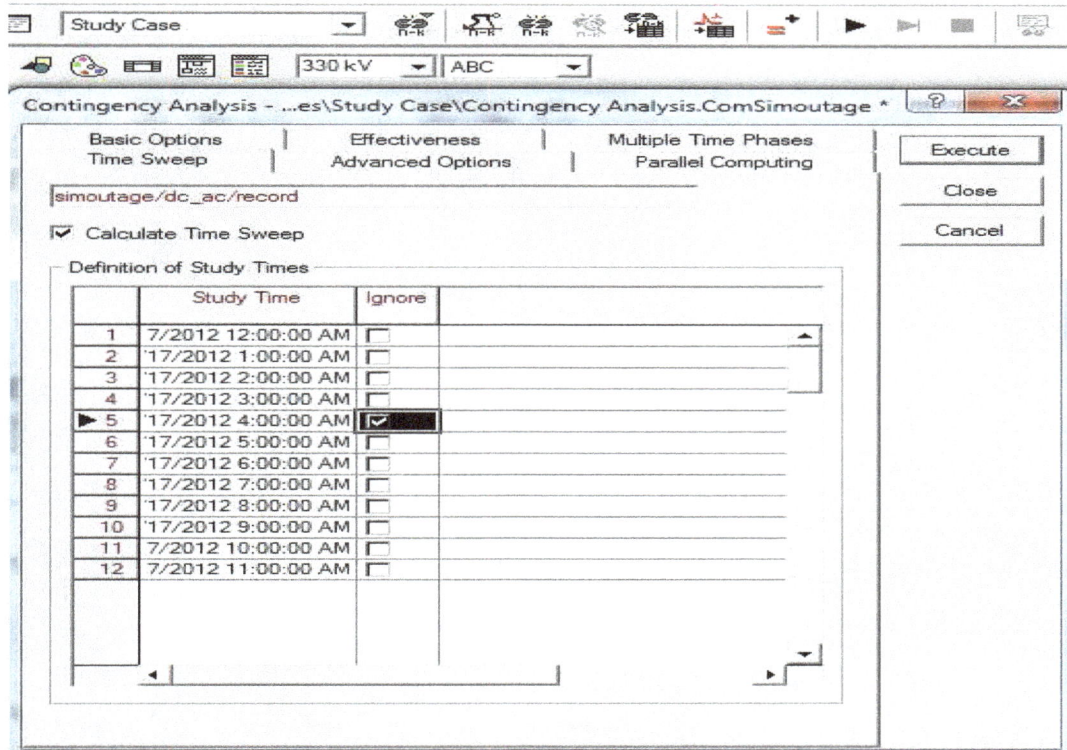

Figure 17. *Contingency analysis command dialogue showing time sweep.*

In multiple-time phase contingency analysis, contingency analysis for multiple time phases was performed, allowing the definition of post-fault actions. Line 2 (Sokoto – Kebbi), line 1 (Katsina – Kano), line 10 (Osun – Ondo), line 12 (Ogun – Oyo), line 16 (Delta – Edo), line 19 (Bauchi – Gombe), line 2 (Kano – Jigawa), line 21 (Gombe – Borno), line 3 (Zamfara – Kebbi), line 31 (F.C.T. – Niger), line 32 (Akwa – Cross River), line 39 (Imo – Ebonyi), N2 (Zamfara), N16 (Bayelsa), L22 (Ebonyi load) and L30 (Yobe load) were defined as contingency cases.

Table 2. *Single-point-in-time period analysis (connectivity method).*

Fault Clearance Breakers use all circuit breakers

Time to open remote controlled switches	1.00 minutes
Consider Maintenance	Yes
Protection/switching failures	Yes
Double earth faults	Yes
Independent second failures	Yes
SAIFI	0.034831 1/Ca
CAIFI	0.041377 1/Ca
SAIDI	0.348 h/Ca
CAIDI	10.002 h
ASAI	0.999960231
ASUI	3.97693E-05
ENS	149.400 MWh/a
AENS	0.011 MWh/Ca
ACCI	0.033 MWh/Ca
EIC	4.264 M$/a
IEAR	28.538 $/kWh
SES	0.000 MWh/a
ASIFI	0.035112 1/a
ASIDI	0.351190 h/a
MAIFI	0.000000 1/Ca

Table 3. *Complete year period analysis – (connectivity method).*

Fault Clearance Breakers
Use all circuit breakers

Switching procedures Sequential			
Calculation time period	2010		
Consider Maintenance	Yes		
Independent second failure	Yes		
Double earth faults	Yes		
Protection/switching failure	Yes		
Buses name	AIT (h/a)	AIF (1/a)	AID (h)
SingleBusbar(25) /N36	0.96	0.1	10
SingleBusbar(24) /N37	0.69	0.07	10
SingleBusbar(26) /N35	0.69	0.07	10
SingleBusbar(9) /N23	0.69	0.07	10
SingleBusbar(22) /N18	0.64	0.06	10
SingleBusbar(5) /N7	0.54	0.05	10
SingleBusbar /N2	0.32	0.03	10
SingleBusbar(1) /N4	0.32	0.03	10
SingleBusbar(10) /N22	0.32	0.03	10
SingleBusbar(11) /N12	0.32	0.03	10
SingleBusbar(12) /N16	0.32	0.03	10
SingleBusbar(13) /N1	0.32	0.03	10
SingleBusbar(14) /N9	0.32	0.03	10
SingleBusbar(15) /N10	0.32	0.03	10
SingleBusbar(16) /N11	0.32	0.03	10
SingleBusbar(17) /N14	0.32	0.03	10
SingleBusbar(2) /N3	0.32	0.03	10
SingleBusbar(20) /N19	0.32	0.03	10
SingleBusbar(21) /N20	0.32	0.03	10
SingleBusbar(23) /N15	0.32	0.03	10
SingleBusbar(28) /N21	0.32	0.03	10
SingleBusbar(29) /N17	0.32	0.03	10
SingleBusbar(3) /N5	0.32	0.03	10
SingleBusbar(34) /N26	0.32	0.03	10

**Fault Clearance Breakers
Use all circuit breakers**

SingleBusbar(35) /N34	0.32	0.03	10
SingleBusbar(36) /N25	0.32	0.03	10
SingleBusbar(4) /N6	0.32	0.03	10
SingleBusbar(6) /N8	0.32	0.03	10
SingleBusbar(7) /N13	0.32	0.03	10
SingleBusbar(8) /N24	0.32	0.03	10
SingleBusbar(18) /N30	0.32	0.03	10
SingleBusbar(19) /N28	0.32	0.03	10
SingleBusbar(27) /N27	0.32	0.03	10
SingleBusbar(30) /N31	0.32	0.03	10
SingleBusbar(31) /N29	0.32	0.03	10
SingleBusbar(32) /N33	0.32	0.03	10
SingleBusbar(33) /N32	0.32	0.03	10

Table 4. *Single-point-in-time period analysis - (load-flow method).*

Fault Clearance Breakers Use all circuit breakers	
Time to open remote controlled switches	1.00 minutes
Consider Maintenance	Yes
Protection/switching failures	Yes
Double earth faults	Yes
Independent second failures	Yes
SAIFI	0.362857 1/Ca
CAIFI	0.362857 1/Ca
SAIDI	3.603 h/Ca
CAIDI	9.931 h
ASAI	0.999588657
ASUI	0.000411343
ENS	1653.312 MWh/a
AENS	0.118 MWh/Ca
ACCI	0.000 MWh/Ca
EIC	4.264 M$/a
IEAR	26.229 $/kWh
SES	206.746 MWh/a
ASIFI	0.391121 1/a
ASIDI	3.886384 h/a
MAIFI	0.004643 1/Ca

Table 5. *Complete year period analysis – (load-flow method).*

Fault Clearance Breakers Use all circuit breakers			
Switching procedures Sequential			
Calculation time period	2010		
Consider Maintenance	Yes		
Independent second failure	Yes		
Double earth faults	Yes		
Protection/switching failure	Yes		
Buses name	AIT (h/a)	AIF (1/a)	AID (h)
SingleBusbar(25) /N36	0.96	0.1	10
SingleBusbar(24) /N37	0.69	0.07	10
SingleBusbar(26) /N35	0.69	0.07	10
SingleBusbar(9) /N23	0.69	0.07	10
SingleBusbar(22) /N18	0.64	0.06	10
SingleBusbar(5) /N7	0.54	0.05	10
SingleBusbar /N2	0.32	0.03	10
SingleBusbar(1) /N4	0.32	0.03	10
SingleBusbar(10) /N22	0.32	0.03	10
SingleBusbar(11) /N12	0.32	0.03	10
SingleBusbar(12) /N16	0.32	0.03	10
SingleBusbar(13) /N1	0.32	0.03	10
SingleBusbar(14) /N9	0.32	0.03	10
SingleBusbar(15) /N10	0.32	0.03	10
SingleBusbar(16) /N11	0.32	0.03	10
SingleBusbar(17) /N14	0.32	0.03	10
SingleBusbar(2) /N3	0.32	0.03	10
SingleBusbar(20) /N19	0.32	0.03	10
SingleBusbar(21) /N20	0.32	0.03	10
SingleBusbar(23) /N15	0.32	0.03	10
SingleBusbar(28) /N21	0.32	0.03	10
SingleBusbar(29) /N17	0.32	0.03	10
SingleBusbar(3) /N5	0.32	0.03	10
SingleBusbar(34) /N26	0.32	0.03	10
SingleBusbar(35) /N34	0.32	0.03	10
SingleBusbar(36) /N25	0.32	0.03	10
SingleBusbar(4) /N6	0.32	0.03	10
SingleBusbar(6) /N8	0.32	0.03	10
SingleBusbar(7) /N13	0.32	0.03	10
SingleBusbar(8) /N24	0.32	0.03	10
SingleBusbar(18) /N30	0.32	0.03	10
SingleBusbar(19) /N28	0.32	0.03	10
SingleBusbar(27) /N27	0.32	0.03	10
SingleBusbar(30) /N31	0.32	0.03	10
SingleBusbar(31) /N29	0.32	0.03	10
SingleBusbar(32) /N33	0.32	0.03	10
SingleBusbar(33) /N32	0.32	0.03	10

The time phases of a contingency analysis are defined in the calculation settings section of the Basic Data tab of the contingency analysis command dialogue, by specifying a 'post contingency time' for each defined time phase. A specified post contingency time defines the end of a time phase and is used to determine which events (actions) from the analyzed contingency are considered. If the time of occurrence of an event from a contingency occurs earlier than or equal to the post contingency time, the event will be considered in the corresponding load flow calculation.

Figure 18 shows the stated study case, while Figures 19 and 20 show minimum voltage violations and maximum voltage violations ASCII report respectively. Figures 21 and 22 ASCII report depicts minimum voltages and voltage steps respectively.

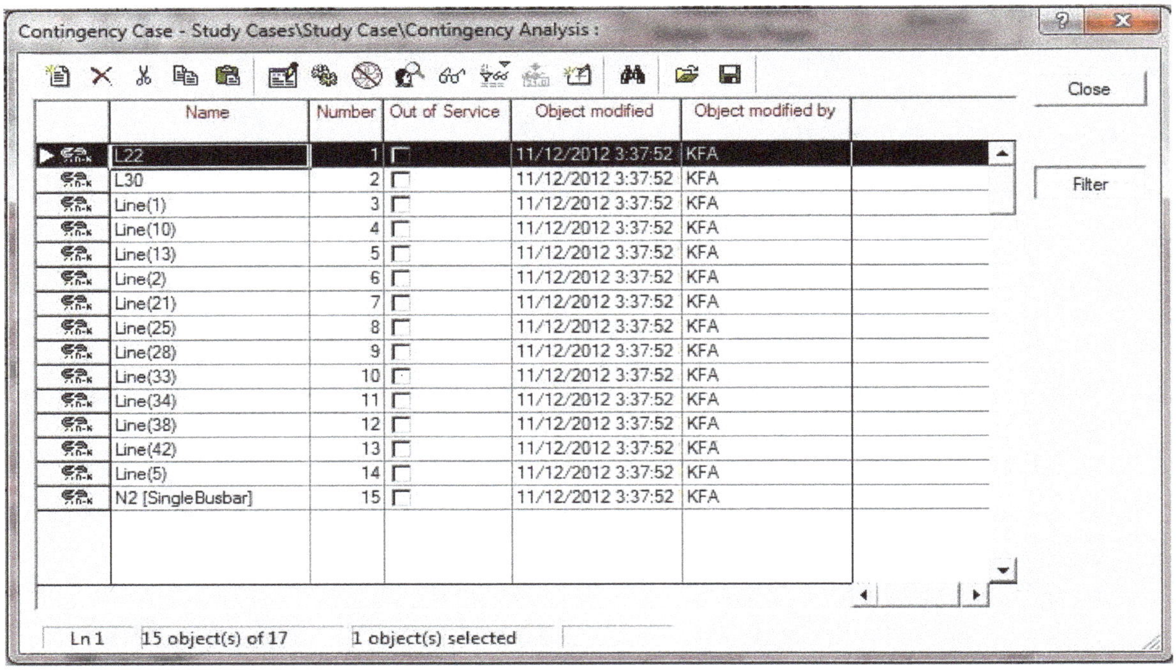

Figure 18. *Contingency analysis – study case*

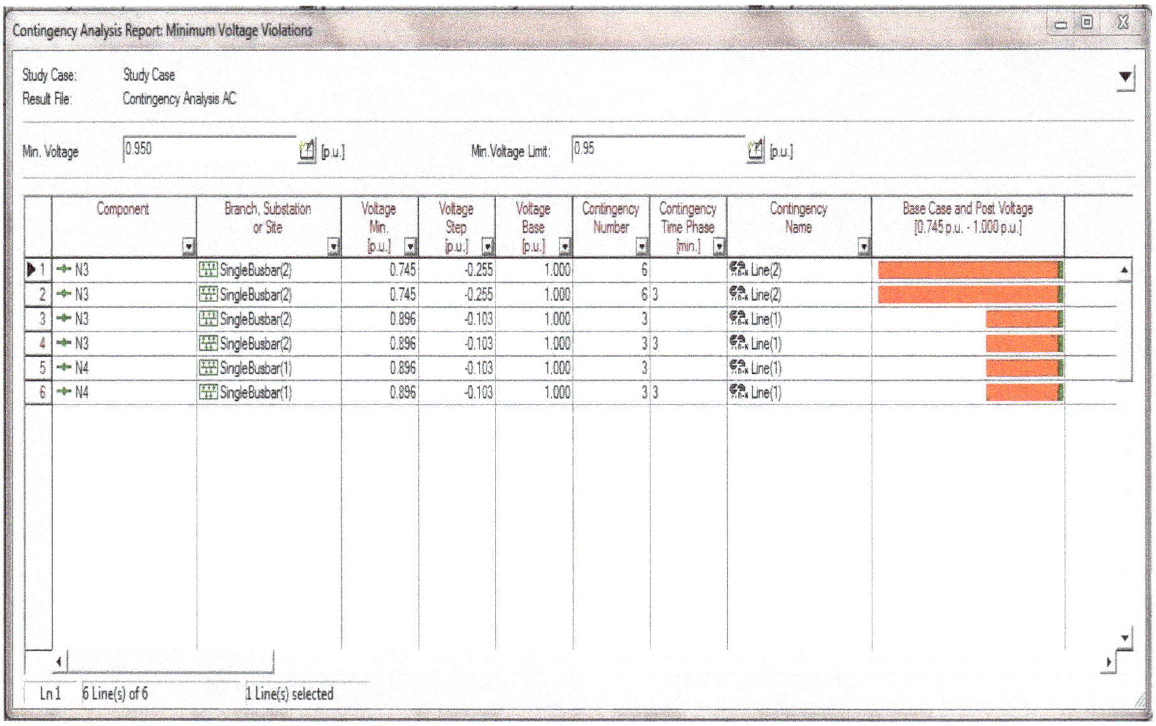

Figure 19. *ASCII report minimum voltage violations.*

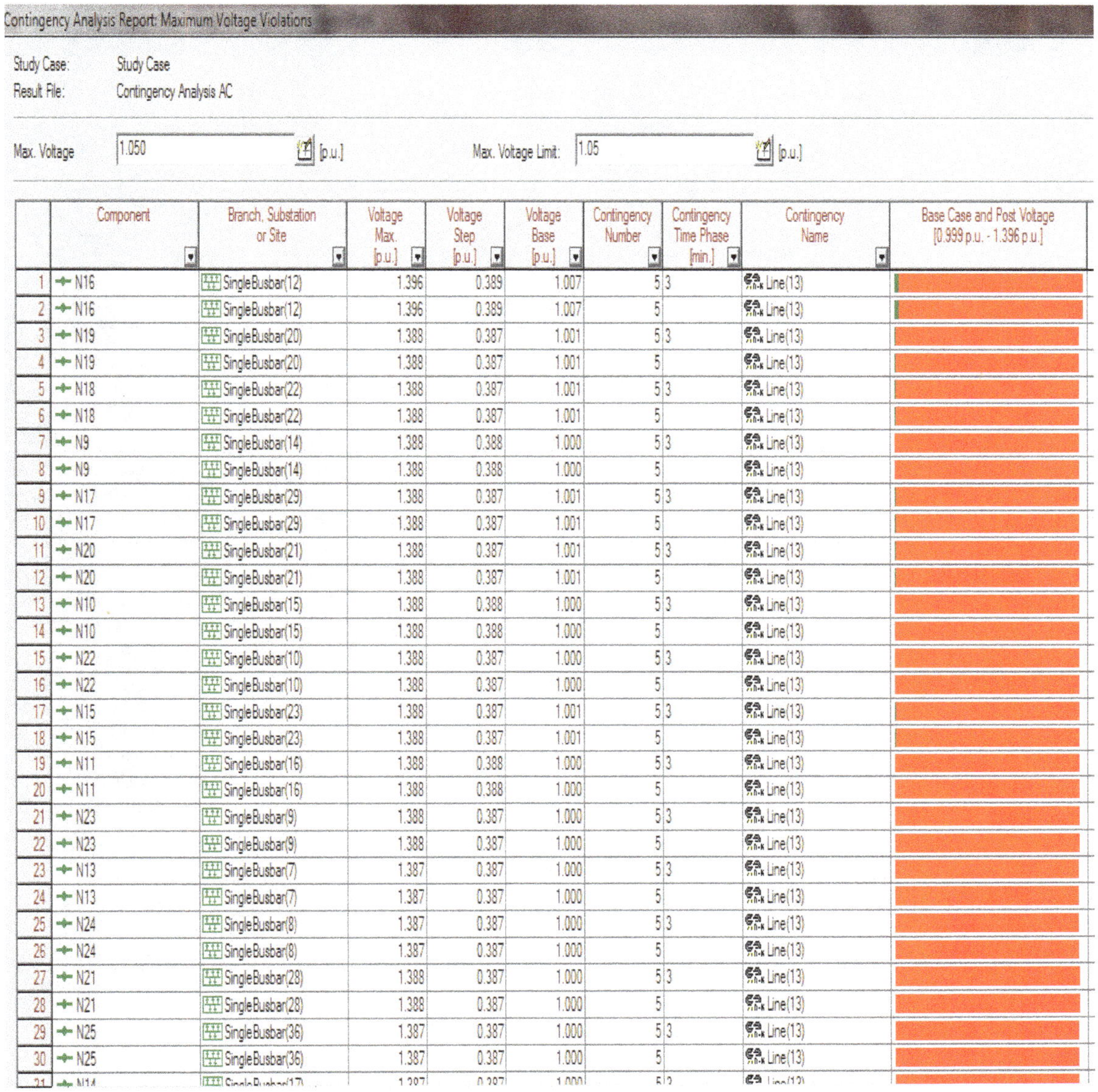

Contingency Analysis Report: Maximum Voltage Violations

Study Case: Study Case
Result File: Contingency Analysis AC

Max. Voltage   1.050   [p.u.]   Max. Voltage Limit: 1.05   [p.u.]

	Component	Branch, Substation or Site	Voltage Max. [p.u.]	Voltage Step [p.u.]	Voltage Base [p.u.]	Contingency Number	Contingency Time Phase [min.]	Contingency Name	Base Case and Post Voltage [0.999 p.u. - 1.396 p.u.]
1	N16	SingleBusbar(12)	1.396	0.389	1.007	5	3	Line(13)	
2	N16	SingleBusbar(12)	1.396	0.389	1.007	5		Line(13)	
3	N19	SingleBusbar(20)	1.388	0.387	1.001	5	3	Line(13)	
4	N19	SingleBusbar(20)	1.388	0.387	1.001	5		Line(13)	
5	N18	SingleBusbar(22)	1.388	0.387	1.001	5	3	Line(13)	
6	N18	SingleBusbar(22)	1.388	0.387	1.001	5		Line(13)	
7	N9	SingleBusbar(14)	1.388	0.388	1.000	5	3	Line(13)	
8	N9	SingleBusbar(14)	1.388	0.388	1.000	5		Line(13)	
9	N17	SingleBusbar(29)	1.388	0.387	1.001	5	3	Line(13)	
10	N17	SingleBusbar(29)	1.388	0.387	1.001	5		Line(13)	
11	N20	SingleBusbar(21)	1.388	0.387	1.001	5	3	Line(13)	
12	N20	SingleBusbar(21)	1.388	0.387	1.001	5		Line(13)	
13	N10	SingleBusbar(15)	1.388	0.388	1.000	5	3	Line(13)	
14	N10	SingleBusbar(15)	1.388	0.388	1.000	5		Line(13)	
15	N22	SingleBusbar(10)	1.388	0.387	1.000	5	3	Line(13)	
16	N22	SingleBusbar(10)	1.388	0.387	1.000	5		Line(13)	
17	N15	SingleBusbar(23)	1.388	0.387	1.001	5	3	Line(13)	
18	N15	SingleBusbar(23)	1.388	0.387	1.001	5		Line(13)	
19	N11	SingleBusbar(16)	1.388	0.388	1.000	5	3	Line(13)	
20	N11	SingleBusbar(16)	1.388	0.388	1.000	5		Line(13)	
21	N23	SingleBusbar(9)	1.388	0.387	1.000	5	3	Line(13)	
22	N23	SingleBusbar(9)	1.388	0.387	1.000	5		Line(13)	
23	N13	SingleBusbar(7)	1.387	0.387	1.000	5	3	Line(13)	
24	N13	SingleBusbar(7)	1.387	0.387	1.000	5		Line(13)	
25	N24	SingleBusbar(8)	1.387	0.387	1.000	5	3	Line(13)	
26	N24	SingleBusbar(8)	1.387	0.387	1.000	5		Line(13)	
27	N21	SingleBusbar(28)	1.388	0.387	1.000	5	3	Line(13)	
28	N21	SingleBusbar(28)	1.388	0.387	1.000	5		Line(13)	
29	N25	SingleBusbar(36)	1.387	0.387	1.000	5	3	Line(13)	
30	N25	SingleBusbar(36)	1.387	0.387	1.000	5		Line(13)	
31	N14	SingleBusbar(17)	1.387	0.387	1.000	5	3	Line(13)	

Figure 20. *ASCII report maximum voltage violations.*

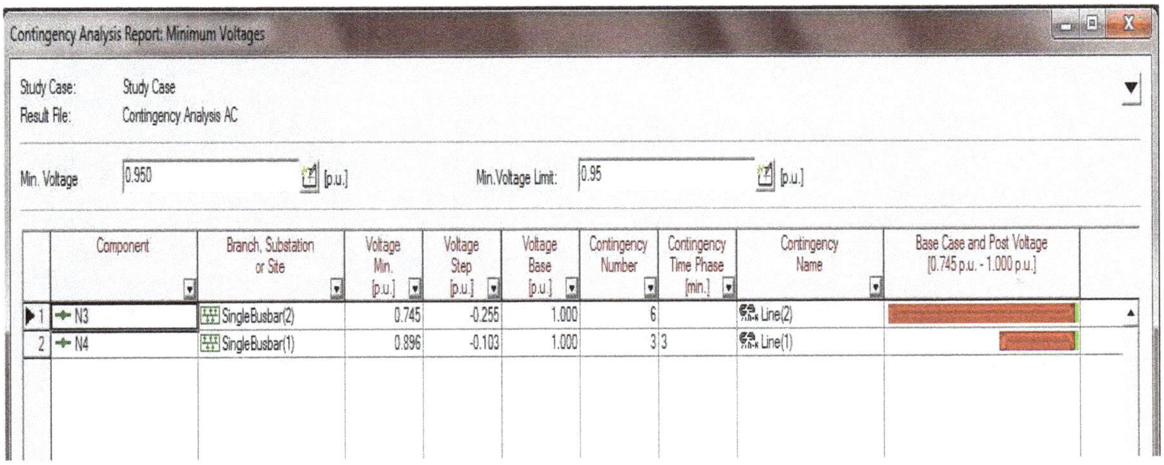

Contingency Analysis Report: Minimum Voltages

Study Case: Study Case
Result File: Contingency Analysis AC

Min. Voltage   0.950   [p.u.]   Min.Voltage Limit: 0.95   [p.u.]

	Component	Branch, Substation or Site	Voltage Min. [p.u.]	Voltage Step [p.u.]	Voltage Base [p.u.]	Contingency Number	Contingency Time Phase [min.]	Contingency Name	Base Case and Post Voltage [0.745 p.u. - 1.000 p.u.]
1	N3	SingleBusbar(2)	0.745	-0.255	1.000	6		Line(2)	
2	N4	SingleBusbar(1)	0.896	-0.103	1.000	3	3	Line(1)	

Figure 21. *ASCII report – minimum voltages.*

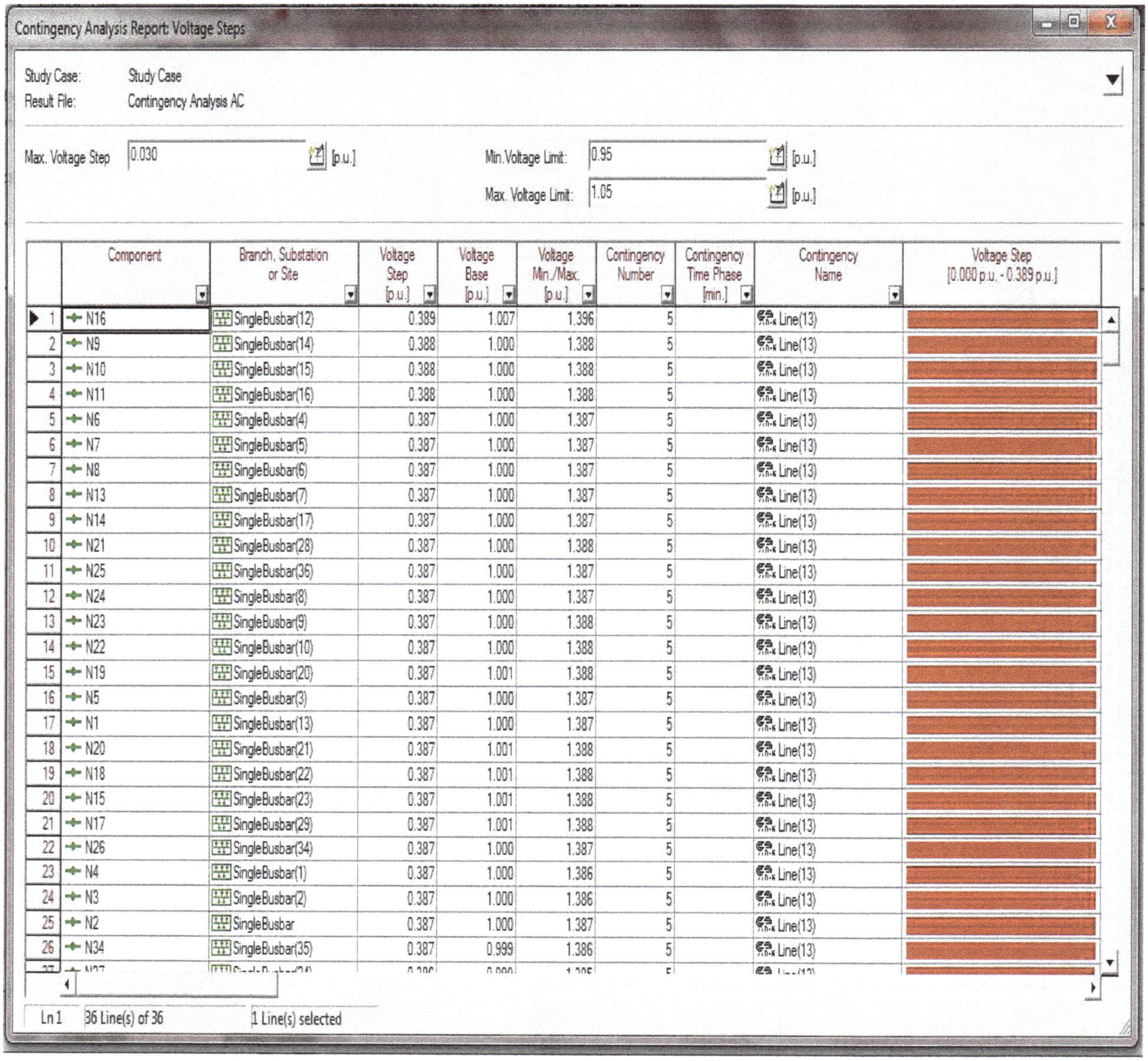

	Component	Branch, Substation or Site	Voltage Step [p.u.]	Voltage Base [p.u.]	Voltage Min./Max. [p.u.]	Contingency Number	Contingency Time Phase [min.]	Contingency Name	Voltage Step [0.000 p.u. - 0.389 p.u.]
1	N16	SingleBusbar(12)	0.389	1.007	1.396	5		Line(13)	
2	N9	SingleBusbar(14)	0.388	1.000	1.388	5		Line(13)	
3	N10	SingleBusbar(15)	0.388	1.000	1.388	5		Line(13)	
4	N11	SingleBusbar(16)	0.388	1.000	1.388	5		Line(13)	
5	N6	SingleBusbar(4)	0.387	1.000	1.387	5		Line(13)	
6	N7	SingleBusbar(5)	0.387	1.000	1.387	5		Line(13)	
7	N8	SingleBusbar(6)	0.387	1.000	1.387	5		Line(13)	
8	N13	SingleBusbar(7)	0.387	1.000	1.387	5		Line(13)	
9	N14	SingleBusbar(17)	0.387	1.000	1.387	5		Line(13)	
10	N21	SingleBusbar(28)	0.387	1.000	1.388	5		Line(13)	
11	N25	SingleBusbar(36)	0.387	1.000	1.387	5		Line(13)	
12	N24	SingleBusbar(8)	0.387	1.000	1.387	5		Line(13)	
13	N23	SingleBusbar(9)	0.387	1.000	1.388	5		Line(13)	
14	N22	SingleBusbar(10)	0.387	1.000	1.388	5		Line(13)	
15	N19	SingleBusbar(20)	0.387	1.001	1.388	5		Line(13)	
16	N5	SingleBusbar(3)	0.387	1.000	1.387	5		Line(13)	
17	N1	SingleBusbar(13)	0.387	1.000	1.387	5		Line(13)	
18	N20	SingleBusbar(21)	0.387	1.001	1.388	5		Line(13)	
19	N18	SingleBusbar(22)	0.387	1.001	1.388	5		Line(13)	
20	N15	SingleBusbar(23)	0.387	1.001	1.388	5		Line(13)	
21	N17	SingleBusbar(29)	0.387	1.001	1.388	5		Line(13)	
22	N26	SingleBusbar(34)	0.387	1.000	1.387	5		Line(13)	
23	N4	SingleBusbar(1)	0.387	1.000	1.386	5		Line(13)	
24	N3	SingleBusbar(2)	0.387	1.000	1.386	5		Line(13)	
25	N2	SingleBusbar	0.387	1.000	1.387	5		Line(13)	
26	N34	SingleBusbar(35)	0.387	0.999	1.386	5		Line(13)	

Figure 22. *ASCII report – voltage steps.*

8. Conclusion

Since objective of electric power systems is to supply electrical energy to customers at low cost, while simultaneously providing acceptable, economically and justifiable service quality, generation adequacy, voltage sag, contingencies analysis and reliability analysis are very important. Deterministic indices reflect postulated conditions. They are not directly indicative of electric system reliability and are not response to most parameters which influence system reliability performance; this is applicable to contingency analysis. Probabilistic indices directly reflect the uncertainty which is inherent in the power system reliability problem and have the capability of reflecting the various parameters which impact reliability [4].

Contingency analysis could be used for blackout prediction in power grid. It simulates and quantifies the results of problems that could occur in the power system in the immediate future. Also, reliability is a key aspect of power system design and planning; giving system interruptions during an operating period, while voltage sag assessment provides frequency and severity of voltage sag during an operation period.

Nomenclature

SAIFI - System Average Interruption Frequency Index
CAIFI - Customer Average Interruption Frequency Index
SAIDI - System Average Interruption Duration Index
CAIDI - Customer Average Interruption Duration Index
ASAI - Average Service Availability Index
ASUI - Average Service Unavailability Index
ENS - Energy Not Supplied
AENS - Average Energy Not Supplied
ACCI - Average Customer Curtailment Index

EIC - Expected Interruption Cost
IEAR - Interrupted Energy Assessment Rate
SES - System energy shed
ASIFI - Average System Interruption Frequency Index
ASIDI - Average System Interruption Duration Index
MAIFI - Momentary Average Interruption Freq. Index
F.C.T. – Federal Capital Territory

References

[1] DIgSILENT PowerFactory Version 14.1 Tutorial, DIgSI-LENT GmbH Heinrich-Hertz-StraBe 9, 72810 Gomaringen, Germany, May, 2011.

[2] D'Annunzio C., "Generation Adequacy Assessment of Power Systems with Significant Wind Generation: A System Planning and Operations Perspective", Unpublished Thesis, the University of Texas at Austin, 2009.

[3] Hegazy Y. G., M. M. A. Salama and A. Y. Chikhani, "Adequacy Assessment of Distributed Generation Systems Using Monte Carlo Simulation", Proc. IEEE Transaction on Power Systems, Vol. 18, No 1, February, 2003.

[4] Amjady N. 'Generation Adequacy Assessment of Power Systems by Time Series and Fuzzy Neural Network', IEEE Transactions on Power Systems, Vol. 21, No. 3, August 2006.

[5] http://www.ece.tamu.edu/People/bios/singh/coursenotes/part2.pdf.

[6] Chen Q. and McCalley J., "Identifying high risk N ¡ k contingencies for online security assessment," IEEE Transactions on Power Systems, vol. 20, no. 2, pp. 823–834, May 2005.

[7] V. Donde, V. L'opez, B. C. Lesieutre, A. Pinar, C. Yang, and J. Meza, "Identification of severe multiple contingencies in electric power networks," in Proceedings of the North American Power Symposium, Ames, IA, October 2005.

[8] Z. Feng, V. Ajjarapu, and D. Maratukulam, "A practical minimum load shedding strategy to mitigate voltage ollapse,"

IEEE Transactions on Power Systems, vol. 13, no. 4, pp. 1285–1291, November 1998.

[9] Zhu D. 'Power System Reliability Analysis with Distributed Generators' Unpublished M.Sc. Thesis, Virginia Polytechnic Institute and State University, Blacksburg, VA, U.S.A., May, 2003.

[10] Wangdee W. 'Bulk Electric System Reliability Simulation and Application', Unpublished PhD Thesis, University of Saskatchewan, Saskatoon, December 2005.

[11] Xie Z., Manimaran G., Vittal V., Phadke A. G., and Centeno V., 'An Information Architecture for Future Power Systems and Its Reliability Analysis', IEEE Transactions on Power Systems, Vol. 17, No. 3, August 2002.

[12] http://triton.elk.itu.edu.tr/~ozdemir/rel.pdf.

[13] Lamoree, J., Smith, J. C., Vinett, P., Duffy, T., and Klein, M., "The Impact of Voltage Sags on Industrial Plant Loads." Paper presented at the First International Conference on Power Quality: End-Use Applications and Perspectives, Paris, France, October 14-16, 1991.

[14] Conrad L., Little K., and Grigg C., 'Predicting and Preventing Problems Associated with Remote Fault-Clearing Voltage Dips," IEEE Transactions on Industry Applications, vol. 27, pp. 167-172, Jan. 1991.

[15] Eddy C. Aeloíza, Prasad N. Enjeti, Luis A. Morán, Oscar C. Montero-Hernandez, and Sangsun Kim 'Analysis and Design of a New Voltage Sag Compensator for Critical Loads in Electrical Power Distribution Systems', IEEE Transactions On Industry Applications, Vol. 39, No. 4, July/August 2003.

[16] Pirjo Heine, and Matti Lehtonen 'Voltage Sag Distributions Caused by Power System Faults', IEEE Transactions on Power Systems, Vol. 18, No. 4, November 2003.

[17] Murta-Vale M.H., Campici P., Menezes T.V., Visacro S. and Nietzsch-Dias R. 'Power System Expansion Planning: Applying LLS Data to Evaluate Lightning-Related Voltage Sags', 19th International Lightning Detection Conference, Tucson, Arizona, U.S.A., 24th – 25th April, 2006.19th International Lightning Detection Conference.

12

Time-dependent exergy analysis of a 120 MW steam turbine unit of sapele power plant

Obodeh, O.[1], Ugwuoke, P. E.[2]

[1]Mechanical Engineering Department Ambrose Alli University, Ekpoma, Edo State, Nigeria
[2]Mechanical Engineering Department Petroleum Training Institute, Effurun, Delta State, Nigeria

Email address:
engobodeh@yahoo.com (Obodeh, O.)

Abstract: Time-dependent exergy model was used to assess the exergy losses that occurred in the major components of a 120 MW steam turbine unit of Sapele power station. Data used for the analysis were both base parameters and measured values recorded in the station operational logbook for the period of January 2007 to December 2011. Component's exergy destruction increments as compared with its base value were highlighted and possible causes of the increment were identified. The boiler section had the highest value. The economiser had a maximum of 4.26 % in 2009 and minimum of 1.25 % in 2007. While the evaporator had a maximium of 5.02 % in 2009 and minimum of 1.50 % in 2008. The superheater had maximum of 4.64 % in 2011 and minimum of 1.48 % in 2007. For the reheater, the maximum was 3.57 % in 2011 while the minimum was 1.71 % in 2007. Tube fouling, defective burners, steam traps and air heater fouling were adduced for the increment. Upgrading components with better designs, optimizing system performance and elimination of conditions that degrade efficiency between maintenance outages were suggested for improving the performance of the boiler section. The analysis showed that for the three turbine stages, HP turbine had the highest increment while the LP turbine had the lowest. The loss in the three turbine stages were attributed to throttling losses at the governor valves and silica deposits at the nozzles and blades. Retrofitting of rotors, diaphragms or complete stator/ rotor modules (inner block) were suggested for improving the situation. The results generally showed that exergy loss increased with increased operation time. It was observed that deterioration and obsolescence may be the major problems and that plant rehabilitation is a feasible solution. It was noted that the suggested modification and refurbishment of Sapele power plant units is an attractive solution to improve the plant economy and keep production cost competitive in a restructured Nigerian power system.

Keywords: Aging Effect, Exergy Destruction, Rehabilitation, Deregulated Market

1. Introduction

Electricity demands in Nigeria far outstrip its supply which is epileptic [1, 2]. Currently, electric energy output is very low, with present installed capacity for energy generation put at 6,200 MW, while actual output hovers between 2,500 MW and 3,200 MW [3]. By the year 2020, the Government's policy objective is that Nigeria should posses a generating capacity of at least 40,00 MW [4]. The investments required to finance an increase in total power station capacity from 12,000 MW to 40,000 MW is huge. Hence the need to incentivize the private sector to partner with government in this endeavour. The unbundling of the Power Holding Company of Nigeria (PHCN) has been an important step

Restructured and liberalized power sectors promote increased competition through unbundling of generation, transmission and privatization of distribution or retailing function [5-8]. Decentralization requires the existing power plants to improve their performance in order to attain high thermal efficiency and reliability, so as to operate at low generation cost. To improve the performance of the plant, first it is necessary to find out the equipment/locations where losses are more [9-12].

Exergy analysis provides the tool for the clear distinction between energy losses to the environment and internal irreversibility of the process. Exergy is defined as the maximum theoretical useful work that can be obtained as a system interacts with an equilibrium state. Exergy is generally not conserved like energy but is destroyed in the system. Exergy can be divided into four distinct classes, viz: physical, Chemical, potential and kinetic exergies. The two important ones are physical exergy and chemical exergy. In thermal power plant, the other two classes are assumed negligible as the elevation and speed have negligible changes. The physical exergy is defined as the maximum theoretical useful work obtained as a system interacts with an equilibrium state. The chemical exergy is associated with the departure of the chemical composition of a system from its chemical equilibrium.

Exergy destruction is the measure of irreversibility that is the source of performance loss. Therefore, an exergy analysis assessing the magnitude of exergy destruction identifies the location, the magnitude and source of thermodynamic inefficiencies in a thermal system. Recent developments in exergy concept have allowed the definition of a new performance criterion which offers some advantages over the tradition ones [13-15]. Hence, exergy analysis can improve resource utilization by determining inefficient, wasteful processes within thermodynamic systems and the results obtained from such analysis can serve as a guide for reducing irreversibilities and performance monitoring giving room for performance improvement.

The objective of this study is to investigate the influence of aging on the exergy destruction in the main components of a 120 MW steam turbine unit of Sapele power station. Physical exergy was used in the analysis. The power plant is strategically located at Ogorode, close to sources of natural gas feedstock and a river for cooling its steam turbine generators. Sapele power plant has an installed capacity of 1020 MW which equates to approximately one-sixth of the nation's installed capacity [5]. It is consists of 6 x 120 MW steam turbines and 4 x 75 MW gas turbines. The steam turbines were commissioned between December, 1978 and April, 1980 while the gas turbines were commissioned between June, 1981 and August, 1981. The steam turbine

designated ST02 used for this study was commissioned on February 2, 1979. The designed power output of the unit is 120 MW but due to operational inefficiencies and other losses, the daily output is about 80 MW. This study therefore seeks to investigate the causes of operational inefficiencies and losses resulting in decrease of expected output of the unit using exergy analysis. The study covers the period between January, 2007 and December, 2011.

Despite many publications on the exergy analysis of power plant [16-22], most of the them applied it to find optimum values for main cycle parameters. Although, these researches are useful to improve the design features of future power plants, they do not suggest any recommendation on how to improve an existing aged power plant. In this study, the calculated parameters based on operating data were compared with the "base" values (these are the values obtainable when the unit was newly installed) to determine the aging effects on the unit performance.

2. Materials and Method

The data used for this study were both base parameters for the steam turbine and measured values recorded in the station operational logbook for the period of January 2007 to December 2011 [23]. Parameters considered during the data collection were the pressures, temperatures and mass flowrates at various points. In the analysis of the data, mean values of daily parameters were computed using statistical methods. This was followed by monthly average and then the yearly average for the period of the research. Fig. 1 shows the schematic diagram of the power plant, demonstrating all its relevant components. The plant unit main thermodynamics data are shown in Table 1.

Using Mollier diagram and thermodynamic tables for steam and thermodynamic equations, the process parameters for the unit are obtained as given in Table 2.

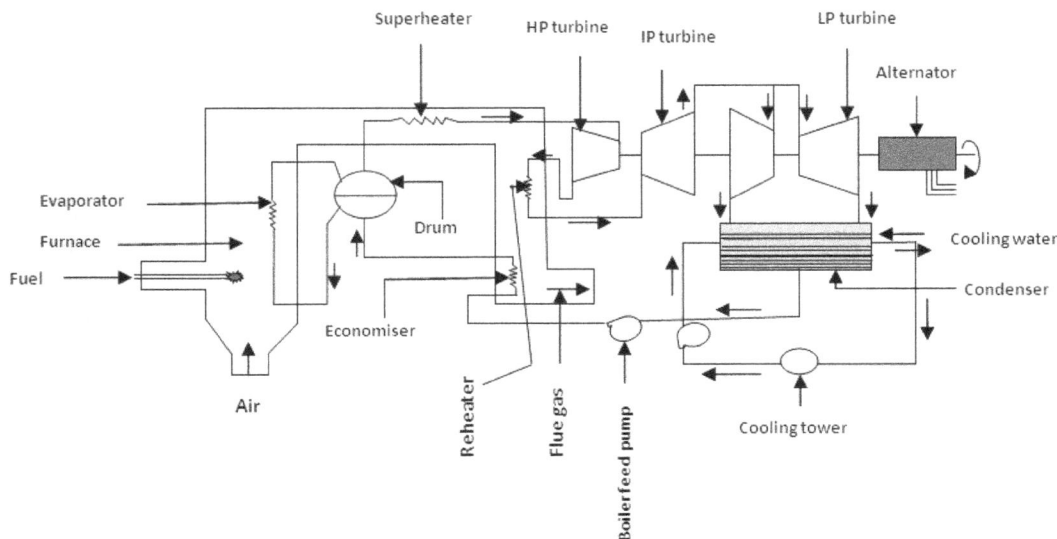

Fig 1. Schematic diagram of the plan

Table 1. Operating conditions of the 120 MW steam turbine unit

| Steam flow | Base parameters | | | Operating parameters | | | | | | | | | | |
| | Temp. (°C) | Press. (bar) | Steam flow rate (kg/s) | Average temperature (°C) | | | | | Average pressure (bar) | | | | |
				2007	2008	2009	2010	2011	2007	2008	2009	2010	2011
Boiler feed pump inlet	128	2.64	87	122	120	118	117	117	2.66	2.67	2.78	2.78	2.84
Boiler feed pump outlet	130	125	87	125	123	121	121	120	124.7	124.5	124	124	123
Economiser inlet	190	123	87	186	185	184	182	182	121.8	121.5	121	120	120
Economiser exhaust to Evaporator	210	122.2	87	207	205	203	201	201	121	120.7	120.1	119.2	119.2
Evaporator exhaust to Superheater	320	122.2	87	317	316	314	313	312	121	120.7	120.1	119.2	119.2
Superheater exhaust to HP turbine	510	120	87	505	502	501	500	498	118.3	118.2	117.6	117.1	117.1
HP turbine exhaust to Reheater	330	24	79	338	340	341	341	345	25.0	25.3	25.4	25.7	25.9
Reheater exhaust to IP turbine	510	24	72.7	505	502	501	500	498	23.8	23.7	23.5	23.4	23.4
IP turbine exhaust to LP turbine	240	2	65	245	247	249	253	252	2.2	2.2	2.4	2.4	2.4
LP turbine exhaust to Condenser	39	0.074	56.4	44	47	49	48	49	0.091	0.101	0.120	0.111	0120
Condenser outlet	38	0.07	58.5	42	46	48	47	47	0.082	0.101	0.111	0.111	0.120

Table 2. Process data for the 120 MW steam turbine unit

| Steam flow | Base parameters | | | Operating parameters | | | | | | | | | | |
| | Enthapy (kJ/kg) | Entropy (kJ/kgK) | Steam flow rate (kg/s) | Enthapy (kJ/kg) | | | | | Entropy (kJ/kgK) | | | | |
				2007	2008	2009	2010	2011	2007	2008	2009	2010	2011
Boiler feed pump inlet	537.8	1.613	87	511.9	504.1	495.5	491.4	491.4	1.549	1.528	1.506	1.495	1.495
Boiler feed pump outlet	546	1.634	87	525.3	516.6	508	508	504.1	1.581	1.56	1.539	1.539	1.528
Economiser inlet	807.4	2.235	87	789.3	784.9	780.6	772	772	2.197	2.188	2.175	2.157	2.157
Economiser exhaust to Evaporator	897.9	2.425	87	884.5	875.3	866.2	851.3	851.3	2.397	2.379	2.361	2.34	2.34
Evaporator exhaust to Superheater	2798	3.449	87	2708	2711	2717	2719	2722	5.562	5.571	5.623	5.687	5.751
Superheater exhaust to HP turbine	3374	6.503	87	3431	3427	3410	3402	3402	6.807	6.791	6.719	6.681	6.681
HP turbine exhaust to Reheater	3103	6.969	79	3122	3127	3130	3131	3140	7.02	7.034	7.039	7.046	7.065
Reheater exhaust to IP turbine	3494	7.537	72.7	3456	3456	3473	3471	3466	7.52	7.51	7.503	7.498	7.492
IP turbine exhaust to LP turbine	2951	7.668	65	2962	2966	2970	2978	2876	7.726	7.734	7.78	7.795	7.792
LP turbine exhaust to Condenser	2383	7.668	56.4	2418	2439	2453	2446	2453	7.557	7.464	7.402	7.432	7.401
Condenser outlet	159.1	0.545	58.5	175.8	192.5	200.9	196.7	196.7	0.599	0.651	0.691	0.6645	0.6645

The mean daily temperature of the region hovers around 27 °C all the year round. The minimum and maximum temperatures are 20 °C and 40 °C respectively [21]. Hence, in this study, 30 °C was used as the mean ambient temperature and 1.013 bar as its pressure.

In analyzing the unit, the cycle was assumed to operate at steady state with no stray heat transfer from any component to its surroundings and negligible kinetic and potential energy effects. Certain components such as boiler stop valves, fuel and oil pumps, induced draught and forced draught fans were neglected in the analysis. Pressure drops along pipelines were assumed to be negligible.

For a control volume, an exergy balance equation is expressed as

$$\sum W = \sum \left(1 - \frac{T_o}{T} \right) Q + \sum \Psi_{in} - \sum \Psi_{out} - \sum \Psi_{des} \quad (1)$$

where

$\sum W$ = sum of ideal work; T_o = reference temperature; T = temperature of system

ΣQ = sum of heat supplied; $\Sigma \dot{\Psi}_{in}$ = sum of exergy inflow; $\Sigma \dot{\Psi}_{out}$ = sum of exergy outflow

$\Sigma \dot{\Psi}_{des}$ = sum of exergy lost in the system due to irreversibilities
where

$$\dot{\Psi} = \dot{m}\left[h - h_o - T_o\left(s - s_o\right)\right] \qquad (2)$$

and

$\dot{m}$ is mass flowrate, h and s represent specific enthalpy and entropy respectively. The subscript o denotes reference condition.

For steam turbine

$$W = W_t = \dot{m}_s\left(h_{out,s} - h_{in,s}\right) \qquad (3)$$

$$Q = \dot{m}_w\left(h_{out,w} - h_{in,w}\right) \qquad (4)$$

where subscript s denotes steam phase and that for water is w.

Exergy of boiler feed pump is given by

$$\dot{\Psi}_{fw} = C_{p_w}\left[\left(T_w - T_o\right) - T_o \ell n\left(\frac{T_w}{T_o}\right)\right] \qquad (5)$$

where C_{pw} is specific heat of water.

And the pump work, W_p is given by

$$W_p = \dot{m}_w v_w\left(P_{out,w} - P_{in,w}\right) \qquad (6)$$

where v is specific volume P, pressure

The exergy efficiency, η_{ex} can be defined, according to Lozano and Valero [24] and Tsatsaronis and Winhold [25] by

$$\eta_{ex} = \frac{product}{input} \qquad (7)$$

This equation establishes a relationship between the desired result (for instance, the heating of a steam flow, or the power in a turbine) and the input (the amount of exergy spent to obtain the result). In some systems there is no universal agreement as to what are an input and an output. Therefore their exergy efficiency must be defined by the expression proposed by Szargut et al. [26] as

$$\eta_{ex} = \frac{outlet}{inlet} \qquad (8)$$

Table 3 summaries the equations used to compute the exergy destruction rate and exergy efficiency of the unit main components

Table 3. *Formulae for the exergy destruction rate and exergy efficiency of the unit's main components*

Component	Exergy destruction rate	Exergy efficiency
Boiler feed pump	$\dot{\Psi}_{des,p} = \dot{\Psi}_{in,p} - \dot{\Psi}_{out,p} + W_p$	$\eta_{ex,p} = 1 - \dfrac{\dot{\Psi}_{des,p}}{W_p}$
Economiser	$\dot{\Psi}_{des,ec} = \dot{\Psi}_{in,ec} - \dot{\Psi}_{out,ec}$	$\eta_{ex,ec} = \dfrac{\dot{\Psi}_{out,ec}}{\dot{\Psi}_{in,ec}}$
Evaporator	$\dot{\Psi}_{des,ev} = \dot{\Psi}_{in,ev} - \dot{\Psi}_{out,ev}$	$\eta_{ex,ev} = \dfrac{\dot{\Psi}_{out,ev}}{\dot{\Psi}_{in,ev}}$
Superheater	$\dot{\Psi}_{des,su} = \dot{\Psi}_{in,su} - \dot{\Psi}_{out,su}$	$\eta_{ex,su} = \dfrac{\dot{\Psi}_{out,su}}{\dot{\Psi}_{in,su}}$
Turbine	$\dot{\Psi}_{des,t} = \dot{\Psi}_{in,t} - \dot{\Psi}_{out,t} - W_t$	$\eta_{ex,t} = 1 - \dfrac{\dot{\Psi}_{des,t}}{\dot{\Psi}_{in,t} - \dot{\Psi}_{out,t}}$
Reheater	$\dot{\Psi}_{des,r} = \dot{\Psi}_{in,r} - \dot{\Psi}_{out,r}$	$\eta_{ex,r} = 1 - \dfrac{\dot{\Psi}_{des,r}}{\dot{\Psi}_{in,r}}$
Condenser	$\dot{\Psi}_{des,c} = \dot{\Psi}_{in,c} - \dot{\Psi}_{out,c}$	$\eta_{ex,c} = \dfrac{\dot{\Psi}_{out,c}}{\dot{\Psi}_{in,c}}$

3. Results and Discussion

When much of the 20 to 25 year old plants (that is, 160,000 to 200,000 operating hours) were originally designed, the plants were expected to run at base load. The only thermal limit applied in the design was creep; thermal fatigue resulting from frequent stops/starts was not anticipated [13]. Due to deterioration, steam power units of more than 200,000 operating hours are facing serious threats in view of their remaining lifetime. Even with proper operation and maintenance, the flow path section in the steam turbine plant will become fouled, eroded, corroded and covered with rust scale. The consequence is increased exergy destruction in their various components.

The base values of exergy destruction and exergy efficiency in boiler feed pump are 0.92 MW and 1.22 % respectively. The increment from the base values for the period under review is presented in Fig. 2.

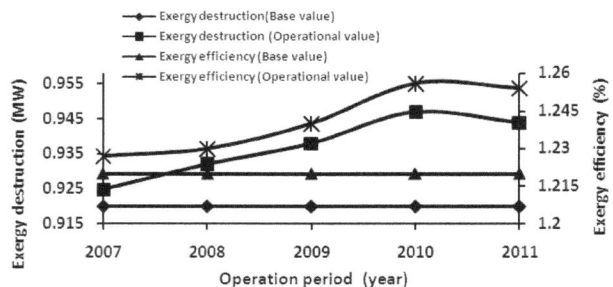

Fig 2. Exergy destruction and efficiency as a function of operation period in boiler feed pump

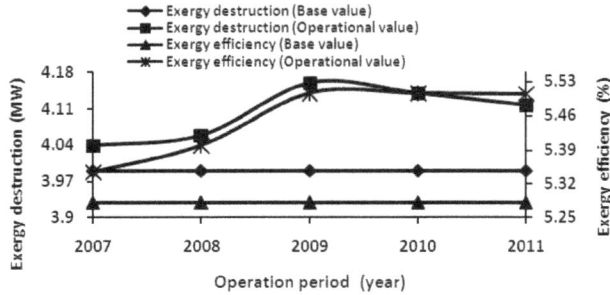

Fig 3. *Variation of exergy destruction and efficiency with operation period in economiser*

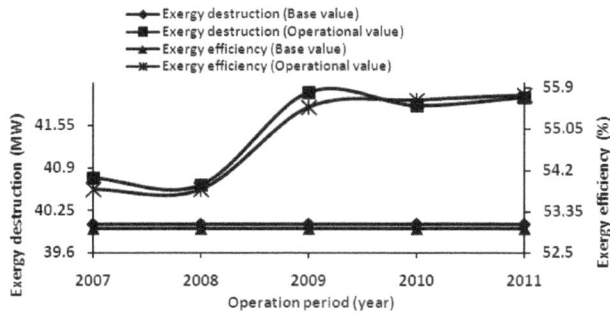

Fig 4. *Exergy destruction and efficiency versus operation period in evaporator*

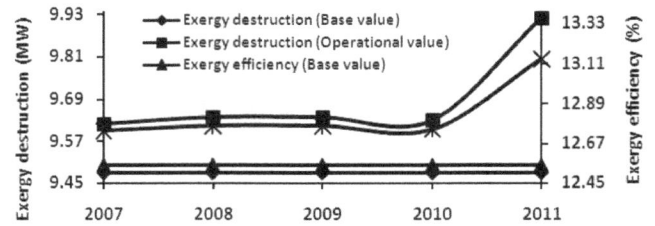

Fig 5. *Exergy destruction and efficiency against operation period in superheater*

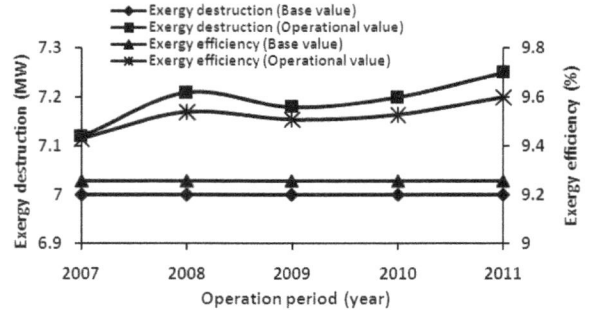

Fig 6. *Exergy destruction and efficiency as a function of operation period in reheater*

Compared with the base values, the boiler feed pump has a minimum increment of 0.54 % in 2007 and a maximum of 2.93 % in 2010 for exergy destruction. While for exergy efficiency, a minimum increment of 0.57 % was obtained in 2007 and maximum of 2.95 % in 2010. These increments may be attributed to deterioration and obsolescence; improvement can be achieved by replacement of major portions or even the complete system.

The boiler section comprises the economiser, evaporator, superheater and reheater. The variation of exergy destruction and efficiency with operation period in the economiser is depicted in Fig. 3.

Its base value for exergy destruction is 3.99 MW while that for exergy efficiency is 5.28 %. Compared with the base value, the exergy destruction had been on the increase ranging from an increment of 1.25 % in 2007 to as high as 4.26 % in 2009. The exergy efficiency had its lowest value of 1.23 % in 2007 and it's highest of 4.31 % in 2009. Fig. 4 depicts the effect of operation period on exergy destruction and efficiency in the evaporator.

The base value of exergy destruction in the evaporator is 40.04 MW and that of exergy efficiency is 53 %. As can be observed, the minimum increment of exergy destruction is 1.50 % in 2008 and a maximum of 5.02 % in 2009. The exergy efficiency had a maximum increment of 5.19 % in 2009 and a minimum of 1.53 % in 2008. Fig. 5 presents the variation of exergy destruction and efficiency with operation period in the superheater.

Exergy destruction and efficiency in the superheater had base values of 9.48 MW and 12.55 % respectively. The exergy destruction increment peaked at 4.64 % in 2011, with its lowest value being 1.48 % in 2007 while exergy efficiency attained its maximum increment of 4.66 % in 2011 and its minimum of 1.51 % in 2007. Effect of operation period on the exergy destruction and efficiency in the reheater is revealed in Fig. 6.

The exergy destruction had a base value of 7.00 MW in the reheater while that of the exergy efficiency is 9.26 %. It showed that exergy destruction had a maximum increment of 3.57 % in 2011 and maximum value of 1.71 % in 2007 while exergy efficiency had a maximum of 3.67 % in 2011 and minimum of 1.84 % in 2007.

The factors contributing to the higher irreversibilities in the boiler section are tube fouling, defective burners, steam traps and air heater fouling. Boiler section efficiency improvements can be realized through upgrading of components with better designs, optimizing system performance and eliminating conditions that degrade efficiency between maintenance outages. Primary areas to be considered are firing system design, furnace wall cleaning, convective heat transfer surface arrangement, air preheater heating element and seal design.

The turbine is a three stage turbine: the HP turbine, the intermediate pressure (IP) turbine and the low pressure (LP) turbine. Fig. 7 shows the plot of exergy destruction and efficiency against the operation period for HP turbine.

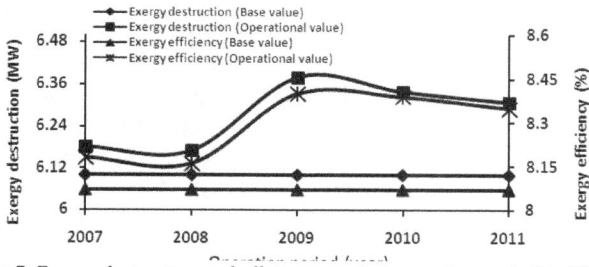

Fig 7. Exergy destruction and efficiency against operation period in HP turbine

The base value of exergy destruction in the HP turbine is 5.29 MW and that of exergy efficiency is 7.01 %. The maximum exergy destruction increment was 3.59 % in 2011 while the minimum was 1.04 % in 2007 and 2008. The exergy efficiency had a maximum of 3.42 % in 2011 and a minimum of 1.07 % in 2008. The variation of exergy destruction and efficiency in IP turbine with operation period is shown in Fig. 8.

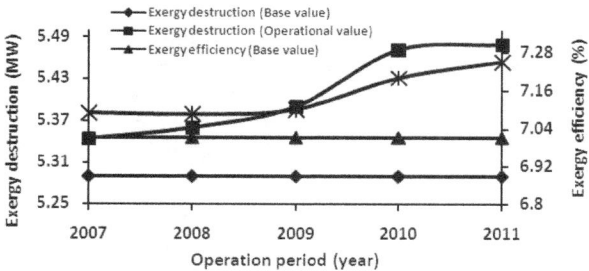

Fig 8. Exergy destruction and efficiency versus operation period in IP turbine

The exergy destruction base value in IP turbine is 6.1 MW and that of the exergy efficiency is 8.07 %. As can be seen, the exergy destruction increment had a maximum value of 4.59 % in 2009 and minimum of 1.15 % in 2008. The exergy efficiency had a maximum increment of 4.09 % in 2009 and minimum of 1.12 % in 2008. Fig. 9 depicts the effect of operation period on exergy destruction and efficiency in LP turbine. The base value of exergy destruction in LP turbine is 3.27 MW and that of the exergy efficiency is 4.33 %.

It was observed that exergy destruction had a maximum increment of 3.43 % in 2010 and minimum of 1.01 % in 2007. In the other hand, the exergy efficiency maximum value is 3.45 % in 2010 and minimum of 1.02 % in 2007.

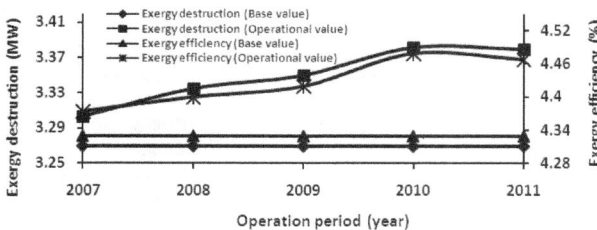

Fig 9. Variation of exergy destruction and efficiency with operation period in LP turbine

The factors that contribute to the irreversibilities in turbine are throttling losses at the turbine governor valves, silica deposits at the nozzles and blades. Amongst the three turbines, the HP turbine produces the highest irreversibility. Retrofitting of rotors, diaphragms or more commonly of complete stator/rotor modules (inner block) is a well proven way of significantly improving the heat rate of a steam turbine.

The variation of exergy destruction and efficiency with operation period in the condenser is shown in Fig. 10

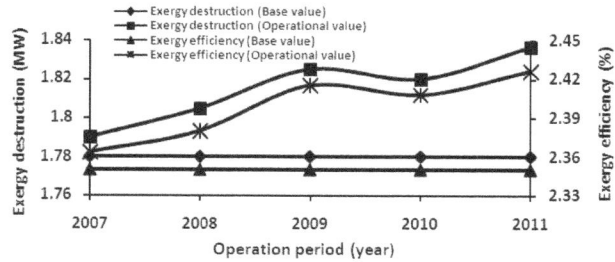

Fig 10. Exergy destruction and efficiency versus operation period in condenser

Exergy destruction and efficiency base values in the condenser are 1.78 MW and 2.35 % respectively. The exergy destruction increment peaked at 3.17 % in 2011 with its lowest value of 0.56 % in 2007 while exergy efficiency attained its maximum increment of 3.23 % in 2011 and its minimum of 0.58 % in 2007.

Sulphur deposits on water distribution plates were adduced for the increments. An effective means of improving the unit efficiency is to improve the vacuum or back pressure. Condenser tube bundle design 25 years ago was largely focused on packing as many tubes into the available volume as possible [22]. As a result some tubes condensed very little steam. A redesigned state-of-the-art tube bundle pattern will allow steam to access all the tubes and an improvement in condensing pressure is achieved. Cooling tower can also be retrofitted with modern packing to give an improvement in cooling water temperature and hence a lower condensing pressure and ultimately a higher power output.

From Figs 2 through 10, the exergy destruction of each component showed an approximately ascending trend with some oscillations. These oscillations might be related to the efforts of the maintenance crew in keeping the plant in good working conditions [10]. As observed in the figures, the exergy destruction and exergy efficiency for each component have similar trend. It is important to note that the civilian administration in Nigeria on its advert in 1999 made concerted efforts in rehabilitation of the plant [3]. This might be the reason why the plant's components had lower irreversibilities in 2007 and 2008. This suggests that steam power plant rehabilitation is a feasible solution for older units that have more than 200,000 operating hours. The results of a successful rehabilitation are reduced electricity production cost achieved by output increase, availability enhancement while at the same time extending lifetime and complying with stricter environmental standards. Implementation of the rehabilitation project is also much shorter with taking typically 1-2 years as opposed to 3-4 years for the construction of a new plant. In addition, it is possible to operate part of the plant whilst one or two units are undergoing the rehabilitation process.

Power plant rehabilitation is a cost-effective method to regain competitive electricity production cost of older power plant units. The suggested necessary modifications and refurbishment of Sapele power plant units is an attractive solution to improve the plant economy and keep its production cost competitive in a restructured Nigerian power system.

4. Conclusion

The analysis revealed that the highest increment in exergy destruction as compared with its base value occurred in the boiler section. The economiser had a maximum of 4.26 % in 2009 and a minimum of 1.25 % in 2007. While for the evaporator, maximum of 5.02 % was obtained in 2009 and minimum of 1.50 % in 2008. The superheater had maximum of 4.64 % in 2011 and minimum of 1.48 % in 2007. For the reheater, the maximum was 3.57 % in 2011 and minimum of 1.71 % in 2007. Tube fouling, defective burners, steam traps and air heater fouling have been adduced for the increments. Upgrading components with better designs, optmizing system performance and elimination of conditions that degrade efficiency between maintenance outages are essential in improving the performance of the boiler section.

The analysis also showed that for the three turbine stages, HP turbine had the highest increment. HP turbine had a maximum of 4.59 % in 2011 and minimum of 1.15 % in 2007 and 2008; the IP turbine had a maximum of 3.59 % in 2009 and minimum of 1.04 % in 2008 while the LP turbine had a maximum of 3.43 % in 2010 and minimum of 1.01 % in 2007. Throttling losses at the governor valves and silica deposits at the nozzles and blades were the bane. Retrofitting of rotors, diaphragms or complete stator/rotor modules (inner block) can improve the situation.

The results of the analysis on boiler feed pump showed a maximum of 2.93 in 2010 and minimum of 0.54 % in 2007. The condenser had a maximum of 3.17 % in 2011 and lowest value of 0.56 % in 2007. Sulphur deposits on water distribution plates of the condenser was advanced for loss of vacuum and power. A redesigned state-of-art-tube bundle will allow steam to access all the tubes resulting in an improvement in condensing pressure. Cooling tower retrofitted with modern packing gives an improvement in cooling water temperature and hence a lower condensing pressure. The results showed that exergy loss increased with increased operation time. These suggest that deterioration and obsolescence may be the major problems; hence plant rehabilitation is a feasible solution. The results of a successful rehabilitation are reduced electricity production cost achieved by output increase, availability

enhancement while at the same time extending lifetime and complying with stricter environmental standards. The suggested necessary modifications and refurbishment of Sapele power plant units is an attractive solution to improve the plant economy and keep production cost competitive in a restructured Nigerian power system.

References

[1]   A. S. Sambo, Achieving the Millennium Development Goals (MDGs): The Implication for Energy Infrastructure in Nigeria, Proceedings of COREN 16th Engineering Assembly, 28-29 August (2007), Abuja, Nigeria, 121-141.

[2]   G. I. Efenedo, I. O. Akalagboro, The Challenges of Power Generation, Utilization and Supply, The Nigeria Vision for the 21st Century, J. Emerging Trends in Engg and Applied Sci., 3 (2012) 886-891.

[3]   O. Obodeh, F. O. Isaac, Performance Indicators for Sapele Thermal Power Station, Nigeria: 1997-2006, Advancd Materials Res. 367(2012) 667-675.

[4]   Presidential Action Committee on Power, Roadmap for Power Sector Reform: A customer-Driven Sector-Wide Plan to Achieve Stable Power (2010), accessed from www.nigeriaelectricityprivatisation.com/wp-content/plugins/download-monitor/download.php?id=43 on December 12, 2012.

[5]   A. O. Cole, Restructing the Electric Power Utility Industry in Nigeria, Proceedings of National Conference of the Nigerian Society of Engineers (Electrical Division), 6-7 October (2004) 1-6.

[6]   K. Jin-ho, P. Jong-Bae, P. Jong-Keun, C. Yeung-Han, Generating Unit Maintenance Scheduling under Competitive Market Environments, Elect. Power and Energy Systems, 27 (2005) 189-194.

[7]   A. Badri, A. N. Niazi, Preventive Generation Maintenance Scheduling Considering System Reliability and Energy Purchase in Restructured Power System Systems, J. Basic and Applied Sci. Res. 2 (2012) 12773-12786.

[8]   O. Obodeh, P. E. Ugwuoke, P. E. (In Press); Optimal Maintenance Scheduling of Generating Units in a Restructured Nigerian Power System, American J. Elect. Power and Energy Systems.

[9]   M. A. Rosen, I. Dincer, M. Kanoglu, Role of Exergy in Increasing Efficiency and Sustainability and Reducing Environmental Impact, Energy Policy, 36 (2008) 128-137.

[10] K. R. Tapan, D. Amitava, G. Amitava, G. Ranjan, Exergy-based Performance Analysis for Proper O & M decisions in a Steam, Energy Conversion and Management, 51 (2010) 1333-1344.

[11] S. Salari, H. Abroshan, Aging Influence on Exergy Destruction in an Operating 320 MW Steam Power Plant, American Institute of Physis Conference Proceedings, Melaka, Malaysia, 3-4 October, 1440 (2011) 444-450.

[12] S. Barari, A. A. Shirazi, M. Keshavarzi, I. Rostamsowlat, Numerical Analysis and Field Study of the Time-Dependent Energy-exergy of a Gas-Steam combined Cycle, J. Serbian Chemical Society, 77 (2012) 945-957.

[13] S. K. Som, A. Datta, Thermodynamic Irreversibilities and Exergy Balance in Combustion Processes, Science Direct Progress in Energy and Combustion Science, 34 (2008) 351-376.

[14] P. Regulagadda, I. Dincer, G. F. Naterer, Exergy Analysis of a Thermal Power Plant with Measured Boiler and Turbine Losses, Appl. Thermal Engg, 30 (2010) 970-976.

[15] B. Kiran, K. Sachdeva, Performance Optimization of Steam Power Plant through Energy and Exergy Analysis, Int. J. Current Engg and Techno., 2 (2012) 285-289.

[16] J. C. Ofodu, D. P. S. Abam, Exergy Analysis of Afam Thermal Power Plant, Nigerian Society of Engineers (NSE) Tech. Trans., 37 (2002) 14-28.

[17] S. Sengupta, A. Datta, S. Duttagupta, Exergy Analysis of a Coal-based 210 MW Thermal Power Plant, Int.l J. Energy Res., 31 (2007) 14-28.

[18] P. Peerapong, B. Limmeechokchai, Exergetic and Thermoeconomic Analysis of the Rice-Husk Power Plant in Thailand, J. Metals, Materials and Minerals, 19 (2009) 9-14.

[19] A. Rashad, A. El Maihy, Energy and Exergy Analysis of a Steam Power Plant in Egypt, 13th International Conference on Aerospace Sciences and Aviation Technology, May 26-28, Military Technical College, Kobry Eelkobbah, Cairo, Egypt, (2009) Paper No. ASAT-13-TH-02.

[20] C. Mborah, E. K. Gbadam, On the Energy and Exergy Analysis of a 500 kW Steam Power Plant at Benso Oil Palm Plantation (BOPP), Res. J. Environmental and Earth Sci., 2 (2010) 239-244.

[21] B. T. Lebele-Alawa, J. M. Asuo, Exergy Analysis of Kolo Creek Gas Turbine Plant, Canadian J. Mechanical Sci. and Engg, 2 (2011) 172-184.

[22] A. Geete, A. I. Khandwawala, Exergy Analysis of 120 MW Thermal Power Plant with Different Condenser Back Pressure and Generate Correction Curves, Int. J. Current Engg and Techno., 3 (2013) 164-167.

[23] PHCN (2012); Sapele Power Plant Logbook, Sapele, Delta State, Nigeria.

[24] M. A. Lozano, A. Valero, Theory of Exergy Cost, Energy, 18 (1993) 939-960.

[25] G. Tsatsaronis, M. Winhold, Exergoeconomic Analysis and Evaluation of Energy Conversion Plants-1: A New General Methodology, Energy, 10 (1985) 69-80.

[26] J. Szargut, D. R. Morris, F. R. Steward, Exergy Analysis of Thermal, Chemical and Metallurgical Process, Hemisphere Publishing Co. New York, USA, (1988) 330 pp.

13

Compensation of voltage flicker by using facts devices

Jyothilal Nayak Bharothu, K Lalitha

Sri Vasavi Institute of Engineering & Technology, Nandamuru, A.P.; India

Email address:

nayakeee@gmail.com (J. N. Bharothu), lalitha.kny@gmail.com (K Lalitha)

Abstract: Voltage flicker occurs when heavy loads are periodically turned on and off in a weak distribution system. If the distribution system's short circuit capacity is not large enough, voltage fluctuations will occur. Voltage flickering can be extremely harmful to sensitive electronic equipment. Computerized equipment requires stable voltage to perform properly. This paper covers the contrasting approaches; dealing with the voltage flicker mitigation in three stages and assessing the related results in details. Initially, the voltage flicker mitigation, using FCTCR (Fixed Capacitor Thyristor Controlled Reactor), was simulated. Secondly, the compensation for the Static Synchronous Compensator (STATCOM) has been performed. In this case, injection of harmonics into the system caused some problems which were later overcome by using 12-pulse assignment of SATCOM and RLC filters. The obtained results show that STATCOM is very efficient and effective for the flicker compensation. All the simulations have been performed on the MATLAB Software.

Keywords: Voltage Flicker, STATCOM, FCTCR, Power Quality, RLC Filters Etc

1. Introduction

Voltage flicker occurs when heavy loads are periodically turned on and off in a weak distribution system. If the distribution system's short circuit capacity is not large enough, voltage fluctuations will occur. Starting large motors require an inrush of current, which causes a decrease in voltage. This voltage depression may cause a visible flicker on lighting circuits connected to the same power system. Voltage flickering can be extremely harmful to sensitive electronic equipment. Computerized equipment requires stable voltage to perform properly. For this reason, voltage flicker is a major power quality problem.

The magnitude of the voltage flicker depends upon the size and type of the electrical load that is producing the disturbance. A sag in voltage can also cause a voltage flicker; sudden voltage drops in the electrical distribution system can generate inrush current which can travel to sensitive equipment.

The relationship between power quality and distribution system has been a subject of interest for several years. The concept of power quality describes the quality of the supplier voltage in relation to the transient breaks, falling voltage, harmonics and voltage flicker. Voltage Flicker is the disturbance of lightning induced by voltage fluctuations. Very small variations are enough to induce lightning

disturbance for human eye for a standard 230V, 60W coiled-coil filament lamp. The disturbance becomes perceptible for voltage variation frequency of 10 Hz and relative magnitude of 0.26%. Huge non-linear industrial loads such as the electrical arc furnaces, pumps, welding machines, rolling mills and others are known as flicker generators. In this respect, the quality of supplied voltage is significantly reduced in an electrical power system and the oscillation of supplied voltage appears to be a major problem. Electric arc furnace, the main generator of voltage flicker, behaves in the form of a constant reactance and a variable resistance.

1.1. Literature Survey

D. Czarkowski et al [1] have analysed about the Voltage Flicker Mitigation Using

PWM-Based Distribution STATCOM in his paper "voltage flicker compensation using statcom"[1] .The author concluded thattheconcept of power quality describes the quality of the suppliervoltage in relation to the transient breaks, falling voltage,harmonics and voltage flicker

J. Mckim et al [2] have analysed about theUIE Flicker-meter Demystified in his paper "voltage flicker compensation using statcom"[2].The author concluded that the concept of The disturbance becomes perceptible for voltagevariation frequency of 10 Hz and relative magnitude of 0.26%

R. Collantes-Bellido et al [3] have analysed about the Identification and Modeling of a

Three Phase Arc Furnace for Voltage Distribution Simulation in his paper "voltage flicker compensationusing statcom"[3].The author concluded that the conceptof Huge non-linear industrial loads such as the electrical arc furnacespumps, welding machines, rolling mills andothers are known as flicker generators.

M. Zouiti et al [4] have analysed about the Electronic Based Equipment for Flicker Mitigation in his paper "voltage flicker compensation using statcom"[4]. The author concluded that the concept ofThe transformer-reactance system is modeled as alumped reactance, a furnace reactance (included connectioncables and busses) and a variable resistancewhich modelsthe arc.

S. Saadate et al al [5] have analysed about the Electronic Based Equipment for Flicker Mitigation in his paper "voltage flicker compensation using statcom"[5]. The author concluded that the concept ofTherefore, voltage flicker mitigation depends on reactivepower control

J. R. Clouston et al [5-14] have analysed about theField Demonstration of aDistribution Static Compensator Used to Mitigate Voltage Flicker in his paper "voltage flicker compensation using statcom"[5-14].In this type of compensation, the reactive power consumed by the compensator is keptconstant at a sufficient value.

M. W. Marshall et al [15-16] have analysed about the sing Series Capacitors to Mitigate VoltageFlicker Problems in his paper "voltage flicker compensation using statcom"[15-16].In this type, all the efforts aredone to decrease the voltage drop mentioned above, andfinally the reactive power is kept constant despite the loadfluctuations by controlling the line reactance.

J. Dolezal et al [17] have analysed about the Topologies and controlof active filters for flicker compensation International Symposiumon Industrial Electronics in his paper"voltage flicker compensation using statcom" [17]. In addition to the aforesaid procedures for thecompensators, the active filters are used for the voltageflickers mitigation as well

L. Gyugi,et al[18] have analysed about the Static Shunt Compensation for VoltageFlicker Reduction and Power Factor Correction in his paper "voltage flicker compensation using statcom"[18].Furthermore, the mitigatingdevices based on Static VAR Compensator (SVC) are the most frequentlyused devices for reduction in the voltage flicking.

Y. Hamachi et al [19]haveanalysed about the Voltage Fluctuation SuppressingSystem Using Thyristor Controlled Capacitors in his paper 'voltage flicker compensation using statcom"[19].Furthermore, the mitigating devices based onThyristor Switched Capacitor TSC are the most frequentlyused devices for reduction in the voltage flicking.

F. FrankTYCAP, et al [20] have analysed about thePower Factor Correction EquipmentUsing Thyristor Controlled Capacitor for Arc Furnacesin his paper 'voltage flicker compensation using statcom"[20].Furthermore, the

mitigating devices based onFCTCR are the most frequentlyused devices for reduction in the voltage flicking.

1.2. Problem Formulation

In this respect, the quality of supplied voltage is significantly reduced in an electrical power system and the oscillation of supplied voltage appears to be a major problem. Electric arc furnace, the main generator of voltage flicker, behaves in the form of a constant reactance and a variable resistance.

The transformer-reactance system is modeled as a lumped reactance, a furnace reactance (included connection cables and busses) and a variable resistancewhich models the arc. Connecting this type of load to the network produces voltage variation at the common point of supply to other consumers. The relative voltage drop is expressed by equation

$$\frac{\Delta U}{U_n} = \frac{R\Delta P + X\Delta Q}{U_n^2} \qquad (1)$$

where ΔP and ΔQ are the variation in active and reactive power; Un is the nominal voltage and R and X are short circuit resistance and reactance. Since R is usually very small in comparison to X, ΔU is proportional to Q (reactive power). Therefore, voltage flicker mitigation depends on reactive power control . Two types of structures can be used for the compensation of the reactive power fluctuations that cause the voltage drop: A: shunt structure : in this type of compensation, the reactive power consumed by the compensator is kept constant at a sufficient value. B: series structure : in this type, all the efforts are done to decrease the voltage drop mentioned above, and finally the reactive power is kept constant despite the load fluctuations by controlling the line reactance.

In addition to the aforesaid procedures for the compensators, the active filters are used for the voltage flickers mitigation as well . Furthermore, the mitigating devices based on Static VAR Compensator (SVC) such as Thyristor Switched Capacitor TSC , Thyristor Controlled Reactor (TCR) , and FCTCR , are the most frequently used devices for reduction in the voltage flicking. SVC devices achieved an acceptable level ofmitigation, but because of their complicated control algorithms, they have problems such as injecting a large amount of current harmonics to the system and causing spikes in voltage waveforms. Advent of FACTS devices make them ideal for use in a power system and especially in the voltage flicker mitigation. In this respect, the FACTS devices based on voltage-source converters have been able to improve the problems related to SVC . A new technique based on a novel control algorithm, which extracts the voltage disturbance to suppress the voltage flicker, is presented in this paper. The technique is to use STATCOM for voltage flicker compensation to overcome the aforementioned problems related to other techniques.

2. Introductions to Power Quality

The power quality problem is defined as any problem manifested in voltage, current or frequency deviations that result in mal-operation of customer equipment. The power quality problem causes the deterioration of performance of various sensitive electronic and electric equipments. The good quality of power can be specified as The supply voltage should be within guaranteed tolerance of declared value. The wave shape should be pure sine wave within allowable limits for distortion. The voltage should be balanced in all three phases. Supply should be reliable i.e. continuous availability without interruption Modern industrial machinery and commercial computer networks are prone to many different failure modes. When the assembly line stops, or the computer network crashes for no apparent reason, very often the electric power quality is suspected. It is a convenient culprit, as it is invisible and not easy to defend. Power quality problems may be very difficult to troubleshoot, and often the electric power may not have any relation to the actual problem. For example, in an industrial plant the faults of an automated assembly machine may ultimately be traced to fluctuations in the compressed air supply or a faulty hydraulic valve. Or in an office building, the problems on a local area network may be find their root cause with coaxial cable tee locations that are too close together, causing reflections and signal loss.

The contemporary container crane industry, like many other industry segments, is often enamoured by the bells and whistles, colourful diagnostic displays, high speed performance, and levels of automation that can be achieved. Although these features and their indirectly related computer based enhancements are key issues to an efficient terminal operation, we must not forget the foundation upon which we are building. Power quality is the mortar which bonds the Foundation blocks. Power quality also affects terminal operating economics, crane reliability, our environment, and initial investment in power distribution systems to support new crane installations.

To quote the utility company newsletter which accompanied the last monthly issue of my home utility billing: 'Using electricity wisely is a good environmental and business practice which saves you money, reduces emissions from generating plants, and conserves our Natural resources.' As we are all aware, container crane performance requirements continue to increase at an astounding rate. Next generation container cranes, already in the bidding process, will require average power demands of 1500 to 2000 kW – almost double the total average Demand three years ago. The rapid increase in power demand levels, an increase in container crane population, SCR converter crane drive retrofits and the large AC and DC drives needed to power and control these cranes will increase awareness of the power quality issue in the very near future.

2.1. Power Quality Problems

For the purpose of this article, we shall define power quality problems as:

'Any power problem that results in failure or disoperation of customer equipment manifests itself as an economic burden to the user, or produces negative impacts on the environment.'

When applied to the container crane industry, the power issues which degrade power quality include:
- Power Factor
- Harmonic Distortion
- Voltage Transients
- Voltage Sags or Dips
- Voltage Swells

The AC and DC variable speed drives utilized on board container cranes are significant contributors to total harmonic current and voltage distortion. Whereas SCR phase control creates the desirable average power factor, DC SCR drives operate at less than this. In addition, line notching occurs when SCR's commutate, creating transient peak recovery voltages that can be 3 to 4 times the nominal line voltage depending upon the system impedance and the size of the drives. The frequency and severity of these power system disturbances varies with the speed of the drive. Harmonic current injection by AC and DC drives will be highest when the drives are operating at slow speeds. Power factor will be lowest when DC drives are operating at slow speeds or during initial acceleration and deceleration periods, increasing to its maximum value when the SCR's are fazed on to produce rated or base speed.

Above base speed, the power factor essentially remains constant. Unfortunately, container cranes can spend considerable time at low speeds as the operator attempts to spot and land containers. Poor power factor places a greater kVA demand burden on the utility or engine-alternator power source. Low power factor loads can also affect the voltage stability which can ultimately result in detrimental effects on the life of sensitive electronic equipment or even intermittent malfunction. Voltage transients created by DC drive SCR line notching, AC drive voltage chopping, and high frequency harmonic voltages and currents are all significant sources of noise and disturbance to sensitive electronic equipment

It has been our experience that end users often do not associate power quality problems with Container cranes, either because they are totally unaware of such issues or there was no economic Consequence if power quality was not addressed. Before the advent of solid-state power supplies, Power factor was reasonable, and harmonic current injection was minimal. Not until the crane Population multiplied, power demands per crane increased, and static power conversion became the way of life, did power quality issues begin to emerge.

Even as harmonic distortion and power Factor issues surfaced, no one was really prepared. Even today, crane builders and electrical drive System vendors avoid the issue during competitive bidding for new cranes. Rather than focus on Awareness and understanding of the potential issues, the power quality issue is intentionally or Unintentionally ignored. Power quality problem solutions are available. Although the solutions are not free, in most cases, they do represent a good return on investment. However, if power quality is not specified, it most likely will not be delivered.

Power quality can be improved through:

- Power factor correction,
- Harmonic filtering,
- Special line notch filtering,
- Transient voltage surge suppression,
- Proper earthing systems.

In most cases, the person specifying and/or buying a container crane may not be fully aware of the potential power quality issues. If this article accomplishes nothing else, we would hope to provide that awareness.

In many cases, those involved with specification and procurement of container cranes may not be cognizant of such issues, do not pay the utility billings, or consider it someone else's concern. As a result, container crane specifications may not include definitive power quality criteria such as power factor correction and/or harmonic filtering. Also, many of those specifications which do require power quality equipment do not properly define the criteria. Early in the process of preparing the crane specification:

- Consult with the utility company to determine regulatory or contract requirements that must be
  satisfied, if any.

- Consult with the electrical drive suppliers and determine the power quality profiles that can be expected based on the drive sizes and technologies proposed for the specific project.

- Evaluate the economics of power quality correction not only on the present situation, but consider the impact of future utility deregulation and the future development plans for the terminal

2.2. The Benefits of Power Quality

Power quality in the container terminal environment impacts the economics of the terminal operation, affects reliability of the terminal equipment, and affects other consumers served by the same utility service. Each of these concerns is explored in the following paragraphs.

2.2.1. Economic Impact

The economic impact of power quality is the foremost incentive to container terminal operators. Economic impact can be significant and manifest itself in several ways:

2.2.1.1. Power Factor Penalties

Many utility companies invoke penalties for low power factor on monthly billings. There is no industry standard followed by utility companies. Methods of metering and calculating power factor penalties vary from one utility

company to the next. Some utility companies actually meter kVAR usage and establish a fixed rate times the number of kVAR-hours consumed. Other utility companies monitor kVAR demands and calculate power factor. If the power factor falls below a fixed limit value over a demand period, a penalty is billed in the form of an adjustment to the peak demand charges. A number of utility companies servicing container terminal equipment do not yet invoke power factor penalties. However, their service contract with the Port may still require that a minimum power factor over a defined demand period be met. The utility company may not continuously monitor power factor or kVAR usage and reflect them in the monthly utility billings; however, they do reserve the right to monitor the Port service at any time. If the power factor criteria set forth in the service contract are not met, the user may be penalized, or required to take corrective actions at the user's expense. One utility company, which supplies power service to several east coast container terminals in the USA, does not reflect power factor penalties in their monthly billings, however, their service contract with the terminal reads as follows:

'The average power factor under operating conditions of customer's load at the point where service is metered shall be not less than 85%. If below 85%, the customer may be required to furnish, install and maintain at its expense corrective apparatus which will increase the Power factor of the entire installation to not less than 85%. The customer shall ensure that no excessive harmonics or transients are introduced on to the [utility] system. This may require special power conditioning equipment or filters. The IEEE Std. 519-1992 is used as a guide in Determining appropriate design requirements.'

The Port or terminal operations personnel, who are responsible for maintaining container cranes, or specifying new container crane equipment, should be aware of these requirements. Utility deregulation will most likely force utilities to enforce requirements such as the example above. Terminal operators who do not deal with penalty issues today may be faced with some rather severe penalties in the future. A sound, future terminal growth plan should include contingencies for addressing the possible economic impact of utility deregulation.

2.2.1.2. System Losses

Harmonic currents and low power factor created by nonlinear loads, not only result in possible power factor penalties, but also increase the power losses in the distribution system. These losses are not visible as a separate item on your monthly utility billing, but you pay for them each month. Container cranes are significant contributors to harmonic currents and low power factor. Based on the typical demands of today's high speed container cranes, correction of power factor alone on a typical state of the art quay crane can result in a reduction of system losses that converts to a 6 to 10% reduction in the monthly utility billing. For most of the larger terminals, this is a significant annual saving in the cost of operation.

2.2.1.3. Power Service Initial Capital Investments

The power distribution system design and installation for new terminals, as well as modification of systems for terminal capacity upgrades, involves high cost, specialized, high and medium voltage equipment. Transformers, switchgear, feeder cables, cable reel trailing cables, collector bars, etc. must be sized based on the kVA demand. Thus cost of the equipment is directly related to the total kVA demand. As the relationship above indicates, kVA demand is inversely proportional to the overall power factor, i.e. a lower power factor demands higher kVA for the same kW load. Container cranes are one of the most significant users of power in the terminal. Since container cranes with DC, 6 pulse, SCR drives operate at relatively low power factor, the total kVA demand is significantly larger than would be the case if power factor correction equipment were supplied on board each crane or at some common bus location in the terminal. In the absence of power quality corrective equipment, transformers are larger, switchgear current ratings must be higher, feeder cable copper sizes are larger, collector system and cable reel cables must be larger, etc. Consequently, the cost of the initial power distribution system equipment for a system which does not address power quality will most likely be higher than the same system which includes power quality equipment.

2.2.2. Equipment Reliability

Poor power quality can affect machine or equipment reliability and reduce the life ofcomponents. Harmonics, voltage transients, and voltage system sags and swells are all power quality problems and are all interdependent. Harmonics affect power factor, voltage transientscan induce harmonics, the same phenomena which create harmonic current injection in DC SCR variable speed drives are responsible for poor power factor, and dynamically varying power factor of the same drives can create voltage sags and swells. The effects of harmonic distortion, harmonic currents, and line notch ringing can be mitigated using specially designed filters.

2.2.3. Power System Adequacy

When considering the installation of additional cranes to an existing power distribution system, a power system analysis should be completed to determine the adequacy of the system to support additional crane loads. Power quality corrective actions may be dictated due to inadequacy of existing power distribution systems to which new or relocated cranes are to be connected. In other words, addition of power quality equipment may render a workable scenario on an existing power distribution system, which would otherwise be inadequate to support additional cranes without high risk of problems.

2.2.4. Environment

No issue might be as important as the effect of power quality on our environment. Reduction in system losses and lower demands equate to a reduction in the consumption of our natural nm resources and reduction in power plant emissions. It is our responsibility as occupants of this planet to encourage conservation of our natural resources and support measures which improve our air quality

3. Introductions to Facts

Flexible AC Transmission Systems, called FACTS, got in the recent years a well-known term for higher controllability in power systems by means of power electronic devices. Several FACTS-devices have been introduced for various applications worldwide. A number of new types of devices are in the stage of being introduced in practice.

In most of the applications the controllability is used to avoid cost intensive or landscape requiring extensions of power systems, for instance like upgrades or additions of substations and power lines. FACTS-devices provide a better adaptation to varying operational conditions and improve the usage of existing installations. The basic applications of FACTS-devices are:

- Power flow control,
- Increase of transmission capability,
- Voltage control,
- Reactive power compensation,
- Stability improvement,
- Power quality improvement,
- Power conditioning,
- Flicker mitigation,
- Interconnection of renewable and distributed generation and storages.

The usage of lines for active power transmission should be ideally up to the thermal limits. Voltage and stability limits shall be shifted with the means of the several different FACTS devices. It can be seen that with growing line length, the opportunity for FACTS devices gets more and more important.

The influence of FACTS-devices is achieved through switched or controlled shunt compensation, series compensation or phase shift control. The devices work electrically as fast current, voltage or impedance controllers. The power electronic allows very short reaction times down to far below one second.

The development of FACTS-devices has started with the growing capabilities of power electronic components. Devices for high power levels have been made available in converters for high and even highest voltage levels. The overall starting points are network elements influencing the reactive power or the impedance of a part of the power system. Figure 3.2 shows a number of basic devices separated into the conventional ones and the FACTS-devices.

For the FACTS side the taxonomy in terms of 'dynamic' and 'static' needs some explanation. The term 'dynamic' is used to express the fast controllability of FACTS-devices provided by the power electronics. This is one of the main differentiation factors from the conventional devices. The term 'static' means that the devices have no moving parts like mechanical switches to perform the dynamic controllability. Therefore most of the FACTS-devices can equally be static and dynamic.

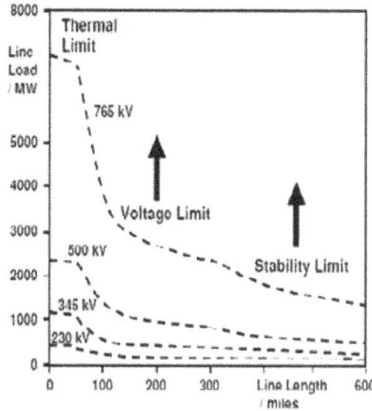

Fig 3.1. Operational limits of transmission lines for different voltage levels

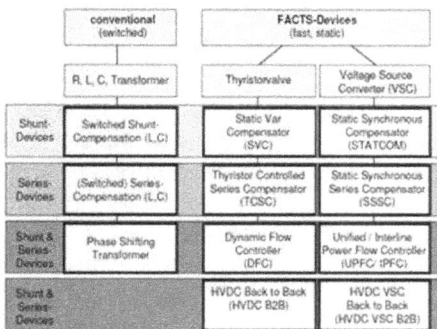

Fig 3.2. Overview of FACTS devices

The left column in Figure 3.2 contains the conventional devices build out of fixed or mechanically switch able components like resistance, inductance or capacitance together with transformers. The FACTS-devices contain these elements as well but use additional power electronic valves or converters to switch the elements in smaller steps or with switching patterns within a cycle of the alternating current. The left column of FACTS-devices uses Thyristor valves or converters. These valves or converters are well known since several years. They have low losses because of their low switching frequency of once a cycle in the converters or the usage of the Thyristors to simply bridge impedances in the valves.

The right column of FACTS-devices contains more advanced technology of voltage source converters based today mainly on Insulated Gate Bipolar Transistors (IGBT) or Insulated Gate Commutated Thyristors (IGCT). Voltage Source Converters provide a free controllable voltage in magnitude and phase due to a pulse width modulation of the IGBTs or IGCTs. High modulation frequencies allow to get low harmonics in the output signal and even to compensate disturbances coming from the network. The disadvantage is that with an increasing switching frequency,

the losses are increasing as well. Therefore special designs of the converters are required to compensate this.

3.1. Shunt Devices

The most used FACTS-device is the SVC or the version with Voltage Source Converter called STATCOM. These shunt devices are operating as reactive power compensators. The main applications in transmission, distribution and industrial networks are:
- Reduction of unwanted reactive power flows and therefore reduced network losses.
- Keeping of contractual power exchanges with balanced reactive power.
- Compensation of consumers and improvement of power quality especially with huge demand fluctuations like industrial machines, metal melting plants, railway or underground train systems.
- Compensation of Thyristor converters e.g. in conventional HVDC lines.
- Improvement of static or transient stability.

Almost half of the SVC and more than half of the STATCOMs are used for industrial applications. Industry as well as commercial and domestic groups of users require power quality. Flickering lamps are no longer accepted, nor are interruptions of industrial processes due to insufficient power quality. Railway or underground systems with huge load variations require SVCs or STATCOMs.

3.1.1. SVC

Electrical loads both generate and absorb reactive power. Since the transmitted load varies considerably from one hour to another, the reactive power balance in a grid varies as well. The result can be unacceptable voltage amplitude variations or even a voltage depression, at the extreme a voltage collapse.

A rapidly operating Static Var Compensator (SVC) can continuously provide the reactive power required to control dynamic voltage oscillations under various system conditions and thereby improve the power system transmission and distribution stability.

Applications of the SVC systems in transmission systems:
a. To increase active power transfer capacity and transient stability margin
b. To damp power oscillations
c. To achieve effective voltage control
In addition, SVCs are also used

3.1.1.1. In Transmission System
a. To reduce temporary over voltages
b. To damp sub synchronous resonances
c. To damp power oscillations in interconnected power systems

3.1.1.2. In Traction Systems
a. To balance loads
b. To improve power factor

c. To improve voltage regulation

3.1.1.3. *In HVDC Systems*

a. To provide reactive power to ac–dc converters

3.1.1.4. *In Arc Furnaces*

a. To reduce voltage variations and associated light flicker

Installing an SVC at one or more suitable points in the network can increase transfer capability and reduce losses while maintaining a smooth voltage profile under different network conditions. In addition an SVC can mitigate active power oscillations through voltage amplitude modulation.

SVC installations consist of a number of building blocks. The most important is the Thyristor valve, i.e. stack assemblies of series connected anti-parallel Thyristors to provide controllability. Air core reactors and high voltage AC capacitors are the reactive power elements used together with the Thyristor valves. The step up connection of this equipment to the transmission voltage is achieved through a power transformer.

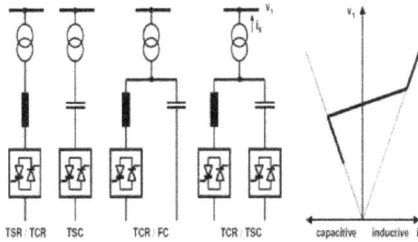

Fig 3.3. SVC building blocks and voltage / current characteristic

In principle the SVC consists of Thyristor Switched Capacitors (TSC) and Thyristor Switched or Controlled Reactors (TSR / TCR). The coordinated control of a combination of these branches varies the reactive power as shown in Figure. The first commercial SVC was installed in 1972 for an electric arc furnace. On transmission level the first SVC was used in 1979. Since then it is widely used and the most accepted FACTS-device.

Fig 3.4. SVC using a TCR and an FC

In this arrangement, two or more FC (fixed capacitor) banks are connected to a TCR (thyristor controlled reactor)

through a step-down transformer. The rating of the reactor is chosen larger than the rating of the capacitor by an amount to provide the maximum lagging vars that have to be absorbed from the system. By changing the firing angle of the thyristor controlling the reactor from 90° to 180°, the reactive power can be varied over the entire range from maximum lagging vars to leading vars that can be absorbed from the system by this compensator.

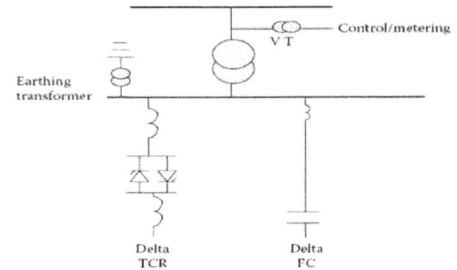

Fig 3.5. SVC using a TCR and an FC

3.1.2. *SVC of the FC/TCR Type*

The main disadvantage of this configuration is the significant harmonics that will be generated because of the partial conduction of the large reactor under normal sinusoidal steady-state operating condition when the SVC is absorbing zero MVAr. These harmonics are filtered in the following manner. Triplex harmonics are canceled by arranging the TCR and the secondary windings of the step-down transformer in delta connection. The capacitor banks with the help of series reactors are tuned to filter fifth, seventh, and other higher-order harmonics as a high-pass filter. Further losses are high due to the circulating current between the reactor and capacitor banks.

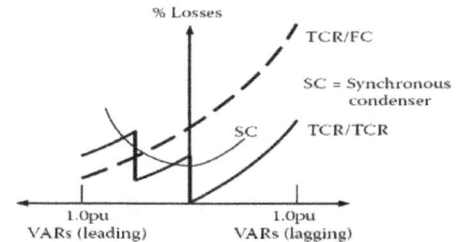

Fig 3.6. Comparison of the loss characteristics of TSC–TCR, TCFC

Compensators and synchronous condenser

These SVCs do not have a short-time overload capability because the reactors are usually of the air-core type. In applications requiring overload capability, TCR must be designed for short-time overloading, or separate thyristor-switched overload reactors must be employed.

3.1.3. *Svc Using a TCR and TSC*

This compensator overcomes two major shortcomings of the earlier compensators by reducing losses under operating conditions and better performance under large system disturbances. In view of the smaller rating of each capacitor bank, the rating of the reactor bank will be 1/n times the maximum output of the SVC, thus reducing the harmonics generated by the reactor. In those situations where harmonics have to be reduced further, a small amount of FCs tuned as filters may be connected in parallel with the TCR.

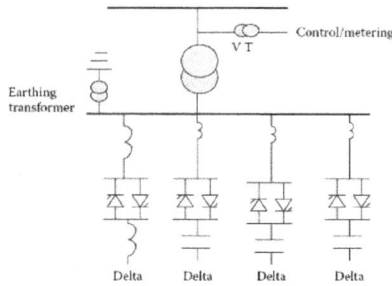

Fig 3.7. SVC of combined TSC and TCR type

When large disturbances occur in a power system due to load rejection, there is a possibility for large voltage transients because of oscillatory interaction between system and the SVC capacitor bank or the parallel. The LC circuit of the SVC in the FC compensator. In the TSC–TCR scheme, due to the flexibility of rapid switching of capacitor banks without appreciable disturbance to the power system, oscillations can be avoided, and hence the transients in the system can also be avoided. The capital cost of this SVC is higher than that of the earlier one due to the increased number of capacitor switches and increased control complexity.

3.1.4. Statcom

In 1999 the first SVC with Voltage Source Converter called STATCOM (STATic COMpensator) went into operation. The STATCOM has a characteristic similar to the synchronous condenser, but as an electronic device it has no inertia and is superior to the synchronous condenser in several ways, such as better dynamics, a lower investment cost and lower operating and maintenance costs. A STATCOM is build with Thyristors with turn-off capability like GTO or today IGCT or with more and more IGBTs. The static line between the current limitations has a certain steepness determining the control characteristic for the voltage.

The advantage of a STATCOM is that the reactive power provision is independent from the actual voltage on the connection point. This can be seen in the diagram for the maximum currents being independent of the voltage in comparison to the SVC. This means, that even during most severe contingencies, the STATCOM keeps its full capability.

In the distributed energy sector the usage of Voltage Source Converters for grid interconnection is common practice today. The next step in STATCOM development is the combination with energy storages on the DC-side. The performance for power quality and balanced network operation can be improved much more with the combination of active and reactive power.

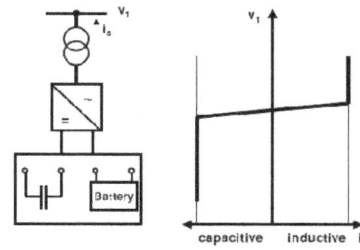

Fig 3.8. STATCOM structure and voltage / current characteristic

STATCOMs are based on Voltage Sourced Converter (VSC) topology and utilize either Gate-Turn-off Thyristors (GTO) or Isolated Gate Bipolar Transistors (IGBT) devices. The STATCOM is a very fast acting, electronic equivalent of a synchronous condenser. If the STATCOM voltage, Vs, (which is proportional to the dc bus voltage Vc) is larger than bus voltage, Es, then leading or capacitive VARS are produced. If Vs is smaller thenEs then lagging or inductive VARS are produced.

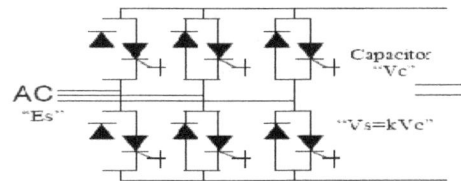

Fig 3.8 6. Pulses STATCOM

The three phases STATCOM makes use of the fact that on a three phase, fundamental frequency, steady state basis, and the instantaneous power entering a purely reactive device must be zero. The reactive power in each phase is supplied by circulating the instantaneous real power between the phases. This is achieved by firing the GTO/diode switches in a manner that maintains the phase difference between the ac bus voltage ES and the STATCOM generated voltage VS. Ideally it is possible to construct a device based on circulating instantaneous power which has no energy storage device (ie no dc capacitor).

A practical STATCOM requires some amount of energy storage to accommodate harmonic power and ac system unbalances, when the instantaneous real power is non-zero. The maximum energy storage required for the STATCOM is much less than for a TCR/TSC type of SVC compensator of comparable rating.

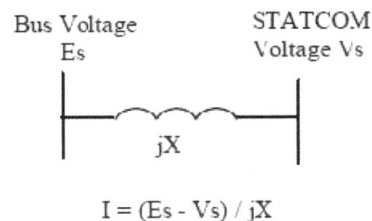

$$I = (Es - Vs) / jX$$

Fig 3.9. STATCOM Equivalent Circuit

Several different control techniques can be used for the

firing control of the STATCOM. Fund.amental switching of the GTO/diode once per cycle can be used. This approach will minimize switching losses, but will generally utilize more complex transformer topologies. As an alternative, Pulse Width Modulated (PWM) techniques, which turn on and off the GTO or IGBT switch more than once per cycle, can be used. This approach allows for simpler transformer topologies at the expense of higher switching losses.

The 6 PulseSTATCOM using 1 harmonics. There are a fundamental switching will of course produce the 6 N variety of methods to decrease the harmonics. These methods include the basic 12 pulse configuration with parallel star / delta transformer connections, a complete elimination of 5th and 7th harmonic current using series connection of star/star and star/delta transformers and a quasi 12 pulse method with a single star-star transformer, and two secondary windings, using control of firing phase shift between the two 6 pulse bridges. This angle to produce a 30 method can be extended to produce a 24 pulse and a 48 pulse STATCOM, thus eliminating harmonics even further. Another possible approach for harmonic cancellation is a multi-level configuration which allows for more than one switching element per level and therefore more than one switching in each bridge arm. The ac voltage derived has a staircase effect, dependent on the number of levels. This staircase voltage can be controlled to eliminate harmonics.

Fig 3.10. Substation with a STATCOM

3.2. Series Devices

Series devices have been further developed from fixed or mechanically switched compensations to the Thyristor Controlled Series Compensation (TCSC) or even Voltage Source Converter based devices.

The main applications are:
- Reduction of series voltage decline in magnitude and angle over apower line,
- Reduction of voltage fluctuations within defined limits during changing power transmissions,
- Improvement of system damping resp. damping of oscillations,
- Limitation of short circuit currents in networks or substations,
- Avoidance of loop flows resp. power flow adjustments.

3.2.1. Tcsc

Thyristor Controlled Series Capacitors (TCSC) address specific dynamical problems in transmission systems. Firstly it increases damping when large electrical systems are interconnected. Secondly it can overcome the problem of Sub Synchronous Resonance (SSR), a phenomenon that involves an interaction between large thermal generating units and series compensated transmission systems.

The TCSC's high speed switching capability provides a mechanism for controlling line power flow, which permits increased loading of existing transmission lines, and allows for rapid readjustment of line power flow in response to various contingencies. The TCSC also can regulate steady-state power flow within its rating limits.

From a principal technology point of view, the TCSC resembles the conventional series capacitor. All the power equipment is located on an isolated steel platform, including the Thyristor valve that is used to control the behavior of the main capacitor bank. Likewise the control and protection is located on ground potential together with other auxiliary systems. Figure shows the principle setup of a TCSC and its operational diagram. The firing angle and the thermal limits of the Thyristors determine the boundaries of the operational diagram.

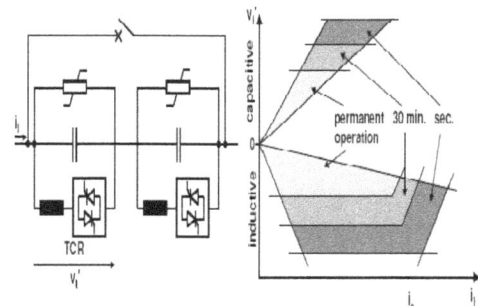

Fig 3.11. principle setup of a TCSC and its operational diagram

Advantages
- Continuous control of desired compensation level
- Direct smooth control of power flow within the network
- Improved capacitor bank protection
- Local mitigation of sub synchronous resonance (SSR). This permits higher levels of compensation in networks where interactions with turbine-generator torsional vibrations or with other control or measuring systems are of concern.
- Damping of electromechanical (0.5-2 Hz) power oscillations which often arise between areas in a large interconnected power network. These oscillations are due to the dynamics of inter area power transfer and often exhibit poor damping when the aggregate power tranfer over a corridor is high relative to the transmission strength.

3.3. Shunt and Series Devices

Dynamic Power Flow Controller

A new device in the area of power flow control is the

Dynamic Power Flow Controller (DFC). The DFC is a hybrid device between a Phase Shifting Transformer (PST) and switched series compensation.

A functional single line diagram of the Dynamic Flow Controller is shown in Figure 1.19. The Dynamic Flow Controller consists of the following components:

• a standard phase shifting transformer with tap-changer (PST)

• series-connected Thyristor Switched Capacitors and Reactors

(TSC / TSR)

• A mechanically switched shunt capacitor (MSC). (This is optional depending on the system reactive power requirements)

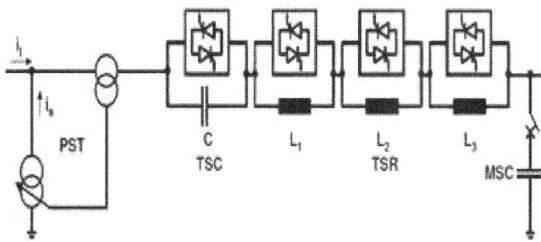

Fig 3.12. Principle configuration of DFC

Based on the system requirements, a DFC might consist of a number of series TSC or TSR. The mechanically switched shunt capacitor (MSC) will provide voltage support in case of overload and other conditions. Normally the reactance of reactors and the capacitors are selected based on a binary basis to result in a desired stepped reactance variation. If a higher power flow resolution is needed, a reactance equivalent to the half of the smallest one can be added.

The switching of series reactors occurs at zero current to avoid any harmonics. However, in general, the principle of phase-angle control used in TCSC can be applied for a continuous control as well. The operation of a DFC is based on the following rules:

• TSC / TSR are switched when a fast response is required.

• The relieve of overload and work in stressed situations is handled by the TSC / TSR.

• The switching of the PST tap-changer should be minimized particularly for the currents higher than normal loading.

• The total reactive power consumption of the device can be optimized by the operation of the MSC, tap changer and the switched capacities and reactors.

In order to visualize the steady state operating range of the DFC, we assume an inductance in parallel representing parallel transmission paths. The overall control objective in steady state would be to control the distribution of power flow between the branch with the DFC and the parallel path. This control is accomplished by control of the injected series voltage.

The PST (assuming a quadrature booster) will inject a voltage in quadrature with the node voltage. The controllable reactance will inject a voltage in quadrature with the throughput current. Assuming that the power flow has a load factor close to one, the two parts of the series voltage will be close to collinear. However, in terms of speed of control, influence on reactive power balance and effectiveness at high/low loading the two parts of the series voltage has quite different characteristics. The steady state control range for loadings up to rated current is illustrated in Figure 3.5.2, where the x-axis corresponds to the throughput current and the y-axis corresponds to the injected series voltage.

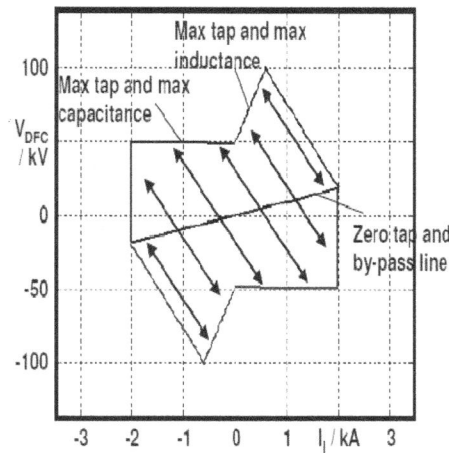

Fig3.13 Operational diagram of a DFC

Operation in the first and third quadrants corresponds to reduction of power through the DFC, whereas operation in the second and fourth quadrants corresponds to increasing the power flow through the DFC. The slope of the line passing through the origin (at which the tap is at zero and TSC / TSR are bypassed) depends on the short circuit reactance of the PST.

Starting at rated current (2 kA) the short circuit reactance by itself provides an injected voltage (approximately 20 kV in this case). If more inductance is switched in and/or the tap is increased, the series voltage increases and the current through the DFC decreases (and the flow on parallel branches increases). The operating point moves along lines parallel to the arrows in the figure. The slope of these arrows depends on the size of the parallel reactance. The maximum series voltage in the first quadrant is obtained when all inductive steps are switched in and the tap is at its maximum.

Now, assuming maximum tap and inductance, if the throughput current decreases (due e.g. to changing loading of the system) the series voltage will decrease. At zero current, it will not matter whether the TSC / TSR steps are in or out, they will not contribute to the series voltage. Consequently, the series voltage at zero current corresponds to rated PST series voltage. Next, moving into the second quadrant, the operating range will be limited by the line corresponding to maximum tap and the capacitive step being switched in (and the inductive steps by-passed). In this case, the capacitive step is approximately as large as the short circuit reactance of the PST, giving an almost constant maximum voltage in the second quadrant.

4. Voltage Flicker

Flicker is a difficult problem to quantify and to solve. The untimely combination of the following factors is required for flicker to be a problem: 1) some deviation in voltage supplying lighting circuits and 2) a person being present to view the possible change in light intensity due to the voltage deviation. The human factor significantly complicates the issue and for this reason flicker has historically been deemed "a problem of perception." The voltage deviations involved are often much less than the thresholds of susceptibility for electrical equipment, so major operating problems are only experienced in rare cases. To office personnel, on the other hand, voltage deviations on the order of a few tenths of one percent could produce extremely annoying fluctuations in the output of lights, especially if the frequency of repetitive deviations is 5-15 Hz. Due to the clear relationship between voltage deviation and light response, the term "flicker" often means different things to different people with the interpretation primarily governed by the concerns of a particular discussion.

Flicker Measurement Introduction

The power supply network voltage varies over time due to perturbations that occur in the processes of electricity generation, transmission and distribution. Interaction of electrical loads with the network causes further deterioration of the electrical power quality. High power loads that draw fluctuating current, such as large motor drives and arc furnaces, cause low frequency cyclic voltage variations that result in: flickering of light sources which can cause significant physiological discomfort, physical and psychological tiredness, and even pathological effects for human beings, problems with the stability of electrical devices and electronic circuits.

4.1 Controlling System

The concept of instantaneous reactive power is used for the controlling system. Following this, the 3-phase voltage upon the use of the park presented by Akagi [24] has been transformed to the synchronous reference frame (Park or dq0 transformation). This transformation leads to the appearances of three instantaneous space vectors: Vd on the d-axis (real or direct axis), Vq on the q-axis (imaginary or quadrature axis) and V0, from the 3-phase voltage of Va, Vb and Vc. The related equations of this transformation, expressed in the MATLAB software, are as follows:

$$V_d = \frac{2}{3}(V_a \sin(\omega t) + V_b \sin(\omega t - \frac{2\pi}{3}) - V_c \sin(\omega t + \frac{2\pi}{3})) \tag{2}$$

$$V_q = \frac{2}{3}(V_a \cos(\omega t) + \cos(\omega t - \frac{2\pi}{3}) + \cos(\omega t + \frac{2\pi}{3})) \tag{3}$$

$$V_0 = \frac{1}{3}(V_a + V_b + V_c) \tag{4}$$

A dynamic computation shows that the voltage oscillations in the connecting node of the flicker-generating load to the network are created by 3 vectors: real current (ip), imaginary current (iq) and the derivative of the real current with respect to time ($\frac{di_p}{dt}$). In general, for the complete voltage flicker compensation, the compensating current (ic) regarding the currents converted to the dq0 axis is given as

$$i_c = j(i_q + i_p \frac{R}{X} f + \frac{1}{\omega} \frac{di_p}{d\omega} f + k)$$

where R and X are the synchronous resistance and reactance of the line and f is the correcting coefficient. The constant k is also used to eliminate the average reactive power of the network . If the compensation current of the above equation is injected to the network, the whole voltage flicker existing in the network will be eliminated. Regarding the equation, related to the dq-transformation of the 3-phase-voltages to the instantaneous vectors, it is obvious that under the conditions of accessing an average voltage flicker, Vd and V0, the obtained values are close to zero and Vq is a proper value adapting to the voltage oscillation of the network. This state of the 3-phase voltage flicker is presented in the following figures (simulated in the MATLAB Simulink package):

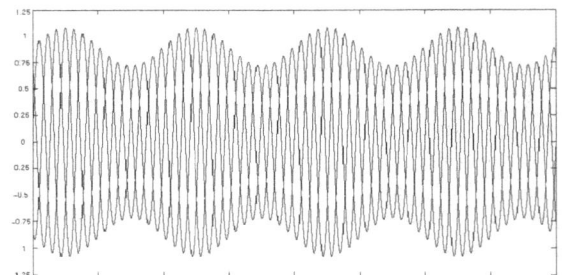

Fig 4.1. voltage flicker extended to circuit

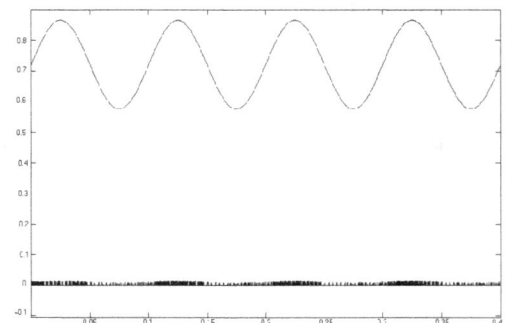

Fig 4.2. instantaneous components of the 3-phase voltage flicker waveform

Then, we may conclude that the decrease of the voltage flicker of the network and the compensating control to decrease the voltage flicker can be limited only based on the amount of the imaginary component of the instantaneous voltage (Vq).

4.2. Compensation System

A typical two-bus power system shown in figure 5.3 is simulated in MATLAB for this study. It can be seen that the voltage oscillation was produced by a 3-phase flicker source connected to the main bus-bar. The complete STATCOM control system scheme implemented on MATLAB is shown in figure 5.3. First, using a 3-phase converter to dq0, the instantaneous vectors Vd, Vq and V0, are evaluated from the output 3-phase voltages whose equations were explained in the previous section. Then, from the obtained instantaneous components, sampling is taken place. Since the controlling system uses just Vq to control the STATCOM, a de-multiplexer is used to extract Vq voltage from Vd and V0. The obtained Vq is then entered as an input to the controlling function upon the MATLAB software. The controlling function generates the amount of conducting angle, needed for the GTOs of the STATCOM. A phase shifting block is designed to control the appropriate phase angle of the exerting pulses upon the GTOs of the STATCOM. The outputs of this unit are entered into the STATCOM as inputs.

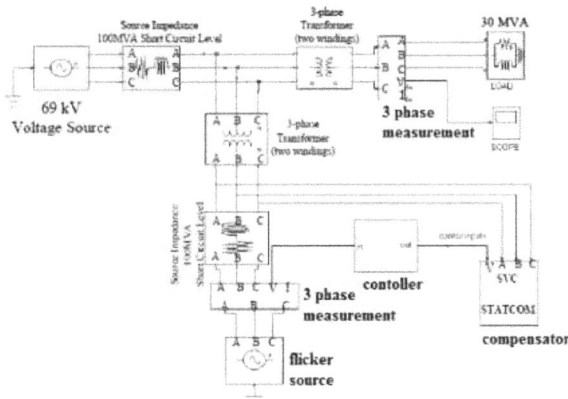

Fig 4.3. The studied power system

Simulation and Analysis of the Results

In order to investigate the influence of the STATCOM as an effective mitigating device for voltage flicker, three types of compensators are simulated in MATLAB. First, the voltage flicker compensation is adopted using FCTCR. Then a 6-pulse voltage-source converter STATCOM is used and finally for a complete voltage flicker mitigation a 12-pulse voltage-source converter STATCOM is designed. The compensation techniques and their results are presented in this section.

4.2.1Compensation Using FCTCR

In this stage a FCTCR; one of the FACTS devices being controlled by a thyristor is used to mitigate the voltage

flicking. In this case, the exerted voltage flicker into the system and the compensated voltage

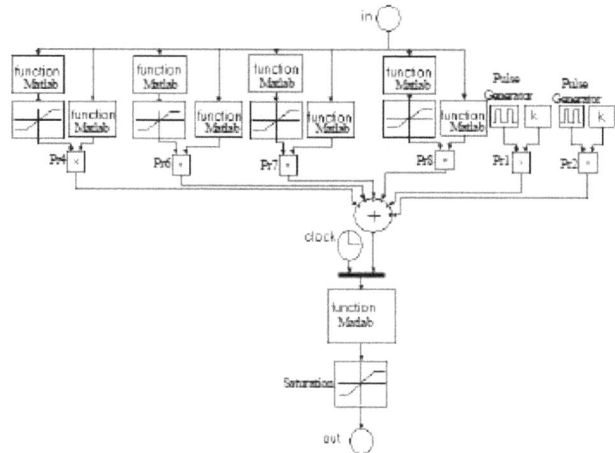

Fig4.4. Compensation using FCTCR

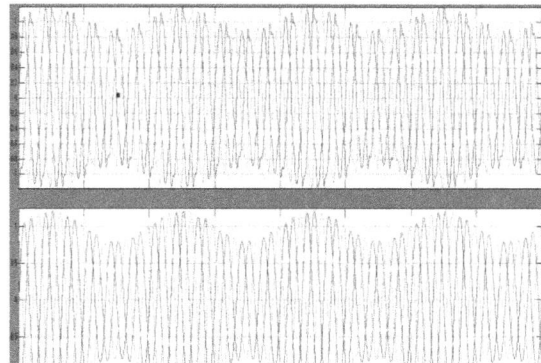

Fig 4.5. output for FCTCR

It is obvious from the output voltage waveform controlled by FCTCR that this technique achieves a reasonable level of mitigation but is incapable to be perfectly successful. Furthermore, in spite of using a snubber circuitto eliminate voltage spikes caused by the huge TCR reactor switching, there are still distortions in the output waveform.

4.2.2. Compensation Using 6-pulse Voltage-source Converter STATCOM

The circuit diagram of a three-phase 6-pulse voltage source converter STATCOM is shown in figure 5.6. Six valves compose the converter and each valve is made up of a GTO with a diode connected in anti-parallel. In this type of STATCOM, each GTO is fired and blocked one time per line voltage cycle. In this case, each GTO in a single branch is conducted during a half-cycle (180 degree) of the fundamental period. The combined pulses of each leg have a 120 degrees phase difference to produce a balanced set of voltages. By adjusting the conducting angle of the GTOs, the generated voltage and then the injected or absorbed power of the STATCOM are controlled. In this respect, the compensated output voltage by 6-pulse voltage-source converter STATCOM is presented in figure 5.12.

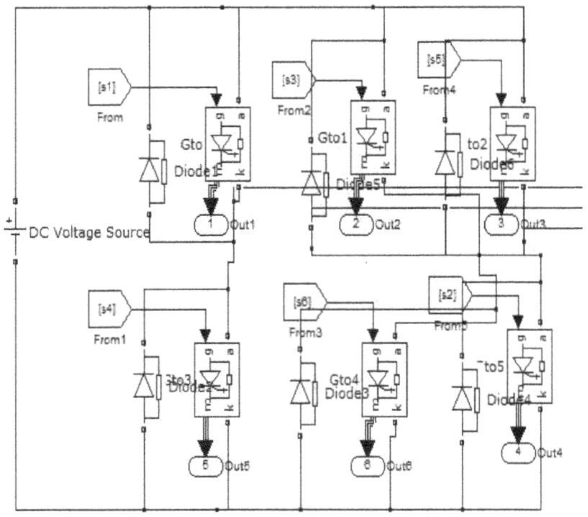

Fig 4.6. pulse VSI

Controlling Unit

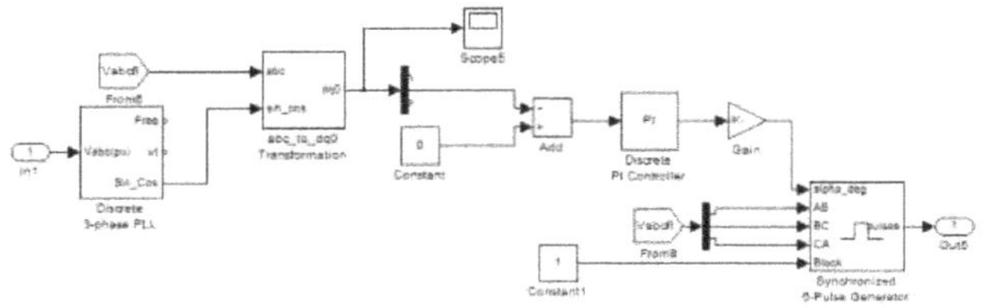

Fig 4.7. controlling block

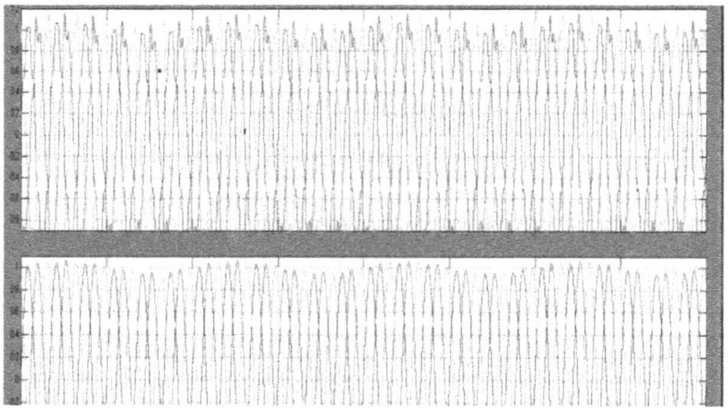

Fig 4.8. output of 6 pulse VSI

It can be seen that the mitigation effects of this compensator is better than that of FCTCR and effectively mitigate the voltage flicker; but the output voltage waveform has some considerable harmonics. The instantaneous output line-to-line voltage (Vab) of the 6-pulse voltage-source converter is as follows

$$V_{ab} = \sum_{n=1,3,5...}^{\infty} \frac{4V_s}{n\pi} \cos \frac{n\pi}{6} \sin n(wt + \frac{\pi}{6}) \quad (6)$$

As we see it is clearly perceptible from the above equation that, the even harmonics in the instantaneous line-to-line voltage has zero value and does not enter the network voltage. Connecting the voltage-source converter with a wye-delta transformer to the network, multiple 3rd Harmonics (3, 9, 15 …) are eliminated from the line

voltages. Therefore, the considerable existing characteristic harmonics in the output voltage waveform in addition to the fundamental component are 5, 7, 11, 13 and higher whose values are shown in the harmonic spectrum of figure 9. It can be observed from the harmonic spectrum that 5th and 7th harmonics have considerable level comparing to the fundamental harmonics. Furthermore, 11th and 13th harmonics are considerable which should be eliminated from the network voltage waveforms. However, higher harmonics (namely 17th, 19th and above) have values very close to zero

4.2.3. Compensation Using 12-pulse Voltage-source Converter Statcom

In order to reduce the harmonic contents at the output voltage, the number of pulses can be increased, forming a multi-pulse configuration. Multi-pulse converters are composed by n (n=2, 4, 8 …), where n is the number of pulses. 6-pulse bridges connected in parallel on the same DC bus and interconnected in series through transformers on the AC side. Depending on the number of pulses, these transformers and their connections can become very complex.

Two 6-pulse bridges are connected, forming a 12-pulse converter for a complete voltage flicker compensation design. In this case, the first converter is connected with a wye-wye transformer and the second one with a wye-delta transformer. These are linked together using a three winding transformer. Moreover, the delta-connected secondary of the second transformer must have 3 times the turns compared to the wye-connected secondary and the pulse train to one converter is shifted by 30 degrees with respect to the other. The 12-pulse voltage-source converter STATCOM circuit diagram is shown in figure 4.10

Pulse generating Unit

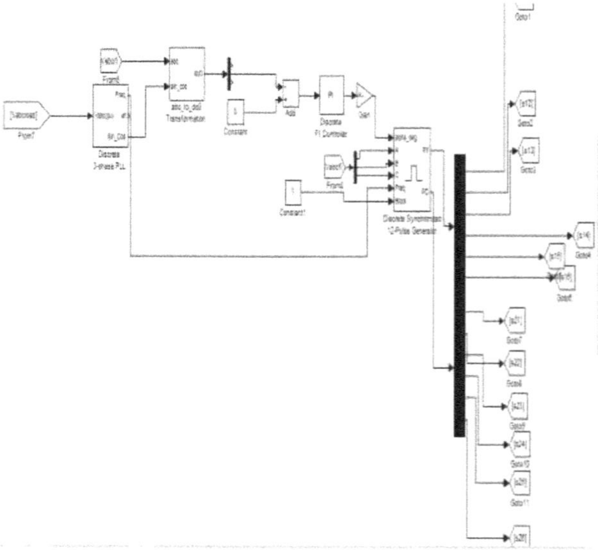

Fig 4.9 pulse generating unit

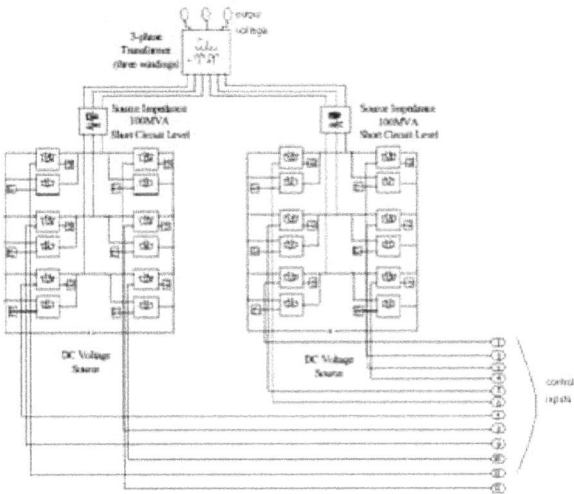

Fig 4.10 12 pulse VSI

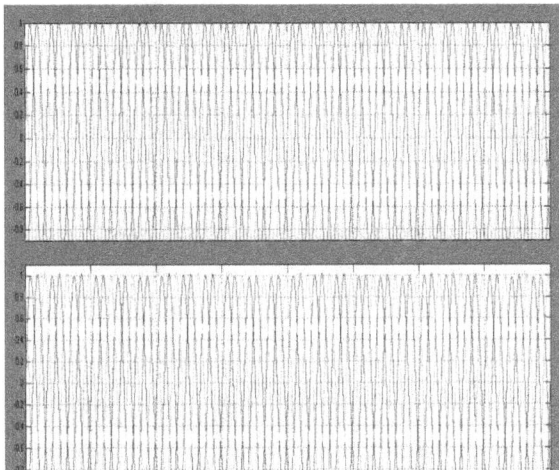

Fig4.11. output of 12 pulse

The complete STATCOM control system scheme is implemented on the power system introduced . The output voltage mitigated by 12-pulse voltage-source converter STATCOM and its harmonic spectrum are depicted in figures 11 and 12 respectively. In this respect, the voltage flicker is completely removed from the output voltage and a sinusoidal waveform is obtained. Furthermore, it is clearly obvious (from the harmonic spectrum) that almost all harmonics are removed from the output voltage. The only injected harmonics to the system are 11 and 13 that are deleted adding an RLC active filter to the designed compensator.

Table 4.1 for total harmonic distortion

model	total harmonic distortion (%)
fctcr	8.03%
6 pulse statcom	5.47%
12 pulse statcom	1.33%

5 Conclusions

The design and application of STATCOM technology based on voltage-source converters for voltage flicker mitigation is discussed in this paper. Mitigation is done in three stages and the results are compared and contrasted. First, FCTCR is used to compensate for the voltage flicker, then a 6-pulse voltage-source converter STATCOM and finally a 12-pulse STATCOM based on voltage-source converter equipped with an RLC filter are designed for complete voltage flicker compensation without harmonics. All the simulated results which have been performed in MATLAB show that a 6-pulse STATCOM is efficiently effective in decreasing the voltage flicker of the generating loads. However, there is injection of the harmonic from STATCOM into the system which can be improved with the increase of the voltage source converters of STATCOM using a 12-pulse STATCOM equipped with an RLC filter. The obtained results clearly demonstrate that 12-pulse STATCOM equipped with an RLC filter can reduce the voltage flicker caused by nonlinear loads such as electric arc furnaces.

Acknowledgements

I Thank to our Institute Executive directors Mr.T Sai kumar &Mr.D Baba for providing creative environment for this work. Also I am very much thankful to our Institute principal Dr. K Ramesh for his kind permission and encouragement to write research paper. I would like to extend my heartfelt thanks to my colleagues. And finally I am very much obliged to my respected parents who inspiring me around the clock.

References

[1]   J. Sun, D. Czarkowski, Z. Zabar, "Voltage Flicker Mitigation Using PWM-Based Distribution STATCOM", IEEE Power Engineering Society Summer Meeting, Vol.1, (21-25 July 2002), pp. 616-621.

[2]   J. Mckim, "The UIE Flicker-meter Demystified", Hewlett-Packard's Power Products Division, 1997.

[3]   R. Collantes-Bellido, T. Gomez, "Identification and Modeling of a Three Phase Arc Furnace for Voltage Distribution Simulation", IEEE Trans. on Power Delivery; Vol.12, No.4, (1997), pp. 1812-1817.

[4]   L. Tang, S. Kolluri, M.F. McGranaghan, "Voltage Flicker Prediction for Two Simultaneously Operated AC Arc Furnaces" IEEE Trans. on Power Delivery; Vol.12, No.2, (1997), pp. 985-991.

[5]   M. Zouiti, S. Saadate, X. Lombard, C. Poumarede, C. Levillain, "Electronic Based Equipment for Flicker Mitigation", Proceedings of International Conference on Harmonics And Quality of Power, Vol.2, (1998), pp. 1182-1187.

[6]   T. Larsson, C. Poumarede, "STATCOM, an efficient means for flicker mitigation" IEEE Power Engineering Society Winter Meeting, Vol.2, (Jan-4Feb 1999), pp. 1208-1213.

[7]   C. S. Chen, H. J. Chuang, C. T. Hsu, S. M. Tscng, "Stochastic Voltage Flicker Analysis and Its Mitigation for Steel Industrial Power Systems", IEEE Power Tech Proceedings, Vol.1, (10-13 Sept. 2001).

[8]   Z. Zhang, N. R. Fahmi, W. T. Norris, "Flicker Analysis and Methods for Electric Arc Furnace Flicker (EAF) Mitigation (A Survey)", IEEE Power Tech Proceedings, Vol.1, (10-13 Sept. 2001).

[9]   J. R. Clouston, J. H. Gurney, "Field Demonstration of a Distribution Static Compensator Used to Mitigate Voltage Flicker", IEEE Power Engineering Society Winter Meeting, Vol.2, (31 Jan-4 Feb 1999), pp. 1138- 1141.

[10]   Elnady, W. El-khattam, M. A. Salama, "Mitigation of AC Arc Furnace Voltage Flicker Using the Unified Power Quality Conditioner", IEEE Power Engineering Society Winter Meeting, Vol.2, (27-31 Jan. 2002), pp. 735-739.

[11]   S. Suzuki, Y. Hara, E. Masada, M. Miyatake, K. Shutoh, "Application of Unified Flow Controller for Power Quality Control at Demand Side", The Third International Power Electronics and Motion Control Conference Proceedings (PIEMC 2000), Vol.3 (15- 18Aug 2000), pp. 1031-1036.

[12]   Y. Hara, E. Masada, M. Miyatake, K. Shutoh, "Application of Unified Flow Controller for Improvement of Power Quality" IEEE Power Engineering Society Winter Meeting, Vol.4, (23-27 Jan. 2000), pp. 2600-2606.

[13]   J. H. R. Enslin, "Unified Approach to Power Quality Mitigation" International Symposium on Industrial Electronics (ISIE '98), IEEE Proceedings, Vol.1, (July1998), pp. 8-20.

[14]   P. Roberts, "Power Quality Solution Alternatives for Low and Medium Voltage Continuous Process Loads", IEEE Rural Electric Power Conference, (5-7 May 2002), pp. C4-C4_7.

[15]   G. C. Montanari, M. Loggini, L. Pitti, E. Tironi, D. Zaninelli, "The effects of series inductors for flicker reduction in electric power systems supplying arc furnaces", IEEE Industry Applications Society Annual Meeting, Vol.2, (2-8 Oct. 1993), pp.1496-1503.

[16]   M. W. Marshall, "sing Series Capacitors to Mitigate Voltage Flicker Problems" IEEE Rural Electric Power Conference, (20-22 April 1997), pp. B3-1-5.

[17]   J. Dolezal, A. G. Castillo, V. Valouch, "Topologies and control of active filters for flicker compensation", International Symposium on Industrial Electronics, IEEE Proceedings, Vol.1, (4-8 Dec, 2000), pp. 90-95.

[18]   L. Gyugi, A. A. Otto, "Static Shunt Compensation for Voltage Flicker Reduction and Power Factor Correction", American Power Conference (1976), pp. 1272-1286.

[19]   Y. Hamachi, M. Takeda, "Voltage Fluctuation Suppressing System Using Thyristor Controlled Capacitors", 8th U.I.E. Congress, (1976).

[20]   F. Frank, S. Ivner, "TYCAP, Power Factor Correction Equipment Using Thyristor Controlled Capacitor for Arc Furnaces", ASEA Journal, No.46, Vol.6, (1973) pp. 147-152.

[21]   R. Mienski, R. Pawelek, I. Wasiak "Shunt Compensation for Power Quality Improvement using a STATCOM controller: Modelling and simulation", IEE Proc.-Gener. Transm. Distrib., No.2, Vol.151, (2004), pp. 274-280.

[22]   Amit K. Jain, AmanBehal, Ximing, Darren M. Dawson, Ned Mohan "Nonlinear Controller for Fast Voltage Regulation Using STATCOMs" IEEE Transaction. On control systems technology, No.6, Vol.12, (2004), pp. 827-842.

[23]   Math Works Company, 'Manual for MATLAB Simulink Software, User's Guide', 2002, Version 6.5.

[24]   H. Akagi, Y. Kanazawa, A. Nabae, "Instantaneous Reactive Power Compensator Comprising Switching Devices Without Energy Storage Components", IEEE Trans. on Industry Applications, No.3, Vol.20, (1984), pp. 625-630.

[25]   Castagnet, T., 'Is the Snubber Circuit Necessary?', STMicroelectronics, Group of Companies, Application Notes, Printed in Italy, 1999.

14

Effective battery charging system by solar energy using C programming and microcontroller

Mohd Tariq[1,*], Sagar Bhardwaj[2], Mohd Rashid[3]

[1]M.Tech Student, Department of Electrical Engineering, Indian Institute of Technology, Kharagpur, India
[2]B.Tech Student, Department of Electrical Engineering, Aligarh Muslim University, Aligarh , India
[3]B.Tech Student, Department of Computer Science, Jamia Hamdard University, New Delhi , India

Email address:

tariq.iitkgp@gmail.com (M. Tariq)

Abstract: Energy is one of the issues that is causing the most controversy as fossil fuels are the greatest pollutants and the greatest contributors to the greenhouse effect .The increasing importance of environmental concern, fuel savings and unavailability of power has led to the renewal of interest in renewable energies. It therefore stands to reason that developing countries whose energy consumption rate is increasing at a very fast rate should be investigating new energy systems based on renewable energies that do not pollute and which are inexhaustible such as the Solar system. In this paper a simple, reliable and effective solar panel charging system has been introduced consisting of a solar panel of desired size and shape. This solar panel is integrated with an embedded system (which contains three modules i.e. dc to ac converter, microcontroller/compiler module and charging output and a battery system).This embedded system regulates the electricity produced (after being converted to ac from dc) between the storage battery and charging output with the help of microcontroller which is programmed to combat the situations in presence and in absence of input supply and able to supply stored energy at night or in unavailability of solar source.

Keywords: Renewable Energy, Solar Energy, Embedded System, Battery, Microcontroller

1. Introduction

The rapid depletion of conventional fossil fuels and environmental concern have resulted in extensive use of renewable energy sources for electrical power generation. Energy is the convertible currency of technology. Without energy the whole fabric of society as we know it would crumble; the effect of a 24 hour cut in electricity supplies to a city shows how totally dependent we are on that particularly useful form of energy. Computers and lifts cease to function, hospitals sink to a care and maintenance level and the lights go out. As populations grow, many faster than the average 2%, the need for more and more energy is exacerbated. Enhanced lifestyle and energy demand rise together and the wealthy industrialized economies which contain 25% of the world's population consume 75% of the world's energy supply [1].

The use of new efficient photovoltaic solar cells (PVSCs) has emerged as an alternative measure of renewable green power, energy conservation and demand-side management

[2]. Renewable energy is the only hope and it is the area of latest research which needs a revolution to make an effective solar panel charging system for the regulation of the flow of current to the desired output and saving the battery from receiving extra voltage and increasing the life.

2. Modelling of PV Cell

PV generators are neither constant voltage sources nor current sources but can be approximated as current generators with dependant voltage sources [3]. During darkness, the solar cell is not an active device. It produces neither a current nor a voltage. However, if it is connected to an external supply (large voltage) it generates a current ID, called diode current or dark current. The diode determines the i-v characteristics of the cell. There are three different models of PV cells generally available. A moderate model of PV cell has been taken in this paper as shown in Figure1.

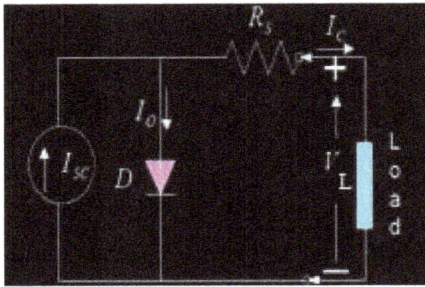

Figure.1. PV cell model.

The model consists of a current source (Isc), a diode (D), and a series resistance (R_S). As the effect of parallel resistance Rp is negligible, it has been omitted in this model[4].

PV arrays are built up with combined series/parallel combinations of PV solar cells, which are usually represented by a simplified equivalent circuit model such as the one given in Fig. 1 and/or by an equation as in (1)

$$V_L = \frac{AKT_C}{q} \ln\left(\frac{I_{sc} + I_o - I_c}{I_o}\right) - I_c\ R_s \qquad (1)$$

Where the symbols are defined as follows:
e: electron charge ($1.602 \times 10\text{-}19$ C).
k: Boltzmann constant ($1.38 \times 10\text{-}23$ J/oK).
Ic: cell output current, A.
I_{SC}: photocurrent, function of irradiation level and junction temperature (5 A).
Io: reverse saturation current of diode (0.0002 A).
Rs: series resistance of cell (0.001 Ω).
Tc: reference cell operating temperature (20 °C).
V_L: cell output voltage, V.

3. Problem Statement

A small effective system comprising of four modules, first the stepping down the dc voltage from Xv to Yv(say) for the microcontroller process to take place, secondly inverting the dc to ac, followed by the relay action of switching and finally passing it to the microcontroller module where it is governed as per the situation of the battery of the module as well as of the system.

4. Set Up

The designed system will solve several of the situations where the solar panel is shown incapable and not worthy for the work. As the system switching the different modes of the battery in the system and in the applied area. Here is the general information describing the overall system, we get the supply from the solar panel system which is step down as per the requirement and inverted if needed which sends us to the next level of relay where the switching takes place as per the command of microcontroller. Considering the three situations in when solar panel is connected:

a) Firstly when the output is connected, in this case the current flows directly to the output, once the output requirements are fulfilled, it automatically switches to the next mode with the help of a zener diode being regulated by microcontroller and commanded to relay, fixed to a certain level and glowing the led for the same.

b) Secondly the switched mode transferred to the charging of the battery placed inside the system for emergency usage, follows the same function of charging and when fully charged to the level of Zener diode given it switches the current to the initial stage.

c) Thirdly the initial stage current not entering the system when both the stages are fulfilled prevent the further depletion of batteries which can be caused if extra current runs through them and increasing the life of the system.

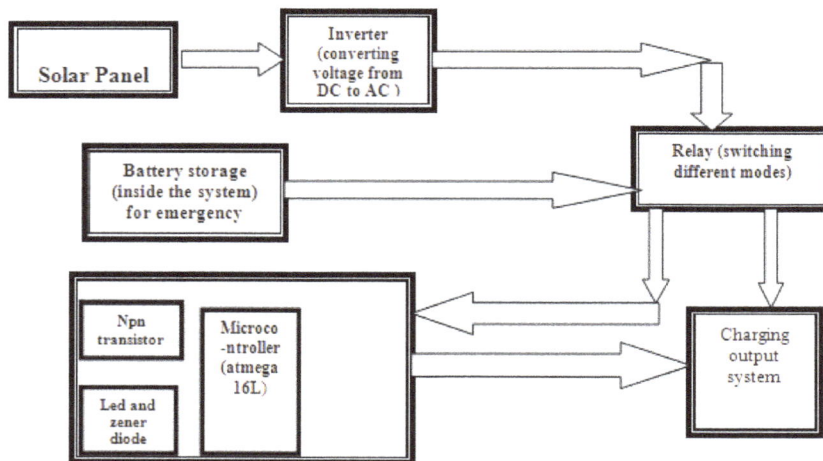

Figure.2. Block Diagram of the Proposed System.

Here the design shows the various connections inside the system taking place.
  ➢   Input from the solar panel

  ➢   Stepping down the voltage
  ➢   Inverting the voltage
  ➢   Relay switching the different modes as per the

commands from microcontroller.

➢ Micro-controller connected with the npn transistor
➢ As soon as the battery of output is filled up the zener diode cut-off the supply and glow the respective led.
➢ And doing same for the system battery and glows the led after the zener diode cut off.
➢ And finally when both the tasks are performed it returns the supply to the input point.

5. Requirements

1. Solar panel:- any solar panel as per the need and requirement.

2. A normal storage battery:- as per the model required

3. A microcontroller (atmel):- preferably atmega16L

4. A relay:- 12V/5V or as per the requirement

5. invertor :- to convert dc to ac voltage

6. voltage chopper- LM317/7805

7. LED and Zener diodes :- signifying the battery is full and zener diode to cut off the supply.

6. Microcontroller Programming

The above algorithm is the key to be built in the micro controller which is regulating the different modes of the charging system:

P1-input port

P2-output port

while(1)//when input supply is there from solar panel
{ If(a==1)//when output battery is not fully charged}
{ P2=0xff;//opening port for the output}
elseif(b==1)//when system battery not fully charged
{P2=0x0f;//opening supply for system battery}
elseif(a=b==0)//when both the batteries are charged
{P2=0x00;//returning the input supply to the initial point}

else
{P2=0x00;//or in any other condition the supply at input and not disturbing the system }
While (0)//when solar panel supply not connected
{ If(b==1)//system battery if charged}
{ P2=0x0f;//supplying the charged system battery output to the output point .}

This is a simple C programming which can be written in the microcontroller.

7. Conclusion

As discussed in the paper the proposed system will be very effective for solving several situations where the solar panel is incapable and not worthy for the work. The proposed effective charging system can be extended to any level, any set-up, which only involves the small embedded kit with the three essential modules empowering the renewable energy

References

[1] Fells I. The problem. In: Dunderdale J,(1990) editor. Energy and the environment. UK: Royal Society of Chemistry.

[2] Abu Tariq, M. Asim, Mohd Tariq (2011) "Simulink based modeling, simulation and Performance Evaluation of an MPPT for maximum power generation on resistive load", 2nd International Conference on Environmental Science and Technology,Singapore.

[3] Khan, B.H., (2006), Renewable energy resources,TataMcGraw-Hill Publishing Company Limited, New Delhi, India.

[4] [Altas. I, A. M. Sharaf, 2007 "A photovoltaic array (PVA) simulation model to use in Matlab Simulink GUI environment." IEEE I-4244- 0632 -03/07.

15

Enhancement of power system stability using self-organized neuro–fuzzy based HVDC controls

Nagu Bhookya, RamanaRao P. V, Sydulu Maheshwarapu

Department of Electrical and Electronics Engineering, National Institute of Technology Warangal, Andhra Pradesh, India

Email address:

nagu.research@gmail.com(N. Bhookya), ramana@nitw.ac.in(Ramanarao P. V), sydulumaheswarapu@yahoo.co.in(S. Maheshwarapu)

Abstract: This paper presents an affective neuro – fuzzy controller (NFC) to improve the transient stability of multi-machine system with HVDC link. Fuzzy rules are used as neurons in artificial neural network (ANN) model. Excellent learning capability of ANN and heuristic fuzzy rules and input/output membership functions of fuzzy logic technique are optimally tuned from training examples by back propagation algorithm (BPA). Considerable time required for fuzzy inference system to match rules is saved using NFC. To illustrate the performance of NFC, transient stability study is carried out on a multi machine system and results are compared with conventional controller as well as fuzzy logic controller.

Keywords: Neuro – Fuzzy Controller, Artificial Neural Network, Transient Stability, Back Propagation Algorithm

1. Introduction

HVDC power transmission system offers several advantages, one of which is rapid control of the transmitted power. Therefore, they have a significant impact on the stability of the associated AC power systems. Moreover, HVDC link effectively uses frequency control and improves the stability of the system using fast load-flow control. The importance of AC-DC power transmission systems in the improvement of stability has been a subject to much research. An HVDC transmission link is highly controllable. It is possible to take advantage of this unique characteristic of the HVDC link to augment the transient stability of the ac systems. In the past, numerous investigations have been carried out to improve transient stability of power system, ranging from theoretical studies to advanced control devices [1-4].

A proper design of the HVDC controls is essential to ensure satisfactory performance of overall AC/DC system [5-6]. The control strategy, traditionally employed for a two-terminal HVDC transmission system is the current margin method, where the rectifier is in current control, and the inverter is in constant extinction angle (CEA) control [4]. Both ends of the dc system rely on PI controllers to provide fast robust control. The conventional methods often require a precise mathematical model of the controlled system.

Because of fixed gains (Kp, Ki, Kd) these controllers perform well over a limited operating range as for power systems in practice, there exists parameter uncertainty in plant modeling and large variations in environmental conditions. Therefore HVDC systems are prone to repetitive commutation failure when connected to a weak AC systems and also when subjected to faults and disturbances. This leads to considerable research in the field of effective control of HVDC systems using adaptive, optimal, intelligent controllers such as neural network, fuzzy logic,neuro – fuzzy controllers etc.

Artificial neural networks and fuzzy logic systems are successfully implemented for improvement of transient stability of power system [7]. The salient features of both techniques are combined to form a hybrid controller i.e. Neuro – fuzzy controller. Self-learning capability of neural network is combined with inference system of fuzzy logic to form self-organizing neuro – fuzzy controller. Thus in this paper, the feasibility of employing a neuro – fuzzy controller for an HVDC transmission system is explored. To demonstrate the effectiveness of proposed NFC, NFC is employed to improve transient stability of a WSCC 9 bus system and the response of the NFC is compared with conventional controller.

2. AC/DC Load Flow Analysis

In transient stability studies it is a prerequisite to do AC/DC load flow calculations in order to obtain system conditions prior to the disturbance [5]. While the conventional approaches are available for conducting the calculations, the eliminated variable method proposed by Anderson, et al[8] is used here which treats the real and reactive powers consumed by the converters as voltage dependent loads. The dc equations are solved analytically or numerically and the dc variables are eliminated from the power flow equations. The method is however unified in the sense that the effect of the dc – link is included in the Jacobian matrix.

2.1. DC System Model

The equations describing the steady state behavior of a mono polar DC link can be summarized as follows [9],

$$V_{dr} = \frac{3\sqrt{2}}{\pi} a_r V_{tr} \cos\alpha_r - \frac{3}{\pi} X_c I_d \qquad (1)$$

$$V_{di} = \frac{3\sqrt{2}}{\pi} a_i V_{ti} \cos\gamma_i - \frac{3}{\pi} X_c I_d \qquad (2)$$

$$V_{dr} = V_{di} + r_d I_d \qquad (3)$$

$$P_{dr} = V_{dr} I_d \qquad (4)$$

$$P_{di} = V_{di} I_d \qquad (5)$$

$$S_{dr} = k \frac{3\sqrt{2}}{\pi} a_r V_{tr} I_d \qquad (6)$$

$$S_{di} = k \frac{3\sqrt{2}}{\pi} a_i V_{ti} I_{d0} \qquad (7)$$

$$Q_{dr} = \sqrt{S_{dr}^2 - P_{dr}^2} \qquad (8)$$

$$Q_{di} = \sqrt{S_{di}^2 - P_{di}^2} \qquad (9)$$

Where,

Vdr, Vdi	voltages at rectifier and inverter end respectively
Vtr, Vti	terminal voltages at rectifier and inverter ends
Id	dc link current
Xc, rd	dc link reactance and resistance
α,□	firing and extinction angle respectively
a	tap ratio
Pdr, Pdi	Real power at rectifier and inverter ends resp.
Qdr, Qdi	Reactive power at rectifier and inverter ends resp.
Sdr, Sdi	Apparent power at rectifier and inverter ends resp.

2.2. The Eliminated Variable Method

The real and reactive powers consumed by the converters are expressed as function of their ac terminal voltages, Vtr and Vti. Their partial derivatives with respect to Vtr and Vti are computed and used in modification of Jacobian

elements of the Newton Raphson power flow solution as shown below,

$$\begin{bmatrix} \Delta P \\ \Delta Q \end{bmatrix} = \begin{bmatrix} H & N \\ M & L \end{bmatrix} \begin{bmatrix} \Delta\delta \\ \Delta V/V \end{bmatrix} \qquad (10)$$

$$N'(tr, tr) = V_{tr} \frac{\partial P_{tr}^{ac}}{\partial V_{tr}} + V_{tr} \frac{\partial P_{dr}(V_{tr}, V_{ti})}{\partial V_{tr}} \qquad (11)$$

$$N'(tr, ti) = V_{ti} \frac{\partial P_{tr}^{ac}}{\partial V_{ti}} + V_{ti} \frac{\partial P_{dr}(V_{tr}, V_{ti})}{\partial V_{ti}} \qquad (12)$$

$$N'(ti, tr) = V_{tr} \frac{\partial P_{tr}^{ac}}{\partial V_{tr}} - V_{tr} \frac{\partial P_{di}(V_{tr}, V_{ti})}{\partial V_{tr}} \qquad (13)$$

$$N'(ti, ti) = V_{ti} \frac{\partial P_{tr}^{ac}}{\partial V_{tr}} - V_{ti} \frac{\partial P_{di}(V_{tr}, V_{ti})}{\partial V_{ti}} \qquad (14)$$

L' is also modified analogously. Thus, in the eliminated variable method, four mismatch equations and up to eight elements of Jacobian have to be modified, but no new variables are added to solution vector, when a dc – link is included in the power flow.

3. Representation of HVDC Systems

Each DC system has unique characteristics tailored to meet the specific needs of its application. Hence, standard models of fixed structures have not been developed for representation of dc systems in stability studies. The current controller employed in this paper is shown in figure 1. It is a proportional integral controller and the auxiliary controller is assumed to be a constant gain controller.

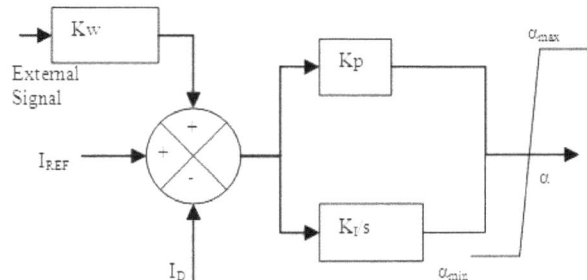

Fig. 1. Block diagram of current controller

The HVDC link can be represented as transfer function model [9] as,

$$I_d = \frac{I_{ref}}{1 + sT_d} \qquad (15)$$

Where,

Id	dc link current
Iref	reference value of current
Td	Time constant of the system.

3.1. Generator Representation

The synchronous machine is represented by a voltage source, in back of a transient reactance, that is constant in magnitude but changes in angular position neglecting the effect of saliency and assumes constant flux linkages and a small change in speed [10]. The classical generator model can be described by following set of differential and algebraic equations,

Differential equations,

$$\frac{d\delta}{dt} = \omega - 2\pi f \qquad (16)$$

$$\frac{d^2\delta}{dt^2} = \frac{d\omega}{dt} = \frac{\pi f}{H}(P_m - P_e) \qquad (17)$$

Algebraic equations,

$$E' = E_t + I_t r_a + jx'_d I_t \qquad (18)$$

Where   E'   Voltage back of transient reactance
     E_t Machine terminal voltage
     I_t Machine terminal current
     r_a Armature resistance
     x'_d Transient reactance
     δ Rotor angle
     ω Speed
     P_m, P_e Mechanical and Electrical Power
     H   Inertia constant

3.2. Load Representation

The static admittance Ypo used to represent the load at bus P, can be obtained from,

$$Y_{po} = \frac{I_{po}}{E_p} \qquad (19)$$

4. Steps of AC-DC Transient Stability Study

The basic structure of transient stability program is given below [14]

1. The initial bus voltages are obtained from the ac/dc load flow solution prior to the disturbance.
2. After the ac/dc load flow solution is obtained, the machine currents and voltages behind transient reactance are calculated.
3. The initial speeds and the initial mechanical powers are obtained for each machine prior to the disturbance.
4. The network data is modified for the new representation. Extra nodes are added to represent the generator internal voltages. Admittance matrix is modified to incorporate the load representation.
5. The time is set as t = 0;
6. If there is any switching operation or change in fault condition, the network data is modified accordingly to run the ac/dc load flow.
7. Using Runge-Kutta method, solution of the machine differential equations are obtained to find the changes in the internal voltage angle and machine speeds.
8. Internal voltage angles and machine speeds are updated.
9. Advance time, t = t + Dt.

10. The time limit is to be checked, if t £ tmax, then the process has to be repeated from step 6,else the process has to be stopped.

In case of multi machine system stability analysis the relative angles are plotted to evaluate the stability of the power system.

5. Conventional Controller

When a multi machine system is subjected to fault, generator closer to location of fault loses synchronism with the system. To stabilize the system, it is necessary to make equal accelerations of all the generators. So an error signal representing average difference in accelerations of the generators is considered. In case of multi machine system, the relative angles are to be maintained within limits to maintain the stability of the system. So, error signals derived from the average difference in the relative angles and average difference in the relative speeds of the generators are considered.

Considering a 3 machine system and first generator falling out of synchronism, the above mentioned errors can be formulated as below.

$$error_1 = \left[\left(\frac{(\omega(2)-\omega(1))+(\omega(3)-\omega(1))}{2}\right) - (\omega(2) - \omega(3))\right] \qquad (20)$$

$$error_2 = \left[\left(\frac{(del(2)-del(1))+(del(3)-del(1))}{2}\right) - (del(2) - del(3))\right] \qquad (21)$$

$$error_3 = \left[\left(\frac{\frac{P\_mis(3)}{H(3)}+\frac{P\_mis(2)}{H(2)}}{2}\right) - \left(\frac{P\_mis(1)}{H(1)}\right)\right] \qquad (22)$$

Combination of the above three signals are considered, in order to improve the stability. Gains of the signals are varied in order to get better transient and dynamic performance. The signal $error_2$ is equivalent to the integration of the signal $error_1$ and the signal $error_3$ is equivalent to the differentiation of the signal $error_1$. Hence, the controller is equivalent to a PID controller . Then the control signal can be represented as,

$$Error = Kp\, e(t) + Ki\, Ie(t) + Kd\, De(t) \qquad (23)$$

$$error = Kp * error1 + Ki * error2 + Kd * error3 \qquad (24)$$

6. Neuro – Fuzzy Controller

The structure of the Neuro – Fuzzy controller is shown in figure 2. The recent direction of research is to design a self-organizing fuzzy logic system that has the capability to create the control strategy by learning [11, 12]. The proposed SONFC is a combination of both neural network and fuzzy logic. The fuzzy method provides a structural control framework to express the input-output relationship of the neural network, and the neural network can embed the salient features of computation power and learning capability into the fuzzy controller.

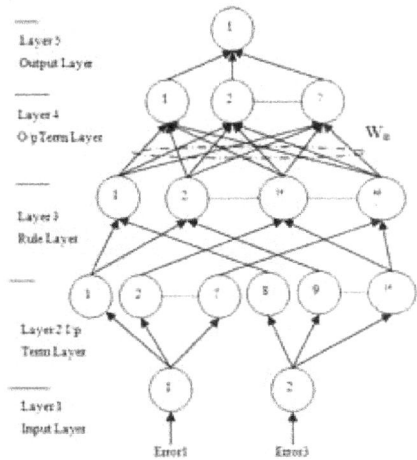

Fig 2. *Topology of Neuro Fuzzy Controller*

6.1. Topology of the Neuro – Fuzzy controller

The proposed NFC is a multilayer neural network-based fuzzy controller. Its overall structure and topology is shown in Fig. 2. The system has a total of five layers. Since two input variables and one output variable are employed in the present work, there are two nodes in layer 1 and one node in layer 5. Nodes in layer 1 are input nodes that directly transmit input signals to the next layer. Layer 5 is the output layer. Nodes in layers 2 and 4 are term nodes that act as membership functions to express the input/output fuzzy linguistic variables. A bell-shaped function is adopted to represent the membership function, in which the mean value m and the variance o will be adapted through the learning process. The fuzzy sets defined for the input/output variables are positive big (PB), positive medium (PM), positive small (PS), zero (ZE), negative small (NS), negative medium (NM), and negative big (NB), which are numbered in descending order in the term nodes. Hence, 14 nodes and 7 nodes are included in layers 2 and 4, respectively, to indicate the input/output linguistic variables. Each node of layer 3 is a rule node that represents one fuzzy control rule. In total, there are 49 nodes in layer 3 to form a fuzzy rule base for two linguistic inputs. Layer 3 links and layer 4 links define the preconditions and consequences of the rule nodes respectively. The NFC adjusts the fuzzy control rules and their membership functions by modifying layer 4 links and the parameters that represent bell – shaped membership function for each node in layer 2 and 4. Following symbols are used to describe various functions:

netiL: the net input value to the i-th node in layer L,

OiL: the output value of the i-th node in layer L,

miL,σiL:the mean and variance of the bell – shaped activation function of the i-th node in layer L,

Wij: the link that connects output layer of j-th node in layer3 with the input to the i-th node in layer 4.

Layer 1: This is a fan – out layer. Inputs are directly transmitted to next layer.

Layer 2: The nodes of this layer act as membership function to express the terms of input linguistic variables.

$$net_i^2 = \begin{cases} O_1^1 & for\ i = 1,2,\dots.7 \\ O_2^2 & for\ i = 8,9\dots14 \end{cases}$$

$$O_i^2 = e^{-\left(\frac{net_i^2 - m_i^2}{\sigma_i^2}\right)^2} \qquad for\ i = 1,2,\dots..14 \qquad (25)$$

Note that layer 2 links are all set to unity.

Layer 3: The links in this layer are used to perform precondition matching of fuzzy rules. Thus, each node has two input values from layer 2. The correlation-minimum inference procedure[15] is utilized here to determine the firing strengths of each rule. The output of nodes in this layer is determined by the fuzzy AND operation. Hence, the functions of the layer are given below:

$$net_i^2 = \min(O_i^2, O_i^2) \qquad (26)$$

The link weights in this layer are also set to unity.

Layer 4: Each node of this layer performs the fuzzy OR operation to integrate the fired rules leading to the same output linguistic variable. Starting with the good initial fuzzy control rules will provide much faster convergence in the learning phase. The functions of this layer are expressed as follows:

$$net_j^4 = \sum_{j=1}^{49} W_{ij} O_j^3 \qquad (27)$$

$$O_j^4 = \min(1, net_j^4) \qquad for\ j = 1,2\dots\dots7 \qquad (28)$$

The link weight Wijin this layer expresses the probability of the j-th rule with the i-th output linguistic variable.

Layer5: The node in this layer computes the control signal of the NFC. The output node together with layer 5 links act as a de-fuzzifier. The de-fuzzification aims at producing anon-fuzzy control action that best represents the possibility distribution of an inferred fuzzy control action. The centre of area de-fuzzificationscheme, in which the fuzzy centroidconstitutes the controller output signal, can be simulated.

$$net_1^5 = \sum_{j=1}^{7} m_j^4 \sigma_j^4 O_j^4 \qquad (29)$$

$$O_j^5 = \frac{net_1^5}{\sum_{j=1}^{7} \sigma_j^4 O_j^4} \qquad (30)$$

6.2. Self-organizing Learning Algorithm

The problem for the self-organized learning can be stated as: Given the training input data xi(t), i = 1, . - . , n , the desired output value yi(t), i = 1,. . . . , m ,the fuzzy partitions|T(x)| and |T(y)|, and the desired shapes of membership functions, we want to locate the membership functions and find the fuzzy logic rules. In this phase, the network works in a two-sided manner; that is, the nodes and links at layer four are in the up-down transmission mode so that the training input and output data are fed into this network from both sides. First, the centres (or means) and the widths (or variances) of the membership functions are determined by self-organized learning techniques analogous to statistical clustering. This serves to allocate network resources efficiently by placing the domains of membership functions covering only those regions of the input/output space where data are present. Kohonen's feature-maps algorithm [20] is adapted here to find the centre mi, of the membership function:

$$\|x(t) - m_{closest}(t)\| = \min_{1 \le i \le k} \{\|x(t) - m_i(t)\|\}$$

$$m_{closest}(t + 1) = m_{closest}(t) + \propto (t)[x(t) - m_{closest}(t)]$$

$$m_i(t + 1) = m_i(t) \, for \, m_i \ne m_{closest}$$

where α (t) is a monotonically decreasing scalar learning rate, and $k = |T(x)|$. This adaptive formulation runs independently for each input and output linguistic variable. The determination of which of the mi's is mclosest can be accomplished in constant time via a winner-take-all circuit. Once the centres of membership functions are found, their widths can be determined by the *N-nearest-neighbors* heuristic by minimizing the following objective function with respect to the widths *(σi's)*

$$\sigma i = \left[\frac{m_i - m_{closest}}{r}\right] \quad for \; i=1,2,....,7 \quad (31)$$

Then the optimal membership functions and fuzzy rules can be found by gradient – descent search techniques. Thus the energy function is defined as,

$$E = \frac{1}{2}(U^d(k) - U(k))^2 \quad (7.10)$$

Now using generalized delta rule [14] to minimize the energy, in standard notations, the delta rule can be expressed as,

$$\chi_i(k + 1) = \chi_i(k) + \eta\left(-\frac{\partial E}{\partial \chi_i}\right) + \lambda\Delta\chi_i(k) \quad (7.11)$$

The error signal term delta produced by the i-th neuron in layer L is defined as,

$$\delta_i^L(k) = -\frac{\partial E}{\partial net_i^L} \quad (7.12)$$

Using above equations, the learning rules of each layer are derived below:
Layer 5: the error signal of the output node is

$$\delta_j^5 = (U^d(k) - U(k)) \quad (7.13)$$

The mean and variance of each output membership function are adapted by,

$$m_i^4(k + 1) = m_i^4(k) + \eta\delta_1^5 \frac{\sigma_i^4 O_i^4}{\sum_{j=1}^{7} \sigma_j^4 O_j^4} + \lambda\Delta m_i^4(k)$$

$$\sigma_i^4(k + 1) = \sigma_i^4(k) + \eta\delta_1^5$$

$$\frac{m_i^4 O_i^4\left(\sum_{j=1}^{7} \sigma_j^4 O_j^4\right) - \sum_{j=1}^{7} m_j^4 \sigma_j^4 O_j^4}{(\sum_{j=1}^{7} \sigma_j^4 O_j^4)^2} + \lambda\Delta\sigma_i^4(k)$$

$$for \; i = 1,2, ... 7 \quad (7.14)$$

Layer 4: the error signal of each node is

$$\delta_i^4 = \delta_1^5 \frac{m_i^4 O_i^4\left(\sum_{j=1}^{7} \sigma_j^4 O_j^4\right) - \sum_{j=1}^{7} m_j^4 \sigma_j^4 O_j^4}{(\sum_{j=1}^{7} \sigma_j^4 O_j^4)^2}$$

$$for \; i = 1,2, ...7; j = 1,2 ... 49$$

The weights between the i-th output linguistic variable and j-th rule is updated by,

$$W_{ij}(k + 1) = W_{ij}(k) + \eta\delta_i^4 O_j^3 + \lambda\Delta W_{ij}(k)$$

$$for \; i = 1,2, ...7$$

Layer 3: No parameter needs to be adjusted in this layer, and only the error signal needs to be computed and propagated backwards. That is,

$$\delta_i^3 = \sum_{j=1}^{7} W_{ij} \, \delta_j^4 \quad (7.15)$$

Layer 2: The mean and variance of the input membership functions can be updated by,

$$m_i^2(k + 1) = m_i^2(k) - \eta\frac{\partial E}{\partial O_i^2} O_i^2 \frac{2(O_1^1 - m_i^2)}{(\sigma_i^2)^2} + \propto \Delta m_i^2(k) \quad (7.16)$$

$$\sigma_i^2(k + 1) = \sigma_i^2(k) - \eta\frac{\partial E}{\partial O_i^2} O_i^2 \frac{2(O_1^1 - m_i^2)}{(\sigma_i^2)^2} + \propto \Delta\sigma_i^2(k)$$

$$for \; i = 1,2, ...14$$

It should be noted that the function of layer 1 is only to distribute the input signal, and hence it is not involved in the learning process. The links connecting layers 4 and 3 can be deleted when the weight is negligibly small or equals zero after learning because it means that this rule node has little or no relationship to the output linguistic variable.

7. Case Study

A WSCC-9 system [13] is taken for stability analysis, it is given in the below figure 4.

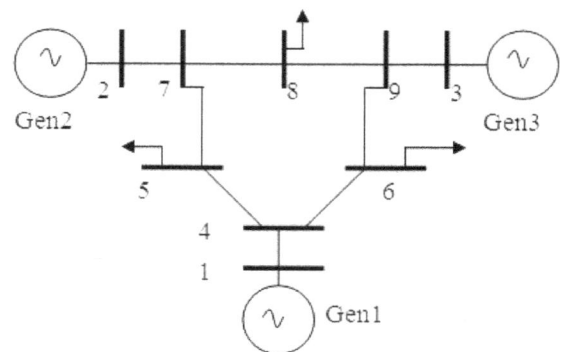

Fig 3. WSCC 9 bus System

To test the effectiveness of the above controllers the HVDC system is subjected to a Three-phase-to-ground fault at the converter end AC bus. Variation of dc link voltage and current, and converter firing angle due to a three-phase-to-ground fault at the rectifier end AC bus. The dc bus voltage completely collapses and results in commutation failure of the converter thyristors. During the fault, the DC link current drops to zero and the firing angle settles at the minimum value. The zero current and zero power condition lead to complete de-energization of the DC link. As soon as the fault is cleared the converter current controller gets activated, and it is in this period when the performance is influenced by the controller actions.

A grounded fault is assumed to occur on Line 4-6, near to Bus 6, at initial time zero and the line from Bus 4 to Bus 6 is removed after 4 cycles. The HVDC line is located between buses 4 –5. Under these conditions, the impact of HVDC on system stability is presented. Initially, a case in which the HVDC line maintains the same control as in the normal state, in which the post-fault HVDC power flow setting remains the same as before, is investigated. It was found that, the system becomes unstable. Then a controller is used to stabilize the system. It is clearly seen from figure 4that the system is becoming unstable, generator 2 and generator 3 are moving together whereas generator 1 falling out of synchronism when no control action is performed.

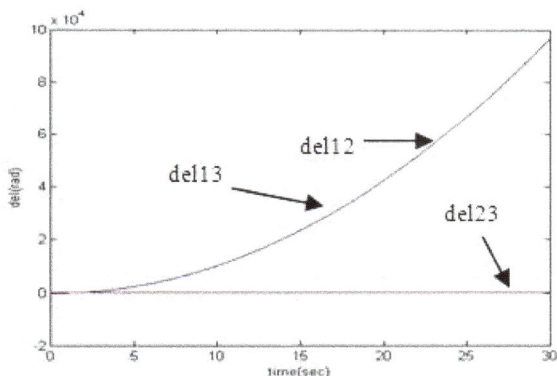

Fig 4. Plot of relative angles without any external control signal applied

Different combinations of the three signals in eq. 20, 21, 22 are considered, in order to improve the stability.

Case 1: Considering the signal error3 as the control input, the plot of relative angles is as shown in the fig. 5.

Case 2: Considering the combination of error2 and error3 signals to generate the control signals, the plot of relative angles will be as shown in figure 6.

Case 3: Considering the combination of all the three signals to generate the control signal, the plots of the relative angles with different gains are as shown in fig.7.

Thus from case 3 it is clear that the plot of relative angles of generators can be improved using conventional PID controller. Therefore Fuzzy logic PID controller is also implemented and the plot of relative angles of generators using fuzzy logic controller is shown in fig. 8.

NFC is trained with the available data. Learned Fuzzy Rule matrix using NFC is shown in table 2. The performance of NFC in figures 8,10 is compared with

conventional controller and fuzzy logic controller and is summarized in table 3. From table 3, it is clear that proposed NFC works satisfactorily for the given system and gives best results. Similar experiment is done by assuming fault on the line between buses 8 and 9 near bus 8 and the HVDC link is in between buses 4 and 5 and the wave forms are plotted.

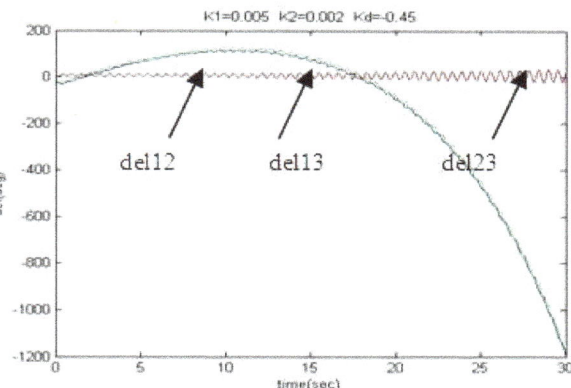

Fig 5. Plot of relative angles with error₃ as the control signal

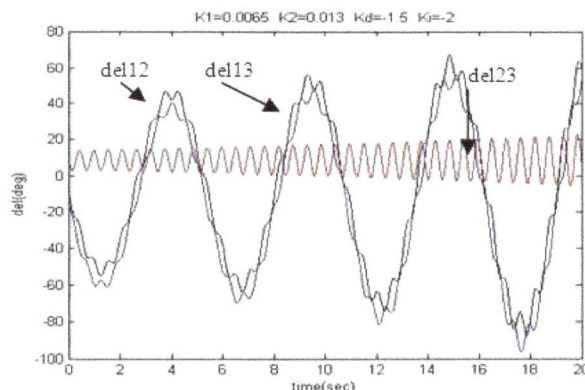

Fig 6. Plot of relative angles with error₂ and error₃ as control signals

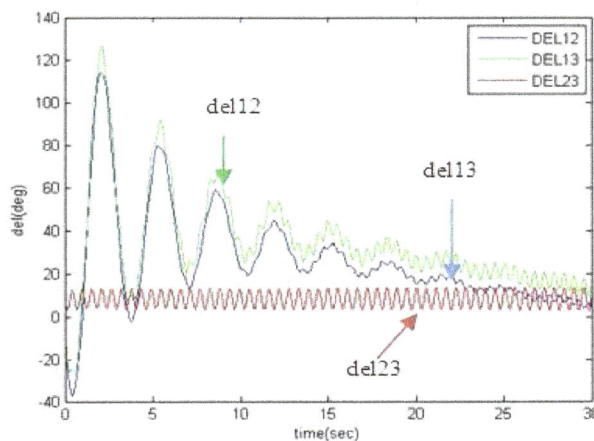

Fig 7. Plot of relative angles with PID controller. When hvdc link is in between buses 4-5and fault is in between buses 4-6.

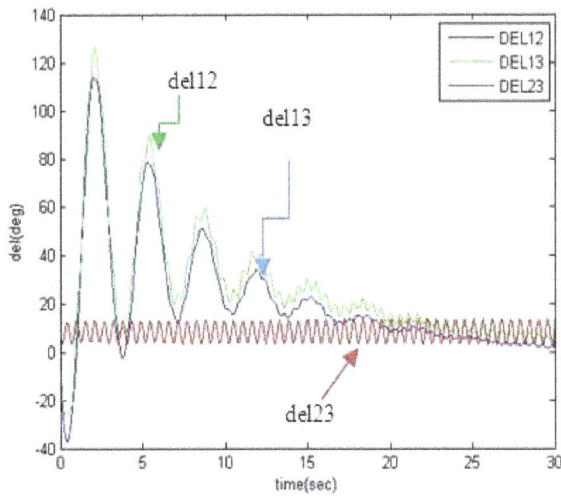

Fig 8. *Plot of relative angles with NeuroFuzzy Controller When hvdc link is in between buses 4-5and fault is in between buses 4-6.*

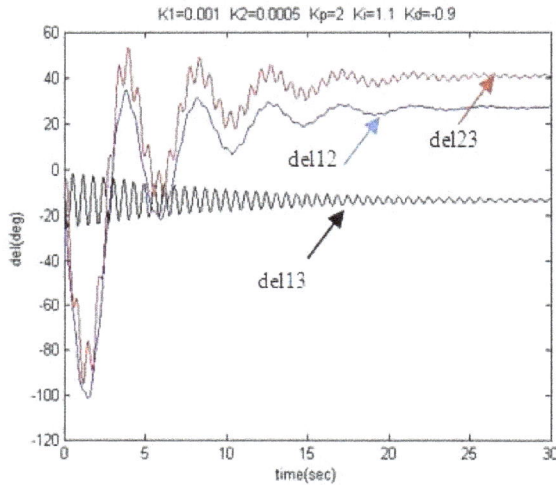

Fig 9. *Plot of relative angles with PID controller. When hvdc link is in between buses 4- 5and ,fault is in between buses 8- 9.*

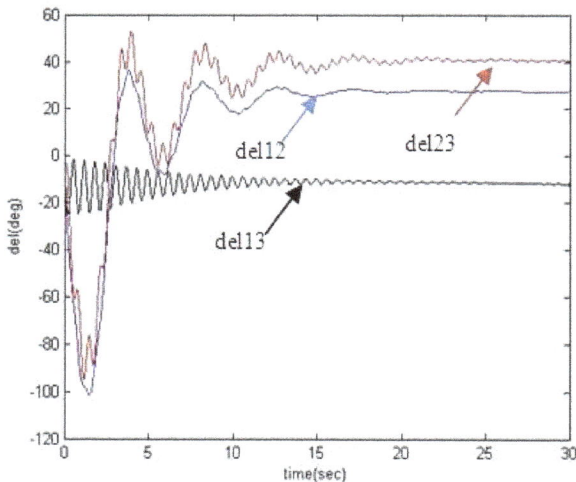

Fig 10. *Plot of relative angles with Neuro-Fuzzy controller. When hvdc link is in between buses 4 and 5 fault is in between buses 8 and 9.*

Table 1. *Initial Fuzzy Rule Matrix*

		Error 1					
		LN	MN	SN	VS	SP	MP	LP
Error3	LP	VS	SP	MP	LP	LP	LP	LP
	MP	SN	VS	SP	MP	MP	LP	LP
	SP	MN	SN	VS	SP	SP	MP	LP
	VS	MN	MN	SN	VS	SP	MP	MP
	SN	LN	MN	SN	SN	VS	SP	MP
	MN	LN	LN	MN	MN	SN	VS	SP
	LN	LN	LN	LN	LN	MN	SN	VS

Table 2. *Learned Fuzzy Rule Matrix*

		Error 1					
		LN	**MN**	**SN**	**VS**	**SP**	**MP**	**LP**
Error3	**LP**				LP			LP
	MP				MP			LP
	SP		LN	MN	SP	LP	LP	LP
	VS	LN	MN	SN	VS	SP	MP	LP
	SN		LN	LN	SN	SN	LP	LP
	MN		LN	LN	MN			
	LN	LN			LN			

Table 3. *Comparison Of Conventional, Fuzzy Logic And Neuro Fuzzy Controller when fault is in between the buses 4 and 6*

Sr. No.	Type of Controller	Generator Controlled	Peak Overshoot	Settling Time
1	Conventional	Gen1	80 deg.	25 sec
2	Fuzzy Logic Based Controller	Gen1	140 deg.	20 sec
3	NFC	Gen1	55 deg.	5 sec

8. Conclusions

Considering the HVDC current controller and line dynamics, it is observed that the transient stability of the multi-machine system is improved only if the combination of all the three signals derived from relative speed (P), rotor angle (D) and average acceleration (I) is used. The paper presents the design of a very simple form of Neuro Fuzzy controller. In this paper, the possibility of replacing a traditional PID controller with a neuro fuzzy controller for the rectifier terminal of an HVDC link is explored.

The neuro fuzzy controller gives better performance than the conventional controller and fuzzy logic controller as expected when compared various aspects of plot of relative angles of the generators such as settling time, peak overshoot etc.

Appendix

Table 4. *Generators Data*

Generator	X_d'	H
1	0.0608	23.64
2	0.1198	6.4
3	0.1813	3.01

Table 5. *Transformers Data*

Transformer	X
1	0.0576
2	0.0625
3	0.0586

Table 6. *Transmission Network Data*

Bus No. P	Bus No. Q	R	X	$y_{pq}/2$
1	4	0.0000	0.0576	0.0000
2	7	0.0000	0.0625	0.0000
3	9	0.0000	0.0586	0.0000
4	6	0.0170	0.0920	0.0790
5	7	0.0320	0.1610	0.1530
6	9	0.0390	0.1700	0.1790
7	8	0.0085	0.0720	0.0745
8	9	0.0119	0.1008	0.1045

Table 7. *Hvdc Line Data*

DC line data		Initial conditions	
r_d=0.017, X_c=0.6,	L_d=0.05.	alfa = 0.2094, I_d=0.3691,	P_{di}=0.406
alfamin=5deg, alfamax=80deg		V_{di}=1.1, $P_{M[1]}$=0.756646	gama=0.3142,
taprmin=0.96, taprmax=1.06		$P_{M[2]}$=1.63, $P_{M[3]}$=0.85, $\delta_{M[1]}$=2.388448deg	
tapimin=0.99, tapimax=1.09		$\delta_{M[2]}$=18.603189deg, $\delta_{M[3]}$=12.314856deg	

Table 8. *Generator Data*

Bus No.	P_{GEN}	P_D	Q_D	V_{sp}
1	0.00	0.00	0.00	1.040
2	1.63	0.00	0.00	1.025
3	0.85	0.00	0.00	1.025
4	0.00	0.00	0.00	--
5	0.00	1.25	0.50	--
6	0.00	0.90	0.30	--
7	0.00	0.00	0.00	--
8	0.00	1.00	0.35	--
9	0.00	0.00	0.00	--

References

[1] H. C. Chang and H. C. Chen, "Fast Generation-Shedding Determination in Transient Emergency State," *IEEE Trans. on Energy Conversion, vol. 8, no. 2, pp. 178-183, 1993.*

[2] M. L. Shelton and P. F. Winkelman, "Bonneville Power Administration 1400-MW Brakine Resistor."*IEEE Trans..vol. PAS- 94, pp. 602-609, 1975.*

[3] C.S. Rao and T. K. Nag Sarkar, "Half Wave Thyristor Controlled Dynamic Brake to Improve Transient Stability," *IEEE Trans., vol. PAS-103, no. 5, pp. 1077-1083, May 1984.*

[4] A. Ekstrom and G. Liss, "A refined HVDC control system," *IEEE Trans. Power Apparatus and Systems, vol. PAS-89, no. 536, May/June 1970.*

[5] P. Kundur, *Power System Stability and Control* McGraw-Hill, Inc., 1994

[6] Garng M. Huang, VikramKrishnaswamy, "HVDC Controls for Power System Stability", *IEEE Power Engineering Society, pp 597- 602, 2002.*

[7] V. K. Sood, N. Kandil, R. V. Patel, K. Khorasani, "Comparative Evaluation of Neural-Network-Based and PI Current Controllers for HVDC Transmission", *IEEE Transactions on Power Electronics, VOL.9, NO.3, May1994.*

[8] T. Smed, G. Anderson, "A New Approach to AC/DC Power Flow", *IEEE Trans. on Power Systems., Vol. 6, No. 3, pp 1238- 1244, Aug. 1991.*

[9] K. R. Padiyar,*HVDC Power Transmission Systems* New Age International (P) Ltd., 2004.

[10] Jos Arrillaga and Bruce Smith, "AC- DC Power System Analysis", *The Institution of Electrical Engineers, 1998.*

[11] M. Sugeno and K. Murakami, "Fuzzy Parking control of Model Car,"*in the 13rd IEEE Conf. on Decision and Control, Las Vegas, 1984.*

[12] R. Tauscheit and E. M. Scharf, "Experiments with the Use of a Rule-Based Self-organizing Controller for Robotics Applications," FuzzySets and System, vol. 26, pp. 195-214, 1988.

[13] P.M.Anderson and A.A.Fouad, *Power System Control and Stability*1[st]ed.,Iowa State University Press, 1977.

[14] Stagg and El- Abiad, *Computer Methods in Power System Analysis* International Student Edition, McGraw- Hill, Book Company, 1968

[15] C. T. Lin and C. S. George Lee, "Neural-Network Based Fuzzy LogicControl and Decision System," *IEEE Trans. on Computers, vol. 40,no.12, pp. 1320-1336, Dec. 1991.*

16

Modeling, analysis and development of "hybrid" manual and solar PV based power generation system

Mohd. Tariq[1,*], Khyzer Shamsi[2], Tabrez Akhtar[3]

[1]M.Tech Student, Department of Electrical Engineering, Indian Institute of Technology, Kharagpur, India
[2]Protection design engineer, Al Jazirah Engineers and consultants, Al-khobar, Kingdom of Saudi Arabia
[3]Assistant Manager, Delhi Metro Rail Corporation Ltd., Delhi, India

Email address:
tariq.iitkgp@gmail.com (M. Tariq)

Abstract: This paper presents the analysis & development of hybrid Manual and Solar PV Based Power Generation System as there is a requirement to supply the rural areas of India and other developing countries with reliable electricity. Most of the rural areas won't have access to reliable electrical power even in the next ten years. Therefore alternate means of generating electrical energy will have to be utilized at local level. One such alternative is to produce electrical energy using hybrid systems utilizing manpower and solar energy. Local institutions like Panchayats might play an important role in the implementation, operation and maintenance of the proposed power generation system. The unregulated voltage generated by the proposed power generation system has been regulated with the help of electronic circuits.

Keywords: DC Generator, Prime Mover, Hybrid Power Generation, Solar PV, Voltage Regulator

1. Introduction

Most of the power generation in India is carried out by conventional energy sources, coal and mineral oil-based power plants which contribute heavily to greenhouse gases emission. Setting up of new power plants is inevitably dependent on import of highly volatile fossil fuels. Thus, it is essential to tackle the energy crisis through judicious utilization of abundant renewable energy resources, such as biomass energy, solar energy, wind energy, geothermal energy and Ocean energy. Last 25 years has been a period of exuberant hunt of activities related to research, development, production and demonstration at India. India has obtained application of a variety of renewable energy technologies for use in different sectors too.

Most of the rural areas of India don't have reliable access to electricity. The cost of installation and maintenance of transmission lines in these areas is quite high due to low population densities. Some of these areas won't have access to reliable electrical power even in the next ten years. Therefore alternate means of generating electrical energy will have to be utilized at local level. One such alternative is to produce electrical energy using Manual Charkha Based Power Generation System. [1]

India has abundant manpower. This can be utilised in the generation of electrical energy. A wheel coupled to an electric generator may be rotated to produce electricity. However, the revolutions per minute (r.p.m.) required by the generator prime-mover is quite high for obtaining the suitable at armature terminals. This problem can be solved by using a gear-system or by using wheels of different radii. However, the gear system leads to

considerable increase in weight of the apparatus. Hence, wheels of different radii are employed [1]. But this system has some limitations as it totally depends upon the manual resources. So for countries having lesser population density, this system /model will not work.

The solar radiation intensity is quite high in India for most of the year [2]. This energy can be harnessed for the generation of electrical energy. The main drawback of solar energy is that it is intermittent in nature. This drawback can be taken care of by using this in tandem with manual power as discussed above.

So, in this paper a Hybrid Manual and Solar PV based power generation system has been proposed to overcome the limitations of Manual & Solar PV based systems installed individually.

This power generation system can be used to provide energy to light loads such as lighting, mobile charging, i-pod charging, charging of flashlights etc. However these equipments require constant voltages. Hence a constant voltage regulator will have to be used such as 78XX.

Mobile-charger will require 7805 to obtain a constant voltage.

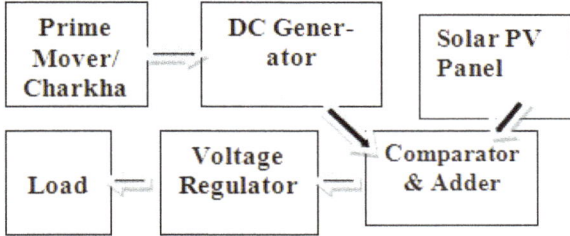

Figure 1. *Block diagram of the Proposed Hybrid Power Generation System.*

The two systems are coordinated such that they supplement each other. When the solar radiations are of low intensity and the power generated is not of desired value, we can run the generator at suitable speed such that the voltage obtained has the desired value and vice-versa.

Rural population constitutes 70% of India's population hence there is approximately 120 to 140 million families or households of which 30%-40% approximately is without electrification, from this it could be inferred that an enhanced living standards at 1kW/household at 4 hours usage per day draws 105MWh per year/house hold which is over 250MW approximately for household consumption sector alone. [3]

The features of rural electricity viz, low and dispersed loads, high T & D costs and seasonality of the load favors decentralized power plants for meeting rural electricity needs in a sustainable manner. Local institutions like Panchayats might play an important role in the implementation, operation and maintenance of the proposed power generation system. This will not only minimize transaction costs but also minimize transmission and distribution costs.[4]

2. Manual Charkha Based Power Generation

A spinning wheel or charkha is a device for spinning thread or yarn from natural or synthetic fibers. The earliest clear illustrations of the spinning wheel come from Baghdad (drawn in 1237), China (1270) and Europe (1280), and there is evidence that spinning wheels had already come into use in both China and the Islamic world during the eleventh century. According to literature, the spinning wheel was introduced into India from Iran in the thirteenth century. The spinning wheel replaced the earlier method of hand spinning with a spindle. The first stage in mechanizing the process was mounting the spindle horizontally so it could be rotated by a cord encircling a large, hand-driven wheel. The great wheel is an example of this type, where the fiber is held in the left hand and the wheel slowly turned with the right. Holding the fiber at a slight angle to

the spindle produced the necessary twist. The spun yarn was then wound onto the spindle by moving it so as to form a right angle with the spindle.

Hand powered spinning wheels are powered by the spinner turning a crank for fly wheel with their hand, as opposed to pressing pedals or using a mechanical engine. The table-top or floor charkha is one of the oldest known forms of the spinning wheel. The charkha works similarly to the great wheel, with a drive wheel being turned by hand, while the yarn is spun off the tip of the spindle.

Figure 2. *Model for Manual Charkha Based Power Generation System [1]*

Table 1. *RPM of Generator and output Voltage [1].*

RPM of Generator	Output Voltage (V)
645	5.00
750	6.28
775	6.50
910	7.50
950	7.85
1030	8.00
1105	8.55
1193	9.00

Observation Table of rpm of Generator and Output Voltage.

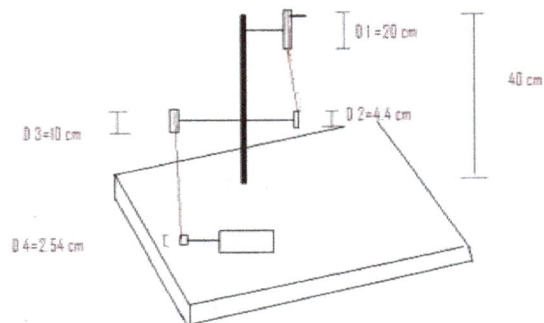

Figure 3. *Diagrammatic Representation Of Multi-Wheel Charkha [1].*

Calculation of Various Radii

D1- Diameter of wheel on which manual power is applied

D2- Diameter of wheel directly below the handled wheel
D3- Diameter of wheel coupled with the second wheel
D4- Diameter of wheel attached with PM dc generator
N1- input mechanical rpm
N2- rpm of wheel having diameter D2
N3- rpm of wheel having diameter D3
N4- required rpm of PM dc generator
Because of same linear velocity

$$\pi D4*N4 = \pi D3*N3$$

Hence, N3=(D4/D3)*N4
Our generator has D4=2.4 cm
For voltage=7.5 V, the reqd. rpm = 910
Therefore, N3=2184/D3
Now, N2=N3
Hence,D1=2184 X (D2/D3)/N1
Assuming N1=50 rpm
D1D3/D2=43.6 cm
We choose,
D1=20 cm, D2=4.6cm, D3=10 cm

3. Solar PV Based Power Generation

Solar energy is the most abundant, inexhaustible and clean of all the renewable energy resources till date. The power from sun intercepted by the earth is about 1.8×10^{11} MW, which is many times larger than the present rate of all the energy consumption. Photovoltaic technology is one of the finest ways to harness the solar power. A photovoltaic power generation system consists of multiple components like cells, mechanical and electrical connections and mountings and means of regulating and/or modifying the electrical output. These systems are rated in peak kilowatts (kWp) which is an amount of electrical power that a system is expected to deliver when the sun is directly overhead on a clear day.[5]

The basic building block of PV technology is the solar cell. Many cells may be wired together to produce a PV module, and many modules are linked together to form a PV array. PV modules sold commercially has range in power output from about 10 watts to 300watts, and produce a direct current. Commercial PV systems are about 7% to 17% efficient.

PV generators are neither constant voltage sources nor current sources but can be approximated as current generators with dependant voltage sources. During darkness, the solar cell is not an active device. It produces neither a current nor a voltage. However, if it is connected to an external supply (large voltage) it generates a current I_d, called diode current or dark current. The diode determines the i-v characteristics of the cell. There are three different models of pv cells generally available.The model consists of a current source (Isc), a diode (D), and a series resistance (Rs) [6]. The equivalent circuit of a PV cell is demonstrated below.

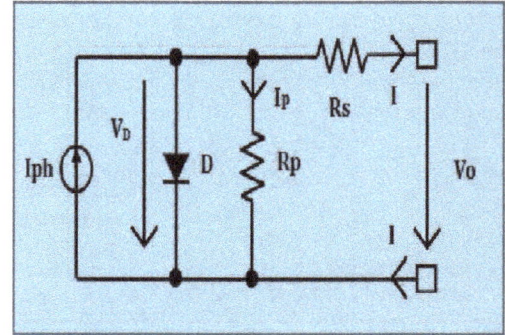

Figure 4. Equivalent circuit of a PV cell.

Derived from Kirchhoff's first law (also referred to as Kirchhoff's current law), the outputcurrent is given by:

$$I = Iph - I_D - Ip$$

$$I = Iph - Isat \left(exp \frac{q.(V_o + I.R_s)}{n.K.T_{cell}.N_s} - 1 \right) - \frac{V_o + I.R_s}{R_p}$$

where
I Output current
IphPhoto current
IsatDiode reverse saturation voltage
VoOutput Voltage
Rs Series resistance (Representing voltage loss on the way to external connectors)
RpParallel resistance (Representing leakage currents)
kBoltzmann's constant
q Charge on electron
Ns Number of cells in series
N Ideality factor
TcellSolar panel temperature

4. Characteristic of the Photovoltaic Array

The current generated in the solar cell by the current source (Iph) is proportional to the amount of light falling on it. When there is no load connected to the output Vo almost all of the generated current flows through diode D. The resistors R and Rp represent small losses due to the connections and leakage respectively. There is very little change in Voc for most instances of load current. However, if a load is connected to the output then the load current draws current away from the diode D. As the load current increases more and more current is diverted away from the diode D. So, as the output load varies so too does the output current, while the output voltage Voc remains largely constant. That is until so much current is being drawn by the load that diode D becomes insufficiently biased and the voltage across it diminishes with increasing load. Hence a number of observations are taken at different insolation levels. These observations are tabulated below in different tables . This results in an I -V characteristic as shown in Figure 5.

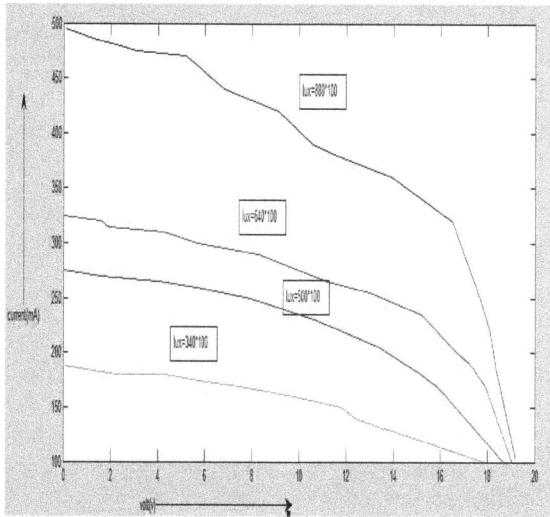

Figure 4. *I-V Characteristic of A Photovoltaic Cell.*

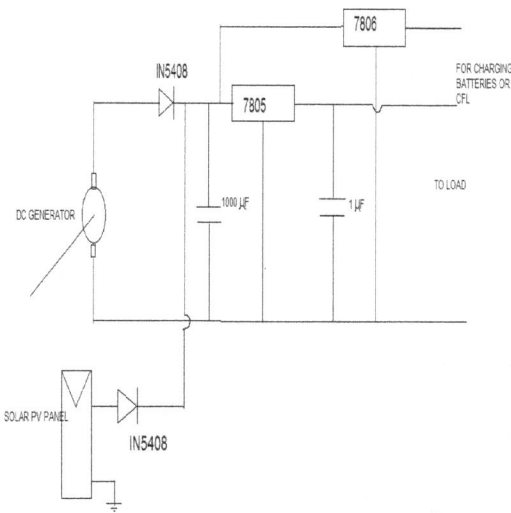

Figure 5. *Circuit Diagram of Proposed Hybrid Power Generation System.*

For Insolation= 340X10² lux

Voltage (V)	Current (mA)	Power (W)
2.22	180	0.39
4.30	180	0.77
5.75	175	1.01
7.14	170	1.21
8.36	165	1.38
9.76	160	1.56
10.74	155	1.66
11.80	150	1.77
12.40	140	1.74
13.00	135	1.72
13.59	130	1.71

For Insolation=500X 10² lux

Voltage (V)	Current (mA)	Power (W)
0.01	275	0.00275
1.59	270	0.43
4.03	265	1.07
5.15	260	1.47
7.96	250	1.99
10.63	230	2.44
11.77	220	2.59
13.43	205	2.75
14.46	190	2.75
15.20	180	2.74
15.79	170	2.68

Table 2. *Voltage, current and power at different insolations.*

For Insolation= 640X10² lux

Voltage (V)	Current (mA)	Power (W)
1.58	325	0.0065
1.58	320	0.50
4.26	310	1.72
4.58	300	1.70
8.28	290	2.40
11.20	265	2.97
13.00	255	3.32
15.20	235	3.57
16.70	200	3.34
17.25	190	3.28
17.90	170	3.046

For Insolation= 800X10² lux

Voltage (V)	Current (mA)	Power (W)
0.1	485	0.48
1.4	480	0.67
2.2	475	1.05
3.1	470	1.46
5.20	440	2.23
6.8	420	2.86
9.1	390	3.55
10.6	380	4.03
11.6	360	4.18
13.9	320	4.45
15.5	250	4.13

5. Hybrid Power Generation

Hybrid power generation system combines a renewable energy.

Source (PV in this case) with other forms of generation, usually a conventional generator powered by diesel or even another renewable form of energy like wind. Such hybrid systems serve to reduce the consumption of non renewable fuel. Barton et al. described a novel method of modelling an energy store used to match the power output from a wind turbine and a solar PV array to a varying electrical load and validated the method against time-stepping methods showing good agreement over a wide range of store power ratings, store efficiencies, wind turbine capacities and solar PV capacities [7]. Katti and Khedkar investigated the application of wind alone, solar-alone, and integrated wind PV generation for utilization as stand-alone generating systems, to be used at the remote areas which was based on the site matching and an energy flow strategy that satisfies the need with optimum unit sizing [8].

The small scale power generation suitable for home purposes has already been discussed in several papers. Bond et al. described current experience and trials in East Timor with solar photovoltaic (PV) technology by introduction of solar home systems (SHS) [9]. Posorski et al. proposed SolarHomeSystems (SHS) that are commercially disseminated and used them cost efficiently to substitute kerosene and dry cell batteries to reduce GHG emissions and thus make a significant contribution to climate protection [10].

Most of the hybrid power generation depends upon Solar with wind or diesel, but in this paper a novel generation system is proposed. Although the solar PV component is providing the sufficient power required for charging the mobile battery or lightning of the lamps, but it has one serious drawback. The sun does not shine all the time. One can collect it only during the day. Also the intensity of solar radiation is quite low in cloudy weather. It further depends on the climate conditions i.e. the solar insolation level varies seasonally. The implication is that power is needed all the time, but the sun and hence, solar energy, is not always available. Therefore, one can say that solar energy is intermittent in nature. So, one cannot totally rely on it. But the solar energy resource is enormous. Solar panels have numerous benefits for the environment. Even if there are some disadvantages of using these systems, the advantages still outweighed them.

Therefore we need a hybrid power generation system which combines the advantages of manual and solar power generation system. When the solar energy is not sufficient to provide the required amount of power for the desired applications, we can generate power manually by the charkha. On the other hand, if the primary purpose of rotating the charkha (i.e. spinning yarn, etc.) ceases to exist, one can revert to solar PV system for providing energy.

Circuit Diagram of The Proposed Hybrid Power Generation System

Two sources of energy are employed here. They are PM DC generator run by manual charkha and solar PV panel. The two diodes IN5408 diode are used to prevent the flow of current from one source of energy to another source of energy. For charging of mobile phone batteries we require 5V supply while 6V supply is required for lighting lamp. The voltage regulators provide the constant required voltage. Two capacitors of 1000 μF and 1 μF are used in the circuit.

6. Conclusion

The analysis and the experimental results of the proposed hybrid power generation system shows that this system can be effectively used in rural/remote areas for low power applications. The proposed system may be installed easily and economically in remote locations with the help of the local committees like panchayats etc. as the system exhibits high performance and low cost to be implemented. It is reliable, simple, and an excellent option to be employed for low power applications. Many countries have enormous manpower in rural areas, so this research is very significant. Though Solar PV cost is high but with the subsidies provided by the governments of several countries, this proposed hybrid power generation system can be of much use in rural areas.

References

[1]   Mohd Tariq, Khyzer Shamsi & Tabrez Akhtar, "Analysis, Modeling and Development of Manual Charkha Based Power Generation System",Energy and Environmental Engineering Journal, 2012 1(3),pp-108-111.

[2]   B.H. Khan, "Non- Conventional Energy Sources", 2nd Edition, Tata McGraw hill Pub. Co., 2009.

[3]   United Nations. "Energy Services for Sustainable Development in Rural Areas in Asia and the Pacific: Policy and Practice". New York: United Nations, 2005.

[4]   Mohd Tariq & Khyzer Shamsi "Application of RET to Develop educational infrastructure in Uttar Pradesh.", International Journal of Recent Trend In engineering, ACEEE (USA), Vol.4 Nov.2010, pp 187-190.

[5]   Bhubaneswari Paridaa, S. Iniyanb & Ranko Goicc, "A review of solar photovoltaic technologies", Renewable and Sustainable Energy Reviews 15 (2011) 1625–1636.

[6]   Abu Tariq, M. Asim, Mohd Tariq (2011) "Simulink based modeling, simulation and Performance Evaluation of an MPPT for maximum power generation on resistive load", 2nd International Conference on Environmental Science and Technology,Singapore.

[7]   John PB, David GI. "A probabilistic method for calculating the usefulness of a store with finite energy capacity for smoothing electricity generation from wind and solar power". Journal of Power Sources 2006;162:943–8.

[8]   Katti PK, Khedkar MK. "Alternative energy facilities based on site matching and generation unit sizing for remote area power supply". Renewable Energy 2007;32:1346–62.

[9]   Bond M, Fuller RJ, Lu Aye. "A policy proposal for the introduction of solar home systems in East Timor". Energy Policy 2007;35:6535–45.

[10] Posorski R, Bussmann M, Menke C. "Does the use of Solar Home Systems (SHS) contribute to climate protection?" Renewable Energy 2003;28:1061–80.

17

Conceptual framework of a Solar PV based high voltage Battery charging strategyfor PHEVs and EVs

Mohammad Saad Alam

Lead Systems Engineer- MagnetiMarelli Holdings of North America Inc (FIAT GROUP SpA)

Email address:

hybridvehicle@gmail.com

Abstract: With the emergence of plugin hybrid electric vehicles (PHEVs) and electric vehicles (EVs) in the automotive market, the rate of the demand for electric power supply may be higher than the projected level. Distributed power generation is a feasible solution to balance the demand and supply of the power industry. In this work, a conceptual model of a solar photovoltaic (PV) based distributed generation system is proposed for charging the PHEVs and EVs. Outline of the overall system is developed with details of the background and feasibility analysis through hardware in the loop experiment of the proposed strategy.

Keywords: Solar PV, Electric Vehicle, High Voltage Battery, Charger

1. Introduction

The electric power supply industry is being deregulated into a "competitive electricity market" wherein the power system is being allocated into competitive generating and distributing groups, which are all linked through the transmission grid [1]. The random nature of consumer demand leads to uncertainty in forecasting demand, and hitches are compounded at peak demand periods. Endeavors to balance the proportion of the power supply and demand in peak hours, strain the power system with a risk of vulnerable to most disturbances [2].The demand load on any power grid must be matched by the supply to it and its ability to transmit that power [1-3]. Peak power is required at times of day when high levels of demand are expected (e.g., hot summer afternoons with huge demand for air conditioning). On top of the peak power management, the electric power industry is expecting another challenge of meeting the charging demand of plug-in hybrid (PHEV) and electric vehicles (EV) [3]. Automotive pundits are forecasting that 10-20% of vehicles on the road will be PHEVs and EVs by 2025 [4]. The contemporary establishment of the power generation is not capable of handling the charging needs of even 5% of vehicles and power demands at peak times can lead to blackouts [5-6]. Smart grid is emerging as a potential solution to manage and enhance the efficiency and reliability of the existing electric power generation,

transmission and distribution set up [6]. However, Smart grid will heavily rely on micro grids to meet the peak power demands locally before rustling the power from the power distribution system [7]. The solar PV System has an appropriate output profile with the peak demand time which can be utilized to meet the charging needs of the growing numbers of PHEVs and EVs [3]. Further, in the absence of the PHEV's and EV's charging requirements, the generated power can be fed into Smart grid to further facilitate the peak power management. Implementing solar technology with the automotive industry willalso be anearnest progression towards preserving the environment. The matter of charging the batteries of the vehicle, however poses major implications for the future [8-10]. Having a solar powered charging station set up at a wide range of locations, however can change the considerations of users that are unwilling to buy a hybrid/electric car as a result of lack of affordable charging options.

The conceptual lay out of the proposed solar PV based charging station entails6 kW micro grid including twenty four 240W solar panels, a compatible MPPT charge controller, a 240/210/280V inverter, and SAE™ J1772 compatible Electric Vehicle Supply Equipment (EVSE). Further details will be elaborated in the following sections of the manuscript.

This paper consists of 7 sections. The background of the existing PHEVs and EVs charging principles, schemes, standards, developments and challenge will be covered in brief in section II. Section III and IV comprises the problem formulation and the conceptual lay out of the solar PV based charging station, respectively. Further, a case study of an on board high voltage battery charging for PHEVs and EVs through the proposed solar PV based charging strategy will be performed in section V, followed by cost analysis in section VI. Finally, the conclusion is outlined in section VII.

2. Background

The Society of Automotive Engineers (SAE) in coordination with the Original Automotive Equipment Manufacturers (OEMs) and National Electrical Manufacturers Association (NEMA) have standardized two levels of charging namely Level 1 and Level 2 and SAE™ J1772 as a standard for EVSE [11-12].

The Level 1 schemeis based on the conventional residential and commercial standard power supply of 120VAC, 15Amp/ 20 Amp (12 Amp/ 16 Amp useable) and is being implemented through an extension cord scheme with an on-board charging system [11-12]. However, Level 1 charging is a time consuming method as it is capable of delivering a maximum of only 1.44kW. This method is not the best solution to the PHEVs and EVs charging needs but definitely is important for emergency situations as it has 'Portable Spare Tire' approach [11-12].On the other hand,

Level 1 allows interface to electrical outlets in compliance with the NEC - Article 625 [13].

Currently, the primary charging method for EVs is Level 2 which requires a 240VAC, single-phase, 40-Amp branch circuit with an on-board charging module. Level 2schematic is hard-wired to home wiring system and requires a permanent private residential installation. Level 2 charging operation can be further categorized based on the mode of charging as inductive and conductive equipment. The popular EVSEs are conductive equipment which have "pin and sleeve" type assembly [11-12]. The inductive scheme is based on the inductive transfer of energytothe high voltage battery of the vehicle without any physical contact and is meant only for EVs and not PHEVs.

For universalizing the EVSEs, SAE™ J1772 standards committee facilitates the guideline forEVSEs manufactured by OEMs or their affiliated supplier companies [11-12]. EVSE can be define as aPhysical link for networking the electric power supply grid to the on-board chargers of PHEVs and EVs. SAE™ J1772 endorse the hardware, electric, andoperational requirements for the conductive charging method use for PHEVs and BEVs with a comprehensive inclusion of everything that associates with the charging system [11-13]. AC Level 2 may be utilized at home, workplace, and public charging facilities. In summary, the charging schemeappropriate for usage with electrical ratings islisted in Table 1.

Table I:Charge method Electrical ratings (North America) [11-15]

Charge Method	Supply AC Voltage (V)	Maximum Current(Amp)	Circuit Breaker rating (Amps)
AC Level 1	120 V, 1-phase	12 A/16 A	15 A (minimum)/20 A
AC Level 2	208-240 V, 1-phase	≤ 80 A	Per NEC 625

As a counterpart of commercial gas stations, Level 3 or fast charging method isbeing installed at few locations in USA. Level 3 has an off board charging system of 60 to 150 kWwith 480VAC, three-phase circuit with an estimated charging time of 10-20 minutes to achieve 40-55% of the state of the charge (SOC) of the on-board high voltage battery [14-15].

3. Emerging Need andSystem Requirements of a Charging Station

The aim of the solar PV based charging strategy for PHEVs and EVs is to explore the potential of solar energy as an alternative and/or as a supplement to address the surge in power demand for the charging provisions of PHEVs and EVs. There exists a choice between theisolated or grid connected charging stations. From feasibility and customer's convenience perspective, a grid connected solar PV charging station is being considered in this work. However, the solar PV based PHEV and EV charging setup is modular in nature and can be implemented as an isolated entity or in conjunction with renewable or existing power

generation sources as hybrid power system.

The fundamental issue to address is to predict the energy requirement of the high voltage battery bank of PHEVs and EVs, the frequency of charges, and the number of vehicles being served. For analysis purpose we are considering an exemplary electric vehicle very similar to the commercially available 2012 Nissan Leaf EV and/or 2012Chevy VoltPHEV. It is also assumed that with Level 1 charging scheme, not more than 2 vehicles will be charged from sun rise to sun set on a typical sunny day.

Before accessing the required energy by the high voltage battery pack of an EV, it is beneficial to perform the following analysis. Assuming the battery is made of 48 modules, and each module contains 4 cells; two of which are parallel, to give a total of about 345V [16-17]. The battery compounds are spinel $LiMn_2O_4$ that can deliver high energy density, and is cheaper and less toxic than$LiCoO_2$which brings the specific energy density of 140 Wh/kg [16-17]. Further, a three minute driving cycle sample is taken at 25°C to get a feel for how much maximum energy would be used by the EV, assuming the vehicle is being driven by a driver weighing 200 pounds or less. The velocity of the vehicle vs. the time is plotted in

Figure 1 and the relevant energy consumed is tabulated in Table II.

Table II: *Energy Required by EV*

Interval (seconds)	Energy (J)	Energy (W-hr)
0 to 5	0	0
5 to 11	229850	63.85
11 to 23	66789	18.55
23 to 27	428720	119.09
27 to 87	995100	276.42
87 to 102	-27518	-7.64
102 to 107	0	0
107 to 111	278580	77.38
111 to 151	273080	75.86
151 to 160	-37193	-10.33
160 to 175	36852	10.24
175 to 180	-4599.7	-1.28
Total Energy	= 2239660 J	= 622 W-hr

In forward motion, the energy used, E_f, can be calculated through equation1, assuming the average velocity over the distance, V_{avg} with parameters as air density, $\rho = 1.18 \text{kg/m}^3$, frontal area, $A_f = 2.4 \text{m}^2$, coefficient of drag for an A segment small car, $C_d = .28$ [16-17].

$$E_f = D[sgn[V_{avg}](mg(C_o + C_1 V_{avg}^2) + \frac{\rho}{2}C_d A_f V_{avg}^2) + m\frac{dV}{dt}] \quad (1)$$

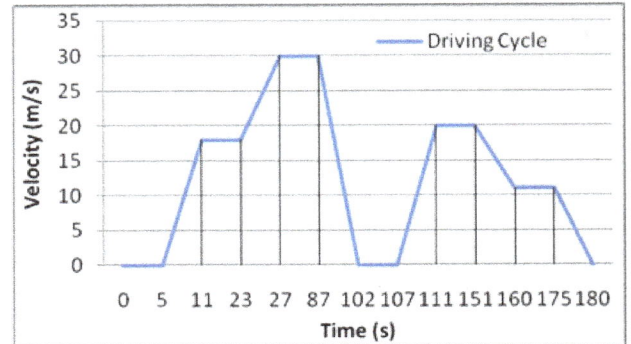

Fig. 1: *The graph illustrates a short drive cycle for a span of 3 minutes.*

This is a well-known fact that, what makes PHEVs and EVsexceptional is the ability to regenerate an average of approximately 30% of the energy that can be recovered during braking and can be calculated as [16-17].

$$E_r(sgn[brakeapplied]) = .3\left(\frac{1}{2}mV_{avg}^2 - DF_{t_{avg}}\right)(sgn[brakeapplied]) \quad (2)$$

Where E_r is the energy recovered in joules and $D$is the distance EV will travel over the course of braking, and $F_{t_{avg}}$is the average of the total drag force minus$m\frac{dV}{dt}$.If we only consider E_f & E_r as our energy loss and gain and neglecting the other energy loss factors, with the illustrated driving cycle amplified to 30 minutes, the consumption of energy would be 6.22kWh used. That is only 26% of the 24kWh battery capacity found in the EV. The other energy consumption factors which were not included could easily have elevated the consumption to over 30% for a 30 minute drive. The highest recommended state of charge (SOC) should be around 80%; thus, dropping the total capacity to around 19kWh. Up to 80% depletion is recommended as maximum use before the EV must be recharged (24kWh*.8 = 19.2 and 19.2kWh*.8=15.36kWh is the practical high voltage battery capacity can deliver). Moreover, to examine the maximum driving range EV can deliver within the safe range of battery usage (using the driving cycle example) of3788 meters (2.35miles) isthe total distance was covered during the trip.For 15.36kWh capacity $\frac{3790m}{.622kWh} =$ $93590m \approx 58miles$ .Note, 15.36kWh/58miles $\frac{15.36kWh}{58miles} \approx$ $264Wh/mile$ Therefore, in order to reach the 100 miles range EV need 26.6kWh usable energy capacity.In order to realize the required energy prediction the discharging and charging power limits vs. the drivable state of the charge (SOC) is plotted for a typical EV High voltage Li-ion battery in Figure 2.Now, a charging set up is need to be developed that can address the requirement of the EV/ PHEV charging system as calculated above. A solar PV charging system is being proposed in the next section.The main charging power source will be Solar PV arrays. In case of the unavailability of the required charging power, necessary power will be pulled from the grid.

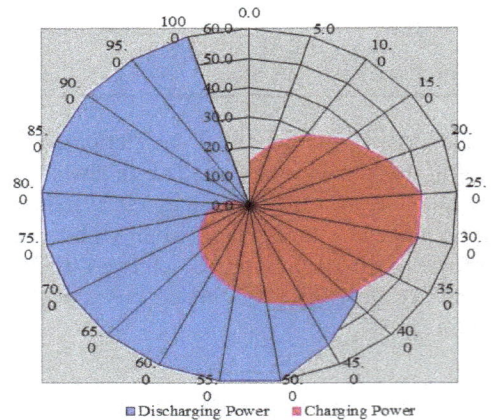

Fig. 2: *Discharging and Charging Power Limits Vs State of the Charge (SOC) of the High Voltage Battery*

4. Conceptual Model of a Solar PV Based Charging Station

In the propose charging system, DC power generated from the PV arrays will be fed to the converter and then the rectified power will be utilized to charge the High Voltage (HV) battery pack of EVs. A generic case study is performed to charge a 16 kW-hr High voltage battery pack, very similar to commercially available PHEV like Chevy Volt [18-19]. Also, possible loads of an EV like the Nissan

Leaf's 24 kW-hr lithium battery were reviewed. It was assumed that the EV or PHEV being charged has not a dead battery or with 0%SOC and thus, determined that this system could handle the load independently.The block diagram of the charging station is shown in Figure 3.

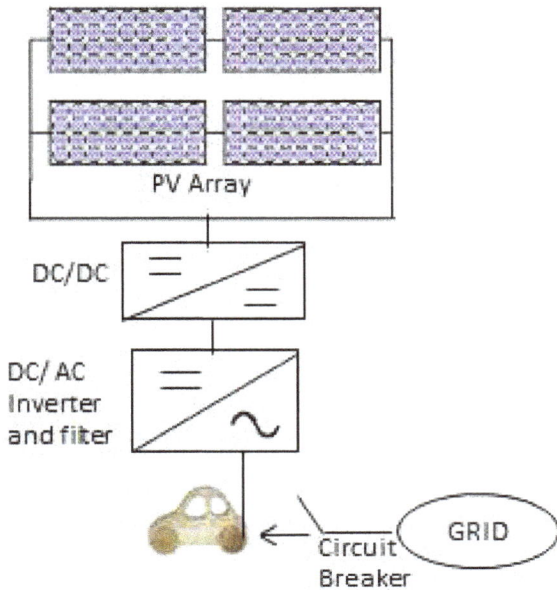

Fig. 3: *Block diagram of the proposed PV based EV charging model*

The EV/PHEV charge process should start as per the sequence described in the following flow chart of Figure 4.

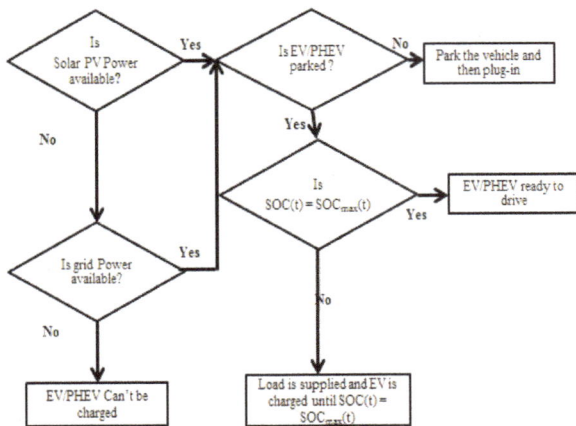

Fig. 4: *Flow chart outlining the overview of the PHEV/EV charging sequence.*

After the EV/PHEV is parked, the Solar PV power or the power from the grid in case of the unavailability of the power from Solar PV arrays like during night time will charge the high voltage battery till the SOC reached to its maximum value or the person unplugged the EVSE. To analyze the proposed lay out a case study is performed and reported in the next section.

5. Case Study ofHV Battery Charging For PHEVs and EVs

For a parking lot location in Michigan, USA, the annual solar radiation profile obtained through weather data analysis[20-22]is plotted in Figure 5 and the daily solar irradiance for a typical day of June is plotted in Figure 6.

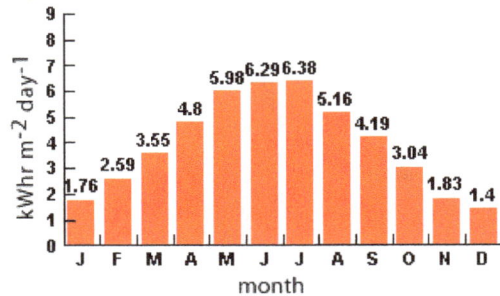

Fig.5: *Annual Solar radiation profile [20-21]*

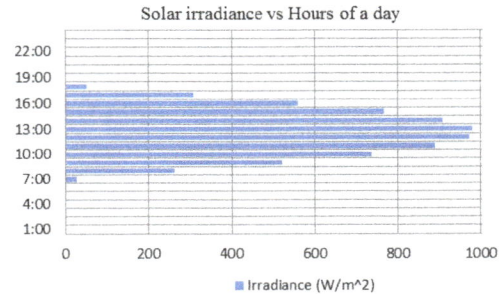

Fig. 6:*Daily solar radiation profile on a typical day of June 2012 [21-22]*

Installing the Solar PV system at an appropriate location is a key factor in determining the output efficiency. It isevident from Figure 6 that the operational solar irradiance is available between 9 AM – 4 PM. Also, as Level1 charge takes 6-7 hours for 50 %SOC for a typical EV, a convenient and optimal choice for an EV charging station would be the parking lot of the driver's work place. The charging set up in the parking lot should be free from tree or building shadows and the panels must be installed facing south with a 42.9 degree tilt angle for maximum achievable efficiency [23-25].

The optimal choice for the solar PV module is achieved through a compromise on the cost and the requirement and Trina Solar TMS-240PA05 modules are being selected for the proposed charging set up [26]. These modules consists of 240W panel, with a maximum voltage and current of 30.6V and 7.84Amps respectively with claimed module efficiency of 14.7%. The Parameters of TMS-240PA05 are tabulated in Table III [26] and the I-V curve is plotted in Figure 7.

Compatible MPPT charge controllers are being installed for achieving maximum power efficiency and two arrays are being designed with 12 modules in each array.The selected inverters areFronius IG Plus 11.4-1 UNI inverters [27]. These inverters have efficiency of 96.2%. The inverters have a Maximum Power Point Tracking input voltage range of 230-500Vdc, a nominal input current of 31.4Amp. The output has a 60Hz nominal frequency output, with a THD of <3%, and an output power factor of 1. Rest of the required electrical components selected for the charging set up are NEC approved and the charging setup meet the entire local, state and federal safety codes.

Table III: *Parameters for TMS-240PA05 PV Module*

Parameter	Values
Peak Power Watts-P_{MAX}(W)	240
Power output Tolerance (%)	0/+3
Maximum Power Voltage-V_{MPP} (V)	30.6
Maximum Power Current- I_{MPP} (A)	7.84
Open Circuit Voltage-V_{OC} (V)	37.5
Short Circuit Current-I_{sc}(A)	8.38
Temperature Coefficient of P_{MAX} (%)	-0.45%/^{0}C
Temperature Coefficient of V_{OC}(%)	-0.35%/^{0}C
Temperature Coefficient of I_{sc}(%)	0.05%/^{0}C

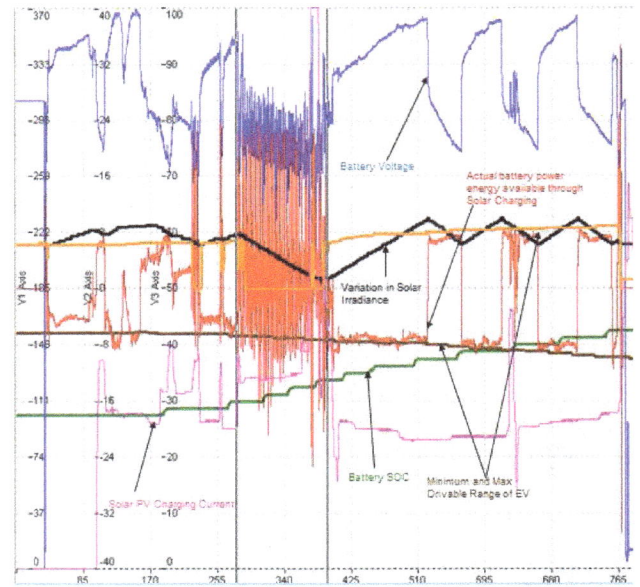

Fig. 8: *Plot of solar irradiance, and various EV charging parameters*

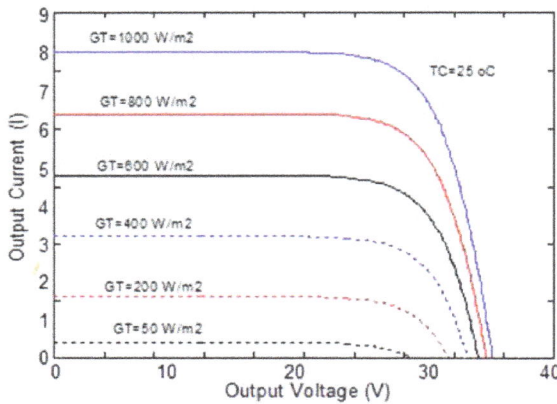

Fig. 7:*I-V curves for TMS-240PA05 PV Module at 25^{0}C*

The charging of the EV for this case study is performed through hardware in the loop (HIL) system and the charging process follow the steps as elaborated in the flow chart in Figure 4.

Level 1 charging set up was performed through HIL at 25^{0}C ambient temperature and varying solar irradiance profile of Figure 6 plotted from real weather data. The EV is charged from an SAE™ J1772 compatibleEVSE for 6 hours from 9:30 AM to 3:30 PM. The charging started at 13% SOC and at 3:30 PM, when EVSE is unplugged, observed SOC was 67%, which will give an approximate 45-50 miles non-highway drive. The battery voltage, actual battery power available, battery SOC, solar PV charging current, variation in solar irradiance including clouding effect and the minimum and maximum drivable range of the EV are plotted in Figure 8. A disturbance in the middle of the charging can be observed which arise due to the switching of Air conditioning and accessories for some time to replicate a real world scenario.

Assuming an average 40-50 miles being driven on a typical week day, the per cent contribution of the proposed solar PV charging scheme and the power drawn from the grid for EV charging for the whole year is calculated through NREL's Hybrid Optimization Model for Electric Renewables [28] tool and the available average annual solar irradiance (as plotted in Figure 5) tabulated in Table III. For a single EV/PHEV, the proposed solar PV charging schemes is capable of providing 76% of the charging demand. This is a significant achievement though at high initial setup cost. The cost analysis, payback period and gain margin is presented in the next section.

Table IV: *Power production from Solar PV and power pulled from grid.*

Charging	kWh/yr	% provision
Solar PV System	7567	76
Grid	2389	24
Total	9956	100

6. Cost analysis

The total solar PV charging system cost includes the 24 solar PV modules;two grid tied inverter, 24 MPPT charge controllers, supplies, mounting racks, and setup costs. The initial cost of the system is approximately $19,560, reduced to $13,692 due to a 30 % incentive. The incentive amountwill be a rebate from Consumers Energy's "Experimental Advanced Renewable" program [29]. With a 12 year contract, through this program, the commercial suppliers can be paid $0.45/kWh or $0.375/kWh. A rate of 10 cents per Kilowatt hour with 2% inflation for the next 20 years is computed. RET-Screen®[30] was used to analyze the overall cost of the proposedsolar PV charging system and the estimated payback period is plotted in Figure 9.

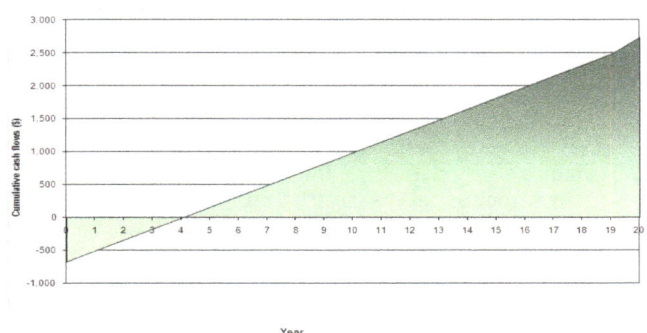

Fig. 9: Payback period of the proposed Solar PV charging System

The analysis estimated that with a capacity factor of 13.8%, the charging system would deliver 7567 Kwh of electricity annually, contributes to 76% of the EV charging requirements. The remaining 24% of the charging demand will be fulfilled from the grid. The base case for electricity rates were determined by the maximum monthly kWh cost of electricity over the last two years, with a 10% annual addendum with an estimated electricity rate of 13cents /kWh and an annual inflation rate of 2.0% for the next 20years.It will take approximately four years to start profiting off the design as shown in the cumulative cash flow diagram in figure 9 as payback period.

Though the initial set up cost is high but with the environmental benefits, state incentives, profitable after 4 years, significantly no maintenance cost and provision of the charging power during peak demand time, this would be the best scenario.

7. Conclusion

Solar PV charging system is a viable technology with initial cost constraints but profitable in the reasonable payback period. With the constant increase in the power demand, strained transmission and distribution systems and power market deregulation and structuring, the added demand due to the growth of EVs/ PHEVs can easily be addressed through the proposed solar PV based charging set up.The case study for a PHEV/ EV vehicle, very similar to commercially available Chevy Volt and Nissan Leaf depicted favorable results. At peak hours the system can charge the battery to approximately 67% SOC, enough for daily commute. Also, the proposed strategy contributes to 76% of the charging demand for the whole year and only 24% of the charging demand is met through grid. Additionally this 76% charging demand is met without causing any environmental disturbance. Thus, the overall proposal deemed effective and successful.

References

[1] M. Shahidehpour, H. Yamin and Z.Y. Li, "Market Operations in Electric Power Systems," John Wiley &Sons, Inc., New York, 2002.

[2] S.M. Amin, "U.S. grid gets less reliable [The Data],"*IEEE Spectrum* vol. 48, Issue: 1, 2011.

[3] Escudero-Garzás et al, "Fair Design of Plug-in Electric VehiclesAggregator for V2G Regulation,"IEEE Transactions On Vehicular Technology, vol. 61, no. 8, Oct. 2012 , pp. 3406-3419.

[4] TJ McCue, "Electric-Drive Vehicle Sales Forecast at 3.8 Million Annually." http://evworld.com/news.cfm?newsid=29463 accessed April 2013

[5] J. Paserba, P. Kundur, "Guest editorial - Power grid blackouts remembering and fighting grid failures," *IEEE Power and Energy Magazine*, vol. 4, Issue: 5, pp. 16-21,2006.

[6] S. M. Amin and B. F. Wollenberg, "Toward a smart grid: Power deliveryfor the 21st century," IEEE Power and Energy Magazine, vol. 3, No. 5, pp. 34-4, 2005.

[7] H. Lund, "Large-scale integration of optimal combinations of PV, wind and wave power into the electricity supply," J. of Renewable Energy Vol. 31 ,2006, pp: 503–515.

[8] N. Watrin, R. Roche, H.Ostermann,B. Blunier, and A. Miraoui, "Multi physical Lithium-Based Battery Model for Use in State-of-Charge Determination,"IEEE Transactions on Vehicular Technology, vol. 61, no. 8, Oct. 2012, pp.3420-3429.

[9] R. Castro, R. E. Araújo, J. P. F. Trovão, P. G. Pereirinha, P. Melo, and D. Freitas, "Robust DC-Link Control in EVs With MultipleEnergy Storage Systems," IEEE Transactions On Vehicular Technology, vol. 61, no. 8, Oct. 2012, pp. 3553-3565.

[10] GByeon et al, "Energy Management Strategy of the DC Distribution System in Buildings Using the EV Service Model," IEEE Transactions   On Power Electronics, vol. 28, no. 4, April 2013, pp.1544-1554.

[11] SAEJ1772, SAE Electric Vehicle and Plug in Hybrid Electric Vehicle Conductive Charge Coupler, published 10/15/2012, Society of Automotive Engineers Standardshttp://standards.sae.org/j1772_201210/accessed April 2013

[12] NIST Special Publication 1108, NIST Framework and Roadmap for Smart Grid Interoperability Standards, Release 1.0∗ January 2010.

[13] NEC Article 625, A_NEC_625_2008.pdf, National Electric Hand Book, 2008, pp.:971

[14] Level 3 charger –NFPA catlog April 2013

[15] M. Kintner-Meyer,"Smart Charger Technology for Customer Convenience and Grid Reliability", May 2009.

[16] A. Balboul, M. Dughlas, H.A. Chisht, "Interconnection Challenges of Plug-in Hybrid / Electric Vehicles to the Smart Grid," project reported submitted for Hybrid Vehicle Course, Wayne State Univ, May 2010

[17] Gene Liao, "Advance Battery Systems for Hybrid Electric Vehicles" January 2010.

[18] NissanLEAF.pdf, Sept 2011.www.Nissan.com accessed April 2013

[19] Chvy Volt.pdf, 2012, http://www.chevrolet.com accessed April 2013

[20] Honsberg; Bowden. PVCDROM<http://pvcdrom.pveducation.org/index.html >. accessed April 2013

[21] *National solar radiation database*<http://rredc.nrel.gov>accessed April 2013

[22] National Climatic Data Center, National Environmental Satellite, Data, and Information Services (NESDIS) [online].accessed April 2013

[23] F. Farret, and M. Simões.*Integration of Alternative Sources of Energy*.Wiley-IEEE Press, 2006.

[24] T. Tan, "A Model of PV Generation Suitable for Stability Analysis," *IEEE Trans.on Energy Conversion*, vol. 19, no. 4, pp. 748-755, 2004.

[25] M.S. Alam, and A.T. Alouani, "Dynamic modeling of photovoltaic module for real-time maximum power tracking," *J. of Renewable Science and Energy*, AIP, NY, vol. 2, no. 3, 2010.

[26] TMS-240PA05Whole Sale Solar. Whole Sale Solar. <http://www.wholesalesolar.com/>.accessed April 2013

[27] Fronius IG Plus 11.1-1 UNi-<www.fronius.com/cps/rde/xchg/SID-177ECB63.../2714_1483.htm>accessed April 2013

[28] Hybrid Optimization Model for Electric Renewables (HOMER) software package : http://www.nrel.gov/homer/accessed April 2013

[29] Experimental Advanced Renewable Program, Consumers Energy: http://www.consumersenergy.com/content.aspx?id=1801.accessed April 2013

[30] RETSCreen software<http://www.retscreen.net/ang/home.php> accessed April 2013

18

Unified power flow controller (UPFC) integrated with electromagnetic energy storage system for system stability enhancement

Saravanan Kandasamy, Dawit Leykuen Berhanu, Getnet Zewde Somanu

Department of Electrical and computer Engineering, JIT, Jimma University, Jimma, Ethiopia

Email address:

Saravanan.me07@gmail.com (S. Kandasamy), dawitleykuen@yahoo.com (D. L. Berhanu), findgetnet@gmail.com (G. Z. Somanu)

Abstract: In this paper analysis the UPFC integrated with energy storage devices for improvement of system stability compensation. Before that, used traditional UPFC is to control all transmission line parameters simultaneously or selectively but don't have appropriate control for throughout system. During large transients; traditional UPFC have restricted capability of power flow control. In this paper to do reduce or eliminate that negative aspect of established UPFC using substantial energy storage devices adapted with UPFC. In this proposed system Integration of Superconducting Magnetic Energy Storage (SMES) into UPFC is described. SMES is connected to UPFC through an interface with DC-DC chopper. UPFC with SMES system will inject or absorb real and reactive power to or from an influence system at really quick rate on a repetitive beginning of stability problem. Here Comparative Analysis of the two types, one is integration of electromagnetic energy storage into UPFC and another is integration of electrochemical energy storage into UPFC is done by means of MATLAB/SIMULINK software package.

Keywords: Transient Stability, Flexible AC transmission System, UPFC, SMES, Battery, DC-DC Chopper

1. Introduction

In sensible arrangement of nonlinear network mustn't maintain the stable condition attributable to explosive changes in load or increment of demand. Whenever load changes within the operating power grid, the system select instability state of affairs. Therein state of affairs compensation is required to keep up system stability. At the current time, usage of compensating device used for FACTS devices. The FACTS devices can give the lager quantity of power transfer capability within the power Network. UPFC is one in all the second generation FACTS devices. It's wont to management real and reactive power autonomously [1].

The existing FACTS device had some issues. One is Dc link electrical condenser of UPFC is capable of charging or discharging for compensating method. At the time of equalization method, the convertor loss is there in UPFC. Energy keep dc-link electrical condenser is lean to finish compensation below massive transient amount [2].Another one is that the great amount of load is disconnected from the road, speed of the generator can increase, and attributable to

this reason additional energy or power can seem within the unit. At this condition the facility is absorbed from power grid to dc link electrical condenser through converters and/or power is provide to power grid through converters. Therefore energy provides to power grid or absorption from power grid is definite restricted price. So as to beat the higher than two disadvantages we are able to use substantial energy storage systems. In this energy storage systems, additional energy is keep and/or energy provide to system in step with needs of power grid [3].

The existing FACTS parts area unit extra to Energy storage devices to boost the facility system stability by active power exchange with power grid, Via Energy storage devices like flywheels, ultra-capacitors, batteries and SMES. These devices area unit differing in charging/discharging rate, energy density, voltage level, potency and economical concerns with one another [2].

If electro chemical battery is employed as a considerable power provide, it'll produce some issues like high ohmic resistance, higher ageing, high heating levels, and chemical process happens on electrodes. Electro chemical batteries aren't able to management high power levels for durable periods, Quick, deep discharges results in high heating levels clearly reduces battery period [4,5,6]. Lead acid (flooded type) battery storage system potency is 72–78%, generation 1000–2000 cycles at 70%depth of discharge [7]. Radical capacitors need to face the issues like low energy density, low voltage, and unable to use full energy spectrum etc [8].

Comparing with all alternative substantial energy provides SMES having the benefits of high energy density, quick response, high potency, minimum energy loss throughout the conversion etc, thus It are thought of because the best substantial energy provide which may be simply integrated with FACTS devices for enhancing transient stability. SMES also can be an honest resolution for raising the facility quality [9, 10, and 11]. Advances in each superconducting technologies and also the necessary power physics interface have created SMES a viable technology that may supply versatile, reliable, and fast-acting power compensation [12].

Very few researches solely administered in FACTS with Energy storage system. During this paper MATLAB/Simulink code for UPFC integrated with SMES is enforced. During this methodology of research scheme is employed for sweetening of transient stability. Planned system to analysis completely different fault condition (Three part fault) shows results of theoretical account. During this paper improvement of transient stability additionally power flow management and power internal control.

2. UPFC Operation

The universal and most versatile FACTS device is that the Unified Power Flow Controller (UPFC). UPFC is that the most promising device within the FACTS. It brings a replacement challenge in power natural philosophy and power grid style. The Unified Power Flow Controller (UPFC) has the flexibility to regulate the ability flow between the two buses of the transmission systems by three vital parameters like, electric resistance of conductor, voltage magnitude and phase angle.

The two basic parts of UPFC are two voltage supply converters with semiconductor devices having turn-off capability. They Series device and Shunt device, these converters are connected to every alternative with a typical dc link and these two voltage supply converters sharing a typical dc electrical device shown in fig. 1. The capability of energy storing is mostly tiny in dc electrical device .The dc electrical device permits the two-way flow of active power between the series output terminals of the SSSC and also the shunt output terminals of the STATCOM. One VSI is connected in shunt with line through the shunt electrical device. The opposite one is connected nonparallel with line through the series electrical device. VSC's are coupled along for a full of life power exchange between the two converters and additionally it will severally exchange reactive power with the AC system. The shunt device is primarily used for

providing the real power demand of the series device at the common dc link terminal from the ac power grid. It will generate or absorb reactive power at its ac terminal, that is autonomous of the active power transfer to (or from) the dc terminal. Active power that was drawn by the shunt device ought to be capable the active power generated by the series device.

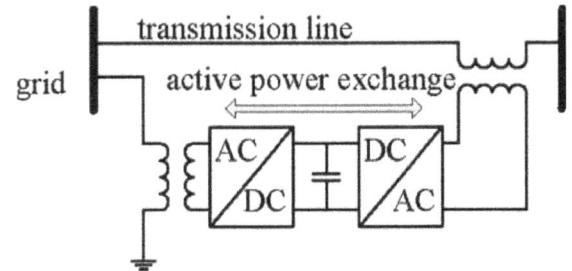

Figure 1. *UPFC operation*

The series-connected electrical converter injects a voltage with governable magnitude and phase nonparallel with the conductor, therefore providing real and reactive power to the conductor. The reactive power is electronically provided by the series electrical converter and then active power is transmitted to the dc terminals. Therefore the shunt electrical converter operated in such the way of keeping voltage across the dc electrical device V_{dc} is constant. Therefore the web power absorbed by the road is capable the losses of the inverters and also the reactive power of their transformers.

The shunt-connected electrical converter provides the active power drawn by the series branch and the losses, and might severally give reactive compensation to the system. The remaining capability of shunt electrical converter is employed to exchange the reactive power among the road for the voltage regulation at the purpose. However, the injected active power ought to be equipped by DC link, successively taken from the AC system throughout the shunt device. Once the sufferers of the converters and also the associated transformers are neglected, then usually active power exchange between the UPFC and also the AC system becomes zero.

The reactive current I by

$$I = \frac{V_L - V_C}{X} \tag{1}$$

The corresponding reactive power Q exchanged can be expressed as follows:

$$Q = \frac{1 - \frac{V_C}{V_L}}{X} V_L^2 \tag{2}$$

Whereas V_L is Transmission line voltage, V_C is converter output voltage and X reactance plus transformer leakage reactance plus system short circuit reactance.

3. Energy Storage System

ESSs is classified in to major three types, based on specific principles: 1) Physical energy storage including compressed air energy storage (CAES), flywheel energy storage (FES) and pumped hydro storage (PHS); 2) Electromagnetic energy storage including super-capacitor energy storage (SCES) and superconducting magnetic energy storage (SMES); 3) Electrochemical energy storage including lead-acid, Nickel Cadmium, lithium-ion, sodium sulfur and fluid flow battery energy storage, etc [19,20,21]. Physical energy storage is doesn't have a large-scale back-up [22]. The technologies of electrochemical energy storage, sodium-sulphur battery can efficiency of energy conversion reach 89% with the energy density three times as lead-acid battery and much longer life cycle. Research works about electromagnetic energy storage efficiency of energy conversion reach 95%. The technology is energy saving and easy to operate with high efficiency [23, 24, 25, 26].

3.1. Integration of Energy Storage Systems into FACTS

An energy storage system (ESS) will play a crucial role in power grid management and supply major enhancements over ancient UPFC performance. SMES in conjunction with UPFC have recently emerged together of the foremost promising near-term storage technologies for power applications. By the addition of an energy storage system to the UPFC it's been potential to manage the active power flow between the UPFC and also the purpose of common coupling (PCC). Thus, the UPFC compensates the reactive power and, additionally stores energy within the storage system once the generated power exceeds the facility limits that might be injected to the distribution grid. Additionally, this resolution provides promotes management of the facility flow at the PCC, by adjusting the direction of power injection, like down or upwards.

Recently, a substantial quantity of attention has been given to developing management ways for a spread of FACTS devices, like static synchronous compensator (STATCOM), the static synchronous series compensator (SSSC), and also the unified power flow controller (UPFC), to be ready to address and mitigate a large vary of potential bulk power transmission issues.

In the absence of energy storage, FACTS devices square measure restricted within the degree of freedom and sustained action during which they will facilitate the facility grid. By the tactic of integration of energy storage system (ESS) into FACTS devices, a freelance real and reactive power absorption or injection into and from the grid is feasible. This integrated system ends up in a lot of economical and versatile power transmission controller for the facility system. Once a line experiences vital power transfer variations in an intermittent manner, a FACTS + ESS combination is put in to manage and change the facility flow inside the loaded line. The improved superior performance of combined FACTS with ESS can have bigger charm to transmission service suppliers. Performance indices were planned for FACTS dynamic performance with energy storage in [27] and management schemes are mentioned in [28, 29 and 30].

Power system liberation, alongside transmission limitations and generation shortage, has modified the ability of grid conditions by making things wherever energy storage technology will play a really important role in maintaining system dependability and power quality. There are multiple edges of energy storage devices like the flexibility to apace damp oscillations, answer sharp load transients, and still offer the load throughout transmission or distribution interruptions. Additionally, this technique will correct load voltage profiles with fast reactive power management, and still enable the generators to balance with the system load at their traditional speed. Fig.1 presents typical design of connected UPFC with SMES to electrical utility system.

The main disadvantage of a conventional UPFC (with no energy storage) is that it's solely two potential steady-states in operation modes, namely, inductive (lagging) and electrical phenomenon (leading). although each the standard UPFC output voltage magnitude and phase will be controlled, they cannot be severally adjusted in steady state because of the shortage of great active power capability of UPFC. Typically, the UPFC device voltage is maintained in section with the PCC voltage, therefore making certain that solely reactive power flows from the UPFC to the system. However, as a result of some losses within the coupling electrical device and device, the device voltage is usually maintained with a little section shift with the PCC voltage. Thus, much, a little quantity of real power flows through the system from PCC to DC bus, to make amends for the losses. However, the \$64000 power capability of the UPFC is extremely restricted because of the absence of any energy storage the DC bus. Compared with the standard UPFC + BESS, the UPFC + SMES provide a lot of flexibility.

In case of UPFC with SMES, the amount of steady-state in operation modes is extended to varied things like inductive mode with DC charge and DC discharge, electrical phenomenon mode with DC charge and discharge. Thus, in steady state, the UPFC with SMES have four in operation modes and may operate at each purpose within the steady-state characteristic circle. Additionally, looking on the energy output of the battery or SMES, the discharge/charge profile is usually adequate to supply enough energy to stabilize the ability regulation within the system and maintain operation till alternative long energy sources are brought into operation.

One of the drawbacks of FACTS + ESS is that for FACTS integration, the scale of the storage systems, significantly battery energy storage (BESS), is also large for sensible use in large-scale transmission-level applications. On bound occasions, giant battery systems tend to exhibit voltage instability once varied cells are placed asynchronous. However, usually it's seen that even giant oscillations will be satisfied with modest power injection from a storage system. To cut back the drawbacks of the UPFC additional to BESS, we tend to use UPFC with SMES system. the flexibility to severally management each active and reactive powers in UPFC + SMES makes them ideal controllers for numerous styles of power regulation system applications, as well as voltage fluctuation mitigation and oscillation damping. Among them, the foremost necessary use of the UPFC + SMES is to stabilize any disturbances occurring within the installation.

3.2. Integration of Energy Storage Systems into UPFC

Comparative Analysis of the Two Types: 1) Electromagnetic energy storage 2) Electrochemical energy storage.

3.3. Integration of Electromagnetic Energy Storage into UPFC

In practical point of view SMES device is very costly compared to battery or ultra capacitor energy storage device. To eliminate these styles of drawback, SMES device connected in parallel with the FACTS. Block Diagram of Integrated UPFC with SMES shown in fig.2. In this system SMES device is generated the magnetic field, when flow of DC source through a superconducting coil. The UPFC model could be a combination of the static synchronous compensator (STATCOM) and static synchronous series compensator (SSSC) models. The currents and square measure the elements of the shunt current. The currents and square measure the elements of the series current. The voltages and square measure the causation finish and receiving finish voltage magnitudes and angles, severally. The UPFC parameters square measure the following:

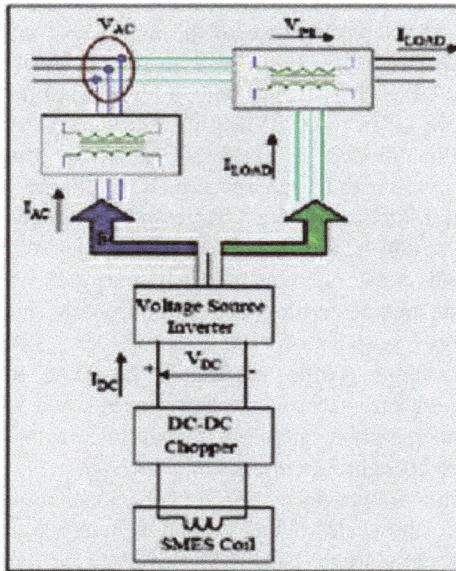

Figure 2. *Block diagram of integrated UPFC with SMES*

The power balance equations at bus 1 are given by

$$0 = V_1\left(\left(i_{d1}-i_{d2}\right)\cos\theta_1 + \left(i_{q1}-i_{q2}\right)\sin\theta_1\right) - V_1\sum_{j=1}^{n} V_j Y_{1j}\cos\left(\theta_1-\theta_j-\phi_{1j}\right) \quad (3)$$

$$0 = V_1\left(\left(i_{d1}-i_{d2}\right)\sin\theta_1 - \left(i_{q1}-i_{q2}\right)\cos\theta_1\right) - V_1\sum_{j=1}^{n} V_j Y_{1j}\sin\left(\theta_1-\theta_j-\phi_{1j}\right) \quad (4)$$

And at bus 2

$$0 = V_2\left(i_{d2}\cos\theta_2 + i_{q2}\sin\theta_2\right)V_2\sum_{j=1}^{n} V_j Y_{2j}\cos\left(\theta_2-\theta_j-\phi_{2j}\right) \quad (5)$$

$$0 = V_2\left(i_{d2}\cos\theta_2 + i_{q2}\sin\theta_2\right) - V_2\sum_{j=1}^{n} V_j Y_{2j}\cos\left(\theta_2-\theta_j-\phi_{2j}\right) \quad (6)$$

$$0 = V_2\left(i_{d2}\sin\theta_2 + i_{q2}\cos\theta_2\right) - V_2\sum_{j=1}^{n} V_j Y_{2j}\sin\left(\theta_2-\theta_j-\phi_{2j}\right) \quad (7)$$

3.4. Basic Principles of the SMES Systems

Electromagnetic ESS could be a reasonably energy storage instrumentation for changing the keep magnetic attraction energy into power directly with none intermediate transferring organizations. Generally, a SMES device consists of the superconducting coil, the refrigerant system, and also the power conversion/conditioning system (PCS) with filtering and protection functions. The coil in SMES doesn't lose Joule heat whereas poured into DC current, therefore the SMES device adopts DC charging supply system, and also the three-phase device transforms ac to dc and reverse. Reckoning on the management loop of its power conversion unit and switch characteristics, varieties of the coupling transformation embody two models: voltage supply device (VSC) and current supply device (CSC), through that SMES will well collaborate with the transmission network. Two-quadrant chopper consists of high power devices, though' that the coil is charged and discharged.

3.5. Power Conditioning System

A power acquisition system (PCS) is needed so as to transfer the energy from the SMES coil into the grid. A PCS consists of a dc-dc chopper and a 3 phase voltage source converter (VSC). Using the voltage–angle control strategy, both the active and reactive power can be controlled. A dc-dc chopper is mainly used to keep the current through the SMES coil constant and to transfer the power to the VSC through the dc-link capacitor. The SMES coil along with a dc-dc chopper is connected to the VSC through a dc-link capacitor. This capacitor acts as a temporary dc voltage source for the VSC to inject active/reactive power into the grid.

3.6. Two Quadrant DC-DC Chopper

In nonlinear grid, want of variable DC provide for improvement of system performance. Thus variable DC supply suggests that DC-DC chopper application. In DC-DC chopper some completely different varieties square measure there like one quadrant, two quadrants and four quadrants chopper. Regarding these varieties, two quadrants chopper was employed in this paper. Two quadrant choppers square measure connected in parallel with DC link electrical condenser of UPFC. SMES device have three modes of operation. 1st mode is charging of superconducting coil, second mode is energy standby in coil, and third mode is discharging of coil.

$$E_{smes} = \frac{1}{2} L_{smes} I_{smes}^2 \quad (8)$$

$$P_{smes} = \frac{dE_{smes}}{dt} = L_{smes} I_{smes} \frac{dI_{smes}}{dt} = V_{smes} I_{smes} \quad (9)$$

3.6.1. SMES Charging Mode

In this mode shown in Fig.3 (a), the SMES coil is charged to its rated capability. Throughout charging mode, GTO2 is often within the ON state. GTO1 are often switched ON or OFF in each cycle. The SMES coil charges once the GTO1 is additionally within the ON state. Once the SMES coil is charging, the connection between the voltage across the SMES coil and also the voltage across the dc link condenser are often given as

$$V_{SMES} = D * V_{DC} \qquad (10)$$

Where V_{SMES} the voltage is across the SMES coil V_{DC} is the voltage across the dc link capacitor and D is the duty cycle of the GTO1 (defined as the ratio of the GTO ON time to the total time for a complete cycle)

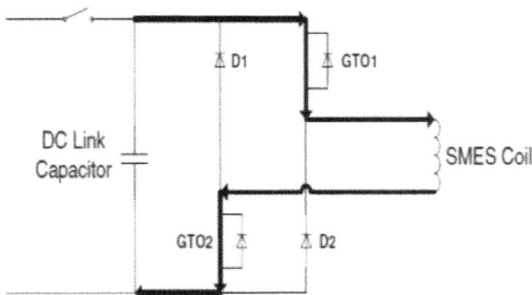

Figure 3 (a). Charging Mode

In this specific case, the duty cycle (D) of the GTO1 is unbroken constant at one second, so the SMES coil charges at the utmost charging rate potential. Within the gift simulation, it takes concerning three seconds to charge the coil to its rated current capability.

3.6.2. SMES Standby Mode

The second mode of operation is named the freewheeling mode. During this mode, the present circulates in a very closed-loop system. This can be additionally referred to as the standby mode. Once the SMES coil is within the freewheel mode, one amongst the 2 GTO's is OFF. In Fig.3 (b), it absolutely was shown that GTO1 is ON and GTO2 is OFF. Throughout this era, there's no vital quantity of loss, because the current through the SMES coil is current in a very closed-loop system. Hence, the present remains fairly constant.

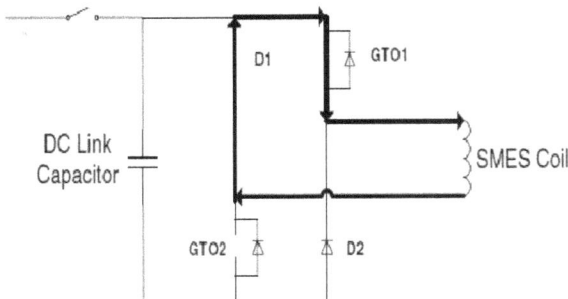

Figure 3 (b). Standby Mode

3.6.3. SMES Discharge Mode

The final mode of operation is that the discharge cycle. The present within the SMES coil discharges into the dc link electrical condenser during this mode of operation. During this mode shown in Fig.3 (c), the GTO2 is often within the OFF state and also the duty cycle of GTO1 will be varied counting on the speed of discharge demand. Throughout the discharge cycle, to own the utmost rate of discharge, each the GTOs square measure unbroken in OFF state. The speed of discharge of the SMES coil will be controlled by creating the duty cycle of 1 of the GTOs to be non-zero. The voltage relationship between the SMES coil and also the dc link electrical condenser throughout the discharge cycle is given as

$$-V_{SMES} = (1 - D) * V_{DC} \qquad (11)$$

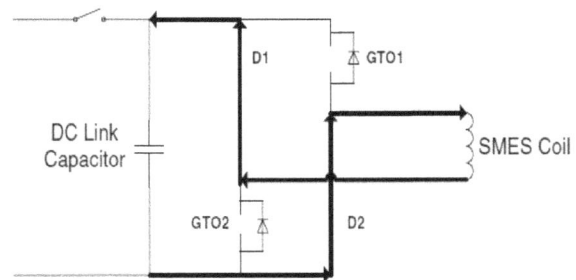

Figure 3(c). Discharge Mode

All the operative mode of DC-DC chopper shown in fig. 3(a), (b), (c), and also the initial mode of operation coil is charging, once giant capability of load is disconnected from the road. At that point of disconnection great amount of power flow in line, that power was absorbed from the road to store in coil. Whenever coil is storing, the voltage of the coil is positive. Second mode is holding on energy in standby mode, once facility is balanced condition. Third mode of operation coil is discharge, once masses square measure adding to the road or when faults clearing time. Throughout the addition of load, need a lot of power injection in line. This injection of additional power transferred from discharging of coil to line. Coil voltage is negative once coil is discharging.

3.7. Integration of Electrochemical Energy Storage into UPFC

UPFC Integration with battery energy device exploitation grid stability improvement. During this system battery is connected across the DC link electrical condenser of UPFC through DC-DC convertor. Whenever power systems lower load or no load condition, the road power flow is raised. The additional power is keep in battery from the line. Once line is want some further energy or power, at the amount of compensation the battery can discharge. Battery energy device isn't applicable for long-standing amount at high power levels. The battery charging and discharging amount chemical change is there. Attributable to chemical change additional thermal drawback in battery, therefore lifespan of the battery is reduced [7].

Table 1. *Battery technologies—characteristics and commercial units used in electric utilities*

Battery type	Comments
Lead acid (flooded type)	η = 72–78%, cost 50–150 Euro/kWh, life span 1000–2000 cycles at 70% depth of discharge, operating temperature −5 to 40∘ C ,25 Wh/kg, self-discharge 2–5%/month, frequent maintenance to replace water lost in operation, deep.
Lead acid (valve regulated)	η = 72–78%, cost 50–150 Euro/kWh, life span 200–300 cycles at 80% depth of discharge, operating temperature −5 to 40∘ C ,30–50 Wh/kg, self-discharge 2–5%/month, less robust, negligible maintenance, more mobile, safe (compared to flooded type)
Nickel Cadmium (NiCd)	η = 72–78%, cost 200–600 Euro/kWh, life span 3000 cycles at 100% depth of discharge, operating temperature −40 to 50 ∘C,45–80 Wh/kg, self-discharge 5–20%/month, high discharge rate, negligible maintenance, NiCd cells are poisonous and heavy
Sodium Sulphur (NaS)	η = 89% (at 325 ∘C), life span 2500 cycles at 100% depth of discharge, operating temperature 325 ∘C, 100Wh/kg, no self-discharge, due to high operating temperature it has to be heated in stand-by mode and this reduces its overall η, have pulse power capability of over 6 times their rating for 30 s
Lithium ion	η ≈100%, cost 700–1000 Euro/kWh, life span 3000 cycle at 80% depth of discharge, operating temperature −30 to 60 ∘C, 90–190 Wh/kg, self-discharge 1%/month, high cost due to special packaging and internal over charge protection.

4. Simulation Results and Discussions

In order to investigate the feasibility of the proposed techniques, UPFC with SMES embedded power flow studies on IEEE 14-bus test system shown in Fig. 4, in this system UPFC with SMES in between bus 1 to 2 or generator 1 to 2 was implemented.

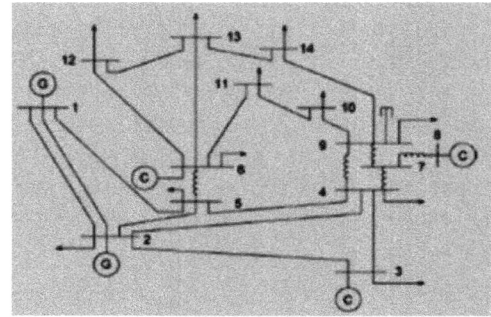

Figure 4. *IEEE 14-bus test system*

Parameters of the given system are mentions below. Fig.4. shown the MatLab/Simulink model for the test case described above. The generator controller will provide the mechanical input Pm, and the field voltage V_f, depending on the electrical load of the system. The controller will also provide damping to the rotor angle during transient condition of the system. The data for various components used in the simulation are as follows (the values are in pu unless stated).

Synchronous generator parameters: 100MVA, V =230KV, f = 60 Hz, X_d = 1.305, X_d^1 = 0.296, X_d'' = 0.255, Xq = 0.474, X_q'' = 0.243, $X1$ = 0.18, X/R=15.

Shunt and series Transformer parameter: 100MVA 230 kV/38 kV, R1=0.002, L1=.08, $R2$ = 0.002, $L2$ = 0.08, Rm = 500, Xm = 500.

Transmission line parameters and Nominal π network parameter (per km): $R1$ = 0.01273, $R0$ = 0.3864, $L1$ = 0.9337 mH, $L0$ = 4.1263 mH, $C1$ = 12.74 nF, $C0$ = 7.751 nF, L=500 km.

UPFC parameters: 500 KV, 100MVAR, Vdc = 38KV, Cdc = 2000e-6F, Vref = 1.0, Kp = 12, Ki = 40.

Receiving end source (infinite bus): 100MVA, V =230KV, f = 60 Hz,X/R=7, L = 1e-3H, R = 0.15ohms.

In the testing system, a symmetrical three phase fault has been applied in transmission line. Transient is created on bus at 0.0167s and has been cleared in 0.1s. In below comparisons shows however stability improvement and the way UPFC with SMES provide higher compensation.

Case (a) is mentioned for the UPFC with battery energy storage system and

Case (b) is mentioned for the UPFC with SMES system.

Compare the Fig.5&fig.6, Fig.6 is Dynamic response for Voltage the successfully created the stability of power system in .12s, although fault cleared in 0.1s.Here stability of power system is faster case (b) than case (a), that is show in figure.

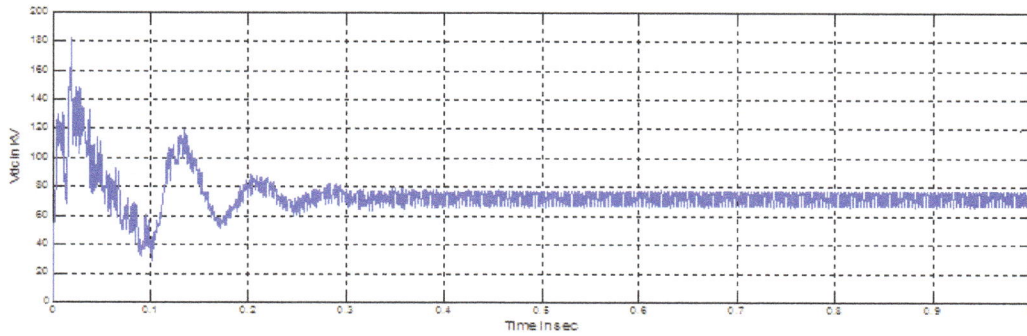

Figure 5. *Voltage across DC link capacitor for case (a)*

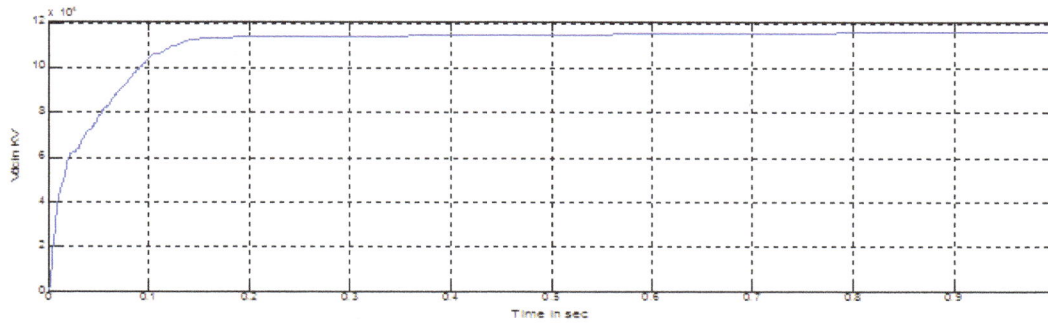

Figure 6. *Voltage across DC link capacitor for case (b)*

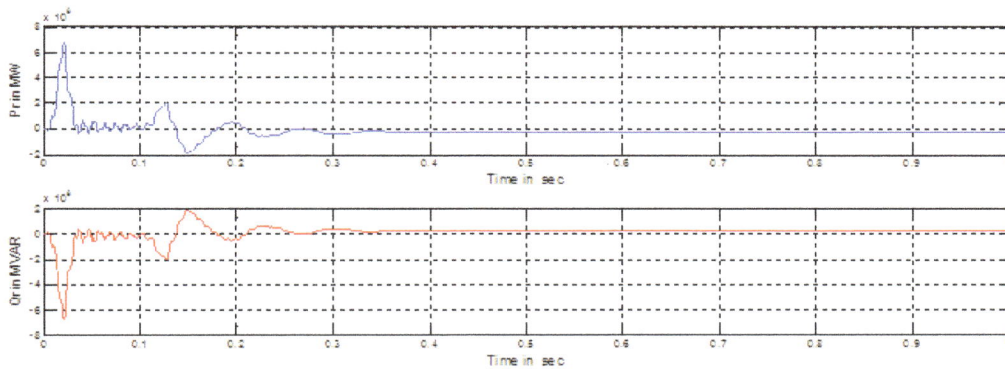

Figure 7. *Sending end Real and Reactive power flow over the transmission line for case (a)*

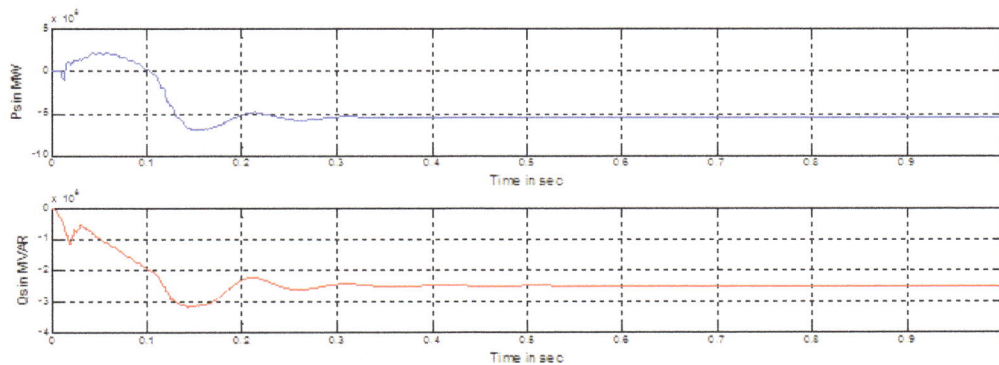

Figure 8. *Sending end Real and Reactive power flow over the transmission line for case (b)*

Fig.7 and fig.8 shows comparison of real and reactive power of the test system. In these figures, negative real and/or reactive power values represent the injected power from the device to the ac system. After fault clearing, amplitude of real and reactive power reduced in case (b)   compared with case (a)

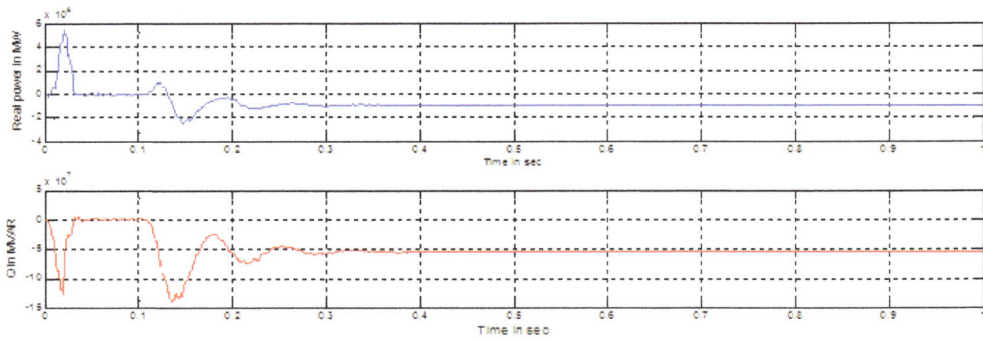

Figure 9. *Receiving end Real and Reactive power flow over the transmission line for case (a)*

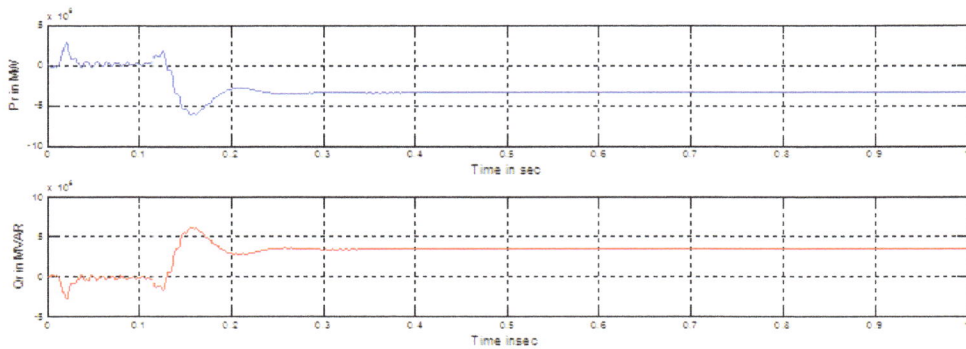

Figure 10. *Receiving end Real and Reactive power flow over the transmission line for case (b)*

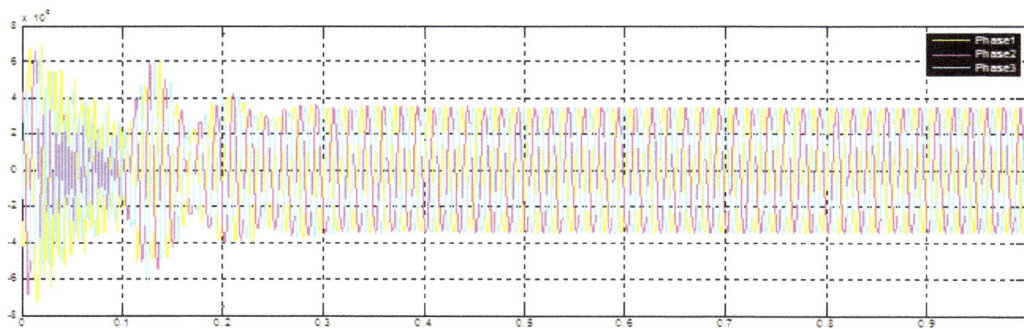

Figure 11. *Receiving end voltage Vs Time for case (a)*

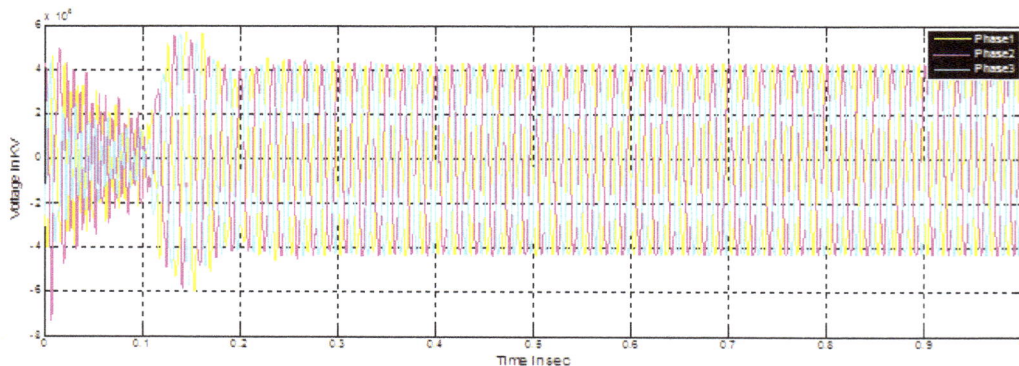

Figure 12. *Receiving end voltage Vs Time for case (b)*

Fig.9 and fig.10 shows comparison of receiving end real   and reactive power of the test system. After fault clearing real and reactive power came to stable state take some time. From the comparisons of case (a) and (b), compensation of real and reactive power better in case (b) then the case (a)

Fig.11 and fig.12 shows comparison of three phase voltage of power system. After the fault cleared better performance of the system case (b) then the case (a).

Shows above simulation results are DC Voltage (Vdc), active power (P) and reactive power (Q) for take a look at system included UPFC with battery and UPFC with SMES. During this UPFC with SMES higher transient stability compensation is provided.

6. Conclusion

In this Paper UPFC integrated with SMES and UPFC integrated with battery square measure analyzed exploitation MATLAB /SIMULINK. Throughout these analysis SMES management exploitation two quadrants DC-DC chopper plays a very important role for stability improvement. This planned UPFC with SMES methodology is extremely economical for transient stability improvement, effective power oscillation damping and to take care of power flow in total distance of the conductor when after disturbances. Conjointly the results shown UPFC-SMES integrated system is effective than UPFC-BATTERY integrated system.

References

[1] N.G.Hingorani, L.Gyugyi. IEEE press, "Understanding FACTS Concepts and Technology of Flexible AC Transmission Systems", 2000.

[2] Paulo F. Ribeiro, Brian K. Johnson, Mariesa L. Crow, AysenArsoy and YILU, "Energy Storage Systems for Advanced Power Applications," proceedings of the IEEE, VOL. 89, NO. 12,2001.

[3] Ali Bidadfar, MehrdadAbedi Mehdi Karari, Chia-Chi Chu. "Power Swings Damping Improvement by Control of UPFC and SMES Based on Direct LyapunovMethodApplication". IEEE Press, 2008.

[4] M. A. Casacca, M. R. Capobianco, and Z. M. Salameh,"Lead-acid battery storage on figurations for improved available capacity,"IEEE Trans. Energy Conversion, vol. 11, pp. 139–145, 1996.

[5] ArindamChakraborty,ShravanaK.Musunuri,Anurag K. Srivastava,and Anil K. Kondabathini,"Integrating STATCOM and Battery Energy Storage System for Power SystemTransient Stability: A Review and Application",Hindawi Publishing Corporation Advances in Power Electronics , Article ID 676010, 12 pages doi:10.1155/2012/676010, 2012.

[6] Z. Yang, C. Shen, L. Zhang, M. L. Crow, and S. Atcitty, "Integration of a StatCom and battery energy storage," IEEE Transactions on Power Systems, vol. 16, no. 2, pp. 254–260, 2001.

[7] K.C. Divya, Jacob Østergaard, "Battery energy storage technology for power systems—an overview", Electric Power Systems Research, Elsevier B.V, 2008.

[8] MahyarZarghami, Mariesa L. Crow, JagannathanSarangapani, Yilu Liu, StanAtcitty, "A Novel Approach to Interarea Oscillation Damping by Unified Power Flow Controllers Utilizing Ultra capacitors ," IEEE Transactions on Power Systems, vol. 25, no. 1, 2010.

[9] P.G. Therond, I. Joly, M. Volker, "Superconducting magnetic energy storage (SMES) for industrial applications-Comparison with battery systems," IEEE Trans. Applied Superconductivity, vol. 3, pp. 250-253, 1993.

[10] M. Parizh, A. K. Kalafala, R. Wilcox, "Superconducting magnetic energy storage for substation applications", IEEE Trans. Applied Superconductivity, vol. 7, no. 2, pp. 849-852, 1997.

[11] C. A. Luongo, "Superconducting storage systems: An overview," IEEE Trans. Magnetics, vol. 32, no. 4, pp. 2214-2223, 1996.

[12] A.B .Arsoy, Y. Liu, P .F .Ribeiro, "Static Synchronous compensatorsand superconducting magnetic energy storage systems in controllingpower system dynamics", IEEE Industry Applications Magazine, 2003.

[13] Antonio Griffo, DavideLauria, "Superconducting Magnetic EnergyStorage Control Strategy for Power System Stability Improvement",Bepress, volume 4, Issue 1, 2005.

[14] IssarachaiNgamroo, "Robust Freqency Stablization By Coordinated Superconducting Magnetic Energy Storage with Static SynchronousSeries Compensator", Bepress, volume 3, Issue 1, Article 1031, 2005.

[15] A. Kazemi, S. Jamali, and H. Shateri, "Effects of SMES Equipped UPFC on Measured Impedance at Relaying Point in Inter Phase Faults", IEEE Electrical Power & Energy Conference, 978-1-4244-2895-3, 2008.

[16] Sandia national laboratory: http://www.sandia.gov/ess.

[17] Renewable energy systems design assistant for storage: http://www.ecn.nl/resdas.

[18] Electricity storage association: http://electricty-storage.org.

[19] J. X. Jin, High temperature superconducting energy storage technologies: principle and application, Beijing: Science Press, 2011.

[20] J. Tian, Y. Q. Zhu, and C. H. Chen, "Application of energy storage technologies in distributed generation," Electrical technology, no. 8, pp.29-32, 2010.

[21] Y. Li, J. Y. Liu, and C. Hu, "Application and prospect of superconducting energy storage technology in the power system,"Sichuan Electric Power Technology, vol. 32, pp. 33-37, December, 2009.

[22] J. M. Miller, "Trends in vehicle energy storage systems: batteries and ultracapacitors to unite," 2008 Vehicle Power and Propulsion Conference, pp. 1-9, Sept. 3-5, 2008.

[23] A. Khamis, Z. M. Badarudin, A. Ahmad, A. A. Rahman, and M. H. Hairi, "Overview of mini scale compressed air energy storage system," 2010 4th International Power Engineering and Optimization Conference (PEOCO), pp. 458-462, June 23-24, 2010.

[24] T. Nanahara, and A. Takimoto, "A study on required reservoir size for pumped hydro storage," IEEE Transactions on Power Systems, vol. 9, no. 1, pp. 318-323, Feb.,1994.

[25] K. J. C. M. Posthumus, R. A. A. Schillemans, and E. C. Kluiters, "Sodium-sulphur batteries for naval applications," Eleventh Annual Battery Conference on Applications and Advances, pp. 301-306, Jan. 9-12, 1996.

[26] T. M. Masaud, L. Keun, and P. K. Sen, "An overview of energy storage technologies in electric power systems: What is the future?," North American Power Symposium (NAPS), pp. 1-6, Sept. 26-28, 2010.

[27] L. Zhang, C. Shen, M. L. Crow, L. Dong, S. Pekarek, and S.Atcitty, "Performance indices for the dynamic performance of FACTS and FACTS with energy storage," Electric Power Components and Systems, vol. 33, no. 3, pp. 299–314, 2005.

[28] Y. Cheng and M. L. Crow, "A diode-clamped multi-level inverter for the StatCom/BESS," in Proceedings of the IEEE Power Engineering SocietyWinter Meeting, vol. 1, pp. 470–475,January, 2002.

[29] R. Kuiava, R. A. Ramos, and N. G. Bretas, "Control design of a STATCOM with energy storage system for stability and power quality improvements," in Proceedings of the IEEE International Conference on IndustrialTechnology (ICIT '09), pp. 1–6, February, 2009.

[30] C. Qian, M. L. Crow, and S. Atcitty, "A multi-processor control system architecture for a cascaded StatCom with energy storage," in Proceedings of the 19th Annual IEEE Applied Power Electronics Conference and Exposition (APEC '04), pp.1757–1763, February, 2004.

19

Design, modeling and simulation of Fuzzy controlled SVC for 750 km (λ/8) Transmission line

Murali. Matcha[1], Sharath Kumar. Papani[1], Vijetha. Killamsetti[2]

[1]Department of Electrical Engineering, N.I.T Warangal, A.P, INDIA-506004
[2]Department of EEE, GMR Institute of Technology, Srikakulam, A.P, INDIA

Email address:

murali233.nitw@gmail.com (Murali. Matcha)

Abstract: Flexible AC transmission system (FACTS) is a technology, which is based on power electronic devices, used to enhance the existing transmission capabilities in order to make the transmission system flexible and independent operation. The FACTS technology is a promising technology to achieve complete deregulation of Power System i.e. Generation, Transmission and Distribution as complete individual units. The loading capability of transmission system can also be enhanced nearer to the thermal limits without affecting the stability. Complete close-loop smooth control of reactive power can be achieved using shunt connected FACTS devices. Static VAR Compensator (SVC) is one of the shunt connected FACTS device, which can be utilized for the purpose of reactive power compensation. Intelligent FACTS devices make them adaptable and hence it is emerging in the present state of art. This paper attempts to design and simulate the Fuzzy logic control of firing angle for SVC in order to achieve better, smooth and adaptive control of reactive power. The design, modeling and simulations are carried out for λ /8 Transmission line and the compensation is placed at the receiving end (load end).

Keywords: Fuzzy Logic, FACTS, SVC

1. Introduction

The reactive power generation and absorption in power system is essential since the reactive power is very precious in keeping the voltage of power system stable. The main elements for generation and absorption of reactive power are transmission line, transformers and alternators. The transmission line distributed parameters throughout the line, on light loads or at no loads become predominant and consequently the line supplies charging VAR (generates reactive power). In order to maintain the terminal voltage at the load bus adequate, reactive reserves are needed. FACTS devices like SVC can supply or absorb the reactive power at receiving end bus or at load end bus in transmission system, which helps in achieving better economy in power transfer.

In this paper Transmission line (λ /8) is simulated using 4π line segments by keeping the sending end voltage constant. The receiving end voltage fluctuations were observed for different loads. In order to maintain the receiving end voltage constant, shunt inductor and capacitor is added for different loading conditions. SVC is simulated

by means of fixed capacitor and thyristor controlled reactor (FC-TCR) which is placed at the receiving end. The firing angle control circuit is designed and the firing angles are varied for various loading conditions to make the receiving end voltage equal to sending end voltage. Fuzzy logic controller is designed to achieve the firing angles for SVC such that it maintains a flat voltage profile. All the results thus obtained, were verified and were utilized in framing of fuzzy rule base in order to achieve better reactive power compensation for the Transmission line (λ/8). Based on observed results for load voltage variations for different values of load resistance, inductance and capacitance a fuzzy controller is designed which controls the firing angle of SVC in order to automatically maintain the receiving end voltage constant.

2. Operating Principles and Modeling of SVC

An elementary single phase thyristor controlled reactor [1] (TCR) shown in Fig.1 consists of a fixed (usually air core) reactor of inductance L and a two anti parallel SCRs.

The device brought into conduction by simultaneous application of gate pulses to SCRs of the same polarity. In addition, it will automatically block immediately after the ac current crosses zero, unless the gate signal is reapplied. The current in the reactor can be controlled from maximum (SCR closed) to zero (SCR open) by the method of firing delay angle control. That is, the SCR conduction delayed with respect to the peak of the applied voltage in each half-cycle, and thus the duration of the current conduction interval is controlled. This method of current control is illustrated separately for the positive and negative current cycles in Fig.2 where the applied voltage V and the reactor current iL(α) at zero delay angle (switch fully closed) and at an arbitrary α delay angle are shown. When α =0, the SCR closes at the crest of the applied voltage and evidently the resulting current in the reactor will be the same as that obtained in steady state with a permanently closed switch. When the gating of the SCR is delayed by an angle α ($0 \leq \alpha \leq \pi/2$) with respect to the crest of the voltage, the current in the reactor can be expressed [1] as follows:

$$V(t) = V \cos \omega t. \tag{1}$$

$$i_L = (1/L) \int_\alpha^{\omega t} V(t)dt = (V/\omega L)(\sin \omega t - \sin \alpha) \tag{2}$$

Since the SCR, by definition, opens as the current reaches zero, is valid for the interval $\alpha \leq \omega t \leq \pi - \alpha$. For subsequent negative half-cycle intervals, the sign of the terms in (1) becomes opposite.

In (1) the term $(V/\omega L) \sin \alpha = 0$ is offset which is shifted down for positive and up for negative current half-cycles obtained at $\alpha = 0$, as illustrated in Fig.2. Since the SCRs automatically turns off at the instant of current zero crossing of SCR this process actually controls the conduction intervals (or angle) of the SCR. That is, the delay angle α defines the prevailing conduction angle σ ($\sigma = \pi - 2\alpha$). Thus, as the delay angle α increases, the corresponding increasing offset results in the reduction of the conduction angle σ of the SCR, and the consequent reduction of the reactor current. At the maximum delay of $\alpha = \pi/2$, the offset also reaches its maximum of $V/\omega L$, at which both the conduction angle and the reactor current becomes zero. The two parameters, delay angle α and conduction angle σ are equivalent and therefore TCR can be characterized by either of them; their use is simply a matter of preference. For this reason, expression related to the TCR can be found in the literature both in terms of α and σ [1].

It is evident that the magnitude of the current in the reactor varied continuously by delay angle control from maximum (α=0) to zero (α=$\pi/2$) shown in Fig.3, where the reactor current iL(α) together with its fundamental component iLF(α) are shown at various delay angles α [1]. However, the adjustment of the current in reactor can take place only once in each-half cycle, in the zero to $\pi/2$ interval [1]. This restriction result in a delay of the attainable current control. The worst-case delay, when changing the current from maximum to zero (or vice versa), is a half-cycle of the applied ac voltage. The amplitude ILF (α) of the fundamental reactor current iLF(α) can be expressed as a function of angle α [1].

$$I_{LF}(\alpha) = V/\omega L (1 - (2/\pi) \alpha - (1/\pi) \sin (2\alpha)) \tag{3}$$

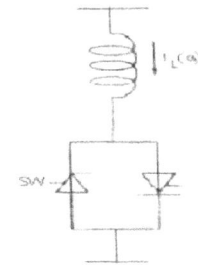

Figure 1. *Basic Thyristor Controlled Reactor*

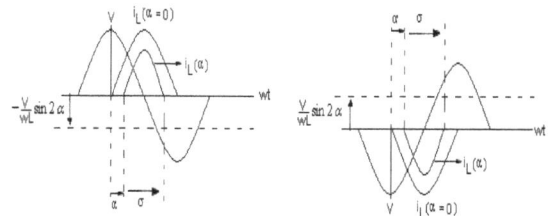

Figure 2. *Firing delay angle*

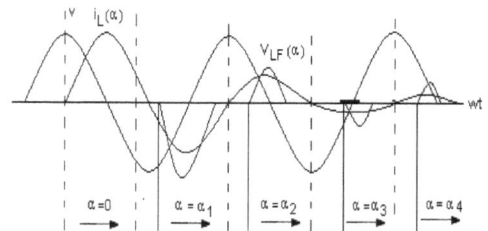

Figure 3. *Operating waveforms*

Where V is the amplitude of the applied voltage, L is the inductance of the thyristor-controlled reactor and ω is the angular frequency of the applied voltage. The variation of the amplitude ILF (α), normalized to the maximum current ILFmax, (ILFmax= V/ωL), is shown plotted against delay angle α shown in Fig.4.

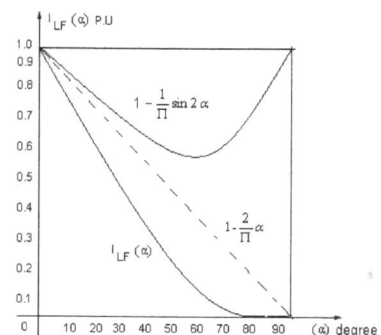

Figure 4. *Amplitude variation of the fundamental TCR current with the delay angle (α)*

It is clear from Fig.4 the TCR can control the fundamental current continuously from zero (SCR open) to a maximum (SCR closed) as if it was a variable reactive admittance. Thus, an effective reactance admittance, BL(α), for the TCR can be defined. This admittance, as a function of angle α is obtained as:

$$B_L(\alpha)=1/\omega L(1-(2/\pi)\alpha-(1/\pi)\sin(2\alpha)) \qquad (4)$$

Evidently, the admittance BL(α) varies with α in the same manner as the fundamental current ILF(α).The meaning of (4) is that at each delay angle α an effective admittance BL(α) can be defined which determines the magnitude of the fundamental current, ILF(α), in the TCR at a given applied voltage V. In practice, the maximal magnitude of the applied voltage and that of the corresponding current limited by the ratings of the power components (reactor and SCRs) used. Thus, a practical TCR can be operated anywhere in a defined V-I area, the boundaries of which are determined by its maximum attainable admittance, voltage and current ratings as illustrated in the Fig.5a. The TCR limits are established by design from actual operating requirements. If the TCR switching is restricted to a fixed delay angle, usually α = 0, then it becomes a thyristor switched reactor (TSR). The TSR provide a fixed inductive admittance and thus, when connected to the ac system, the reactive current in it will be proportion to the applied voltage as the V - I plot in the Fig.5b.

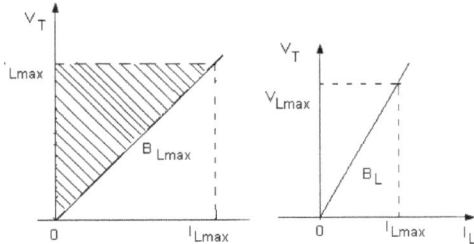

Figure 5. *Operating V-I area of (a) For TCR and (b) For TSR*

VLmax = voltage limit, ILmax = current limit

BLmax = max admittance of TCR,

BL = admittance of reactor

A basic VAR generator arrangement using a fixed capacitor with a thyristor-controlled reactor (FC-TCR) shown in Fig.6 [1].The current in the reactor is varied by the previously discussed method of firing delay angle control. A filter network that has the necessary capacitive impedance at the fundamental frequency to generate the reactive power required usually substitutes the fixed capacitor in practice, fully or partially, but it provides low impedance at selected frequencies to shunt the dominant harmonics produced by the TCR.

The fixed capacitor thyristor-controlled reactor type VAR generator may be considered essentially to consist of a variable reactor (controlled by a delay angle α) and a fixed capacitor. With an overall VAR demand versus VAR output characteristic as shown in Fig.7 in constant capacitive VAR generator (Qc) of the fixed capacitor is opposed by the variable VAR absorption (QL) of the thyristor-controlled reactor, to yield the total VAR output (Q) required. At the maximum capacitive VAR output, the thyristor-controlled

reactor is off (α= 900). To decrease the capacitive output, decreasing delay angle α. At zero VAR output increases the current in the reactor, the capacitive and inductive current becomes equal and thus the capacitive and inductive VARs cancel out. With a further decrease of angle α, the inductive current becomes larger than the capacitive current, resulting in a net inductive VAR output. At zero delay angle, the thyristor-controlled reactor conducts current over the full 180o interval, resulting in maximum inductive VAR output that is equal to the difference between the VARs generated by the capacitor and those absorbed by the fully conducting reactor.

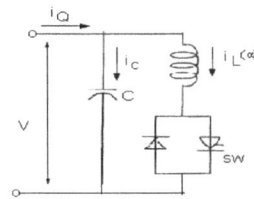

Figure 6. *Basic FC-TCR type static generator*

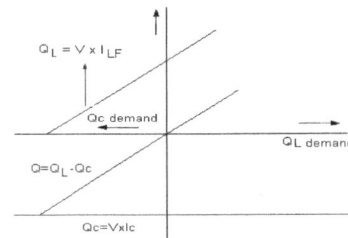

Figure 7. *VAR demand versus VAR output characteristic*

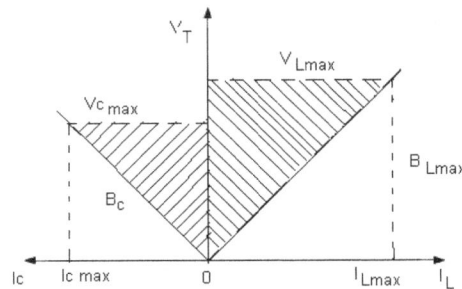

Figure 8. *V-I characteristics of the FC-TCR type VAR Generator*

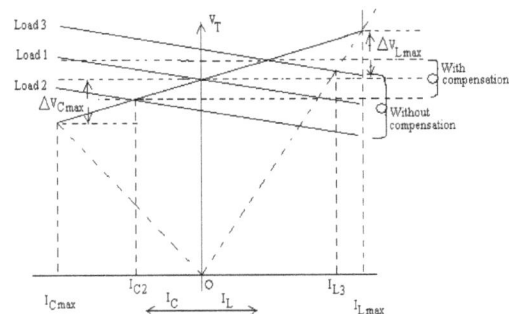

Figure 9. *Dynamic V-I Characteristics of SVC with Load lines*

In Fig.8 [1] the voltage defines the V-I operating area of the FC-TCR VAR generator and current rating of the major power components. In the dynamic V-I Characteristics of SVC along with the Load lines showed in the Fig.9 [1] the load characteristics assumed straight lines for Dynamic studies as easily seen that the voltage improved with compensation when compared without compensation.

V_{Cmax} = voltage limit of capacitor

B_C = admittance of capacitor

V_{Lmax} = voltage limit of TCR

I_{Cmax} = capacitive current limit

I_{Lmax} = inductive current limit

B_{Lmax} = max inductive admittance

3. Fuzzy Logic Controller

Fuzzy logic is a new control approach with great potential for real time applications [2] [3]. Fig.10 shows the structure of the fuzzy logic controller (FIS-Fuzzy inference system) in MATLAB Fuzzy logic toolbox. [5][6]. Load voltage and load current taken as input to fuzzy system. For a closed loop control, error input can be selected as current, voltage or impedance, according to control type [7]. To get the linearity triangular membership function is taken with 50% overlap. The output of fuzzy controller taken as the control signal and the pulse generator provides synchronous firing pulses to thyristors as shown in fig.11. The Fuzzy Logic is a rule based controller, where a set of rules represents a control decision mechanism to correct the effect of certain causes coming from power system [8] [9]. In fuzzy logic, the five linguistic variables expressed by fuzzy sets defined on their respective universes of discourse. Table 1 shows the suggested membership function rules of FC-TCR controller. The rule of this table can be chosen based on practical experience and simulation results observed from the behavior of the system around its stable equilibrium points.

Table 1. *Membership Function Rules*

		Load	voltage			
		NL	NM	P	PM	PB
	NL	PB	PB	NM	NM	NL
	NM	PB	PB	NM	P	NL
Load current	P	P	PM	NM	NM	P
	PM	NM	P	NM	NM	PM
	PB	NL	NM	NM	NL	NL

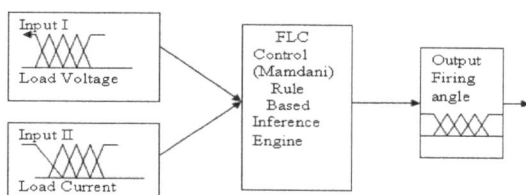

Figure 10. *Structure of fuzzy logic controller*

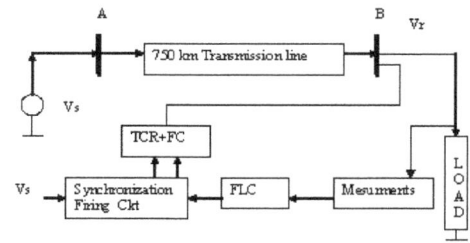

Figure 11. *Single Phase equivalent circuit and fuzzy logic control structure of SVC*

4. Hardware Implementation

An available simple two-bus artificial transmission (λ/8) line model of 4π line segments with 750 km, distributed parameters were used in this study. The line inductance 0.1mH /km, capacitance 0.01µf/km and the line resistance 0.001Ω were used. Each π section is of 187km, 187km, 188km and 188 km. Supply voltage is 230V, 50 Hz having source internal resistance of 1 Ω connected to node A. Static load is connected at receiving end B .The load resistance was varied to obtain the voltage variations at the receiving end. A shunt branch consisting of inductor and capacitor is added to compensate the reactive power of transmission line. With the change of load and due to Ferranti effect, the variations in voltages are observed at receiving end B of transmission line [9] [10]. The practical values of shunt elements are varied for different loading conditions to get both sending and receiving end voltages equal. As shown in Table 2.

Table 2. *Compensated Practical Values of Inductor and Capacitor*

S.No	Load Resistance (Ω)	Compensating Inductance (H)	Compensating Capacitance (µF)
1	500	0.8	1
2	400	0.9	1
3	300	0.19	2
4	200	0.18	5
5	150	0.19	5
6	100	0.22	8
7	50	0.14	8.5
8	40	0.14	9.0
9	30	0.14	10
10	20	0.14	12

4.1. Filter Circuit Design

IC TCA 785 a 16 pin IC shown in Fig.12 is used in this study for firing the SCRs. This IC having output current of 250 mA and a fuzzy logic trainer kit with two input variables and having 5 linguistic sets is used. This can generate 5 X 5 rules. The output of fuzzy logic which varies from DC -10V to +10V is given to IC 785 controller pin11, which controls the comparator voltage VC ,and the firing angle α for one cycle and (180 +α) during negative cycle shown in fig.13.

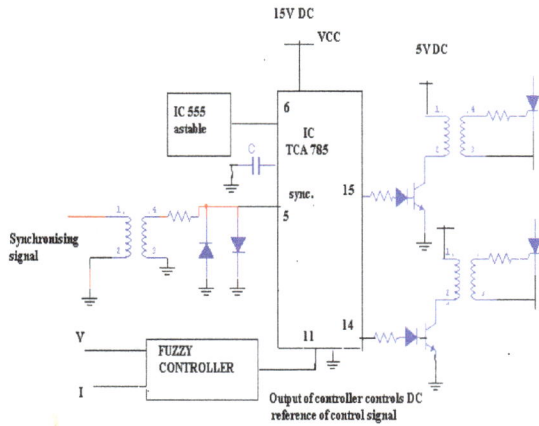

Figure 12. *Firing Scheme with TCA 785 IC*

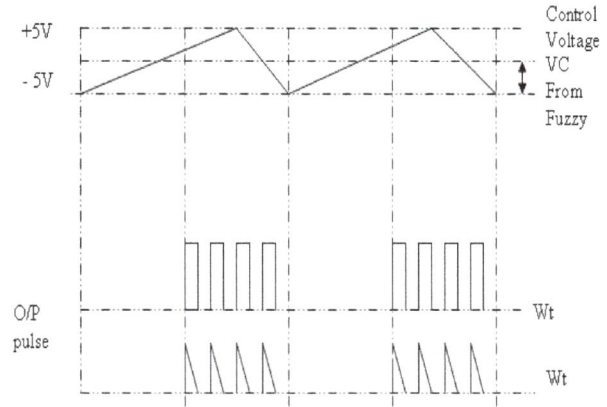

Figure 13. *Generation of wave forms of TCA 785 IC*

Table 3. *Load Voltage Before and After Compensation*

Transmission Line Parameters for Lt = 10mh/km Ct = 0.1μf/km R = 0.001Ω		Before compensation For Vs= 230V (p-p)		After Compensation For Vs= 230 (P-P)		
R (Ω)	V_S (rms) Volts	V_R (rms) Volts	I_R rms Amp	V_R (rms) Volts	I_R (rms) Amp	α
500	162.6	270.80	0.54	162.1	2.032	90
400	162.6	268.10	0.67	162.4	2.036	100
300	162.6	268.00	0.89	162.	2.099	102
200	162.6	261.10	1.30	162.7	2.182	103
180	162.6	258.10	1.43	162.4	2.198	105
160	162.6	256.10	1.59	162.3	2.232	106
140	162.6	250.30	1.78	162.8	2.299	108
120	162.6	243.80	2.03	161.8	2.357	109
100	162.6	234.20	2.34	162.4	2.459	112
80	162.6	219.50	2.74	163.3	2.651	117
60	162.6	195.80	3.26	162.3	3.071	128
50	162.6	156.50	3.91	162.5	4.124	158

5. Test Results

The transmission line without any compensation was not satisfying the essential condition of maintaining the voltage within the reasonable limits. The effect of increasing load was to reduce the voltage level at the load end. At light loads, the load voltage is greater than the sending end voltage as the reactive power generated is greater than absorbed. At higher loads the load voltage drops, as the reactive power absorbed is greater than generated, as shown in Table 3. Fig.14 and Fig.15 indicates unequal voltage profiles. Fig.16 clearly shows the firing angle and inductor current control.

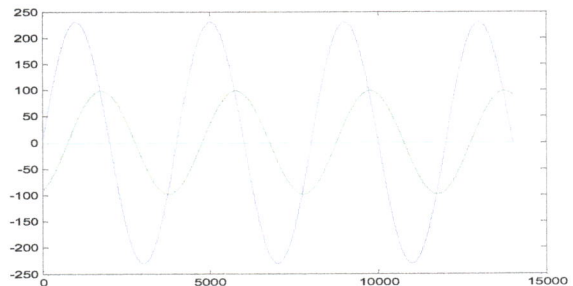

Figure 14. *Uncompensated voltages for heavy loads*

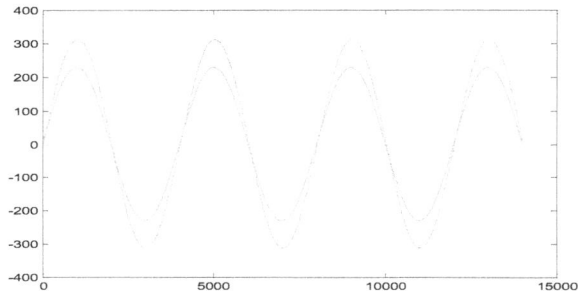

Figure 15. Uncompensated voltage for light load

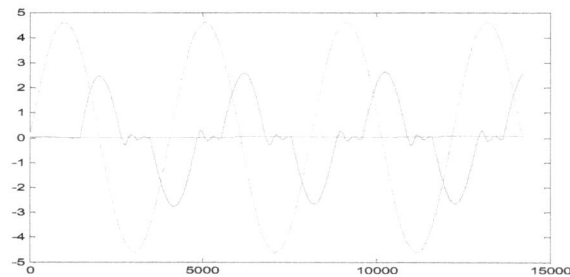

Figure 16. Inductor Current for firing angle 165 deg

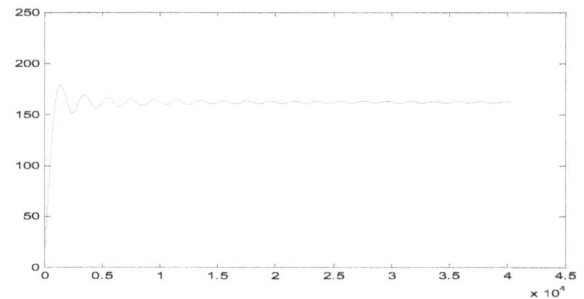

Figure 17. Compensated VR =VS (RMS voltage) for R=200Ω

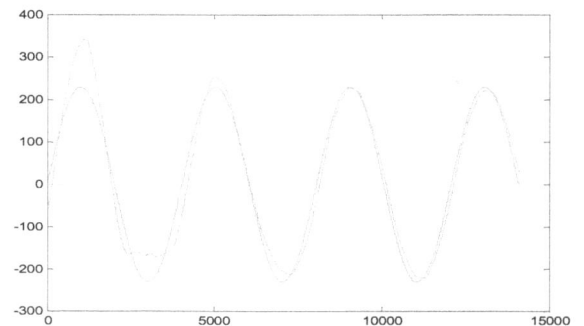

Figure 18. Compensated VR=VS (instantaneous voltage) for R=200 Ω

6. Conclusions

This paper presents an "online Fuzzy control scheme for SVC" and it can be concluded that the use of fuzzy controlled SVC (FC-TCR) compensating device with the firing angle control is continuous, effective and it is a simplest way of controlling the reactive power of transmission line. It is observed that SVC device was able to compensate both over and under voltages. Compensating voltages are shown in Fig.17 and Fig.18. The use of fuzzy logic has facilitated the closed loop control of system, by designing a set of rules, which decides the firing angle given to SVC to attain the required voltage. With MATLAB simulations [4], [5] and actual testing it is observed that SVC (FC-TCR) provides an effective reactive power control irrespective of load variations.

Acknowledgements

The authors would like to thank NIT Warangal for providing required facilities to carry out the simulation of this research work.

References

[1]   Narain. G. Hingorani, "Understanding FACTS, Concepts and Technology Of flexible AC Transmission Systems", by IEEE Press USA.

[2]   Bart Kosko, "Neural Networks and Fuzzy Systems A Dynamical Systems Approach to Machine Intelligence", Prentice-Hall of India New Delhi, June 1994.

[3]   Timothy J Ross, "Fuzzy Logic with Engineering Applications", McGraw-Hill, Inc, New York, 1997.

[4]   Laboratory Manual for Transmission line and fuzzy Trainer Kit of Electrical Engineering Department, NIT Warangal

[5]   SIM Power System User Guide Version 4 MATLAB Manual Periodicals and Conference Proceedings

[6]   S.M.Sadeghzadeh, M. Ehsan " Improvement of Transient Stability Limit in Power System Transmission Lines Using Fuzzy Control of FACTS Devices",IEEE Transactions on Power System Vol.13 No.3 ,August 1998

[7]   Chuen Chien Lee "Fuzzy Logic in Control Systems: Fuzzy Logic Controller", Part I and Part II. IEEE R. IEEE transactions on system, man, and cybernetics, vol.20 March/April11990

[8]   [8]   A.M. Kulkarni, "Design of power system stabilizer for single-machine system using robust periodic output feedback controller", IEE Proceedings Part – C, Vol. 150, No. 2, pp. 211 – 216, March 2003. Technical Reports: Papers from Conference Proceedings unpublished)

[9]   U.Yolac, T.Talcinoz Dept. of Electronic Eng.Nigde 51200, Turkey "Comparison of Fuzzy Logic and PID Controls For TCSC Using MATLAB"

[10]   Jaun Dixon, Luis Moran, Jose Rodrfguz, Ricardo Domke, "Reactive power compensation technology   state- of- art-review" (invited paper)

[11]   Electrical Engineering Dept Pontifica Universidad Catolica De CHILE.

20

Transient stability enhancement of 30 bus multi- machine systems by using PSS & increasing inertia

Nitin Mohan Lal, Arvind Kumar Singh

Electrical Engg. Deptt, B.I.T Sindri, Dhanbad, India

Email address:

nitinsumit009@gmail.com (N. M. Lal), singharvindk67@gmail.com (A. K. Singh)

Abstract: In this paper the transient stability enhancement of a 30- bus multi-machine system by using Power system stabilizer and increasing inertia has been observed, as single method is not sufficient for this purpose. We have created a three phase fault at time 0.04 seconds at bus 7 and cleared at time 0.5 seconds. On implementing PSS and on increasing the inertia of the machine we achieved a better response in terms of power swing when compared with initial condition. The inertia of the machine is kept within a certain limit. And 30 bus multi-machine system maintain its own permissible operating condition.

Keyword: Transient Stability, PSS, Inertia, Three Phase Fault

1. Introduction

Power system stability of modern large inter-connected systems is a major problem for secure operation of the system. Recent major black-outs across the world caused by system instability, even in very sophisticated and secure systems, illustrate the problems facing secure operation of power systems. Stability is defined as the ability of a system to return to its normal or stable operation after having been subjected to some form of disturbance.

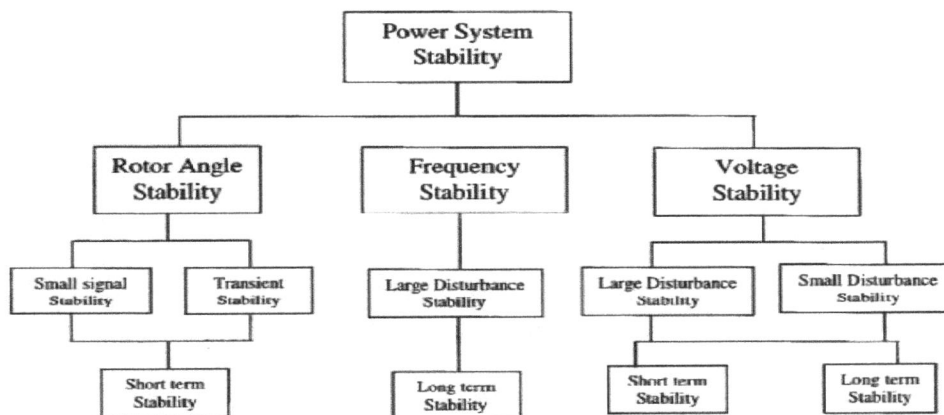

Figure 1. Classification of power system stability [1].

This mainly refers to the ability of the system to remain in synchronism. However, modern power systems operate under complex interconnections, controls and extremely stressed conditions. Further, with increased automation and use of electronic equipment, the quality of power has gained more importance, shifting focus on concepts of voltage stability, frequency stability, inter-area oscillations etc.

For the purpose of analysis the power system stability is sub divided into two major categories

i)   Steady state stability. &

ii) Transient stability.

Steady-state stability refers to the ability of the power system to regain synchronism after small and slow disturbances, such as ground power changes. An extension of the steady-state stability is known as the dynamic stability. Transient stability studies deal with the effects of large, sudden disturbances, such as the occurrence of the fault, the sudden outage of a line

Transient stability of a power system is its ability to maintain synchronous operation of the machines when subjected to a large disturbance. The occurrence of such a disturbance may result in large excursions of the system machine rotor angles and, whenever corrective actions fail, loss of synchronism results among machines. Generally, the loss of synchronism develops in very few seconds after the disturbance inception.[2]

1.1. Transient Stability Analysis

The power system is almost regularly subjected to a variety of disturbances. Even the act of switching on an appliance in the house can be regarded as a disturbance. However, given the size of the system and the scale of the perturbation caused by the switching of an appliance in comparison to the size and capability of the interconnected system, the effects are not measurable. Large disturbance do occur on the system. These include severe lightning strikes, loss of transmission line carrying bulk power due to overloading. The ability of power system to survive the transition following a large disturbance and reach an acceptable operating condition is called transient stability.

Any disruption in the system will cause the imbalance between the mechanical power input to the generator and electrical power output of the generator to be affected. As a result, some of the generators will tend to speed up and some will tend to slow down. If, for a particular generator, this tendency is too great, it will no longer remain in synchronism with the rest of the system and will be automatically disconnected from the system. This phenomenon is referred to as a generator going out of step.

E-Tap allows the user to reduce very large and complex power systems into simple one line diagram and performs operations on it like load the system, subject the system to contingency and study the characteristics of faults. These virtual faults in the simulation model can be compared to the real time system faults.[15] In this paper enhancement of transient stability analysis of 30-bus multi machine system by using power system stabilizer (PSS), and increasing inertia is done. For this purpose we create a three phase fault on specified bus and then analyze the behaviour of the synchronous machine. For this work we have used the licensed version of ETAP software.

1.2. Power System Stabilizer (PSS)

Power system stabilizer provides an additional input signal to the regulator to damp power system oscillations. Some commonly used input signals are rotor speed deviation, accelerating power, and frequency deviation.[14]

The power system stabilizer uses auxiliary stabilizing signals to control the excitation system so as to improve power system dynamic performance this is very effective method of enhancing small signal stability performance.[3]. The main aim of the PSS is to provide damping with the help of a component of electrical torque in phase with the rotor speed variation, to the generator oscillation by using auxiliary stabilizing signal(s)[4].

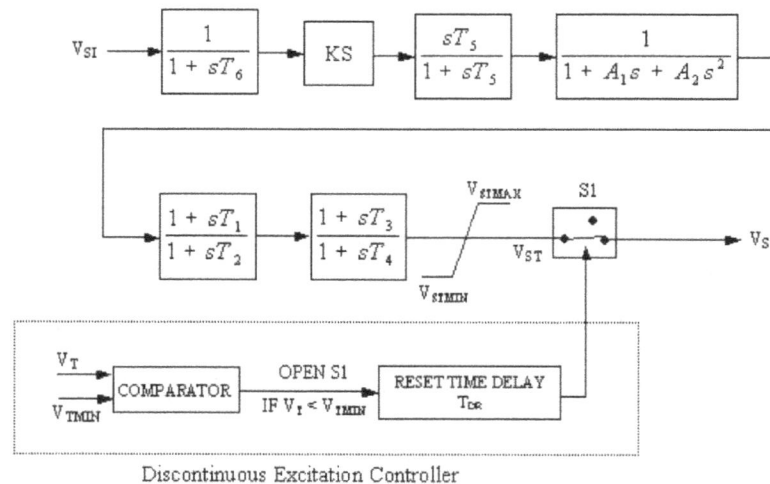

Figure 2. IEEE TYPE-1 PSS(PSS1A).

2. Description of 30 Bus Model

A 30 bus multi- machine system is taken here for the analysis purpose as shown in the figure. it consists of 30 bus , 6 generators, 4 transformers, 20 loads and 37 cables are connected in between the buses. 3 buses are in swing mode, 3 buses are voltage controlled bus and remaining 24 buses are load bus. The length of each cable is 50 km and positive, zero sequence component of impedance is (0.015240+j 0.027432) ohms per conductor per phase. The rating of generators, loads, PSS is given below in the following tables.

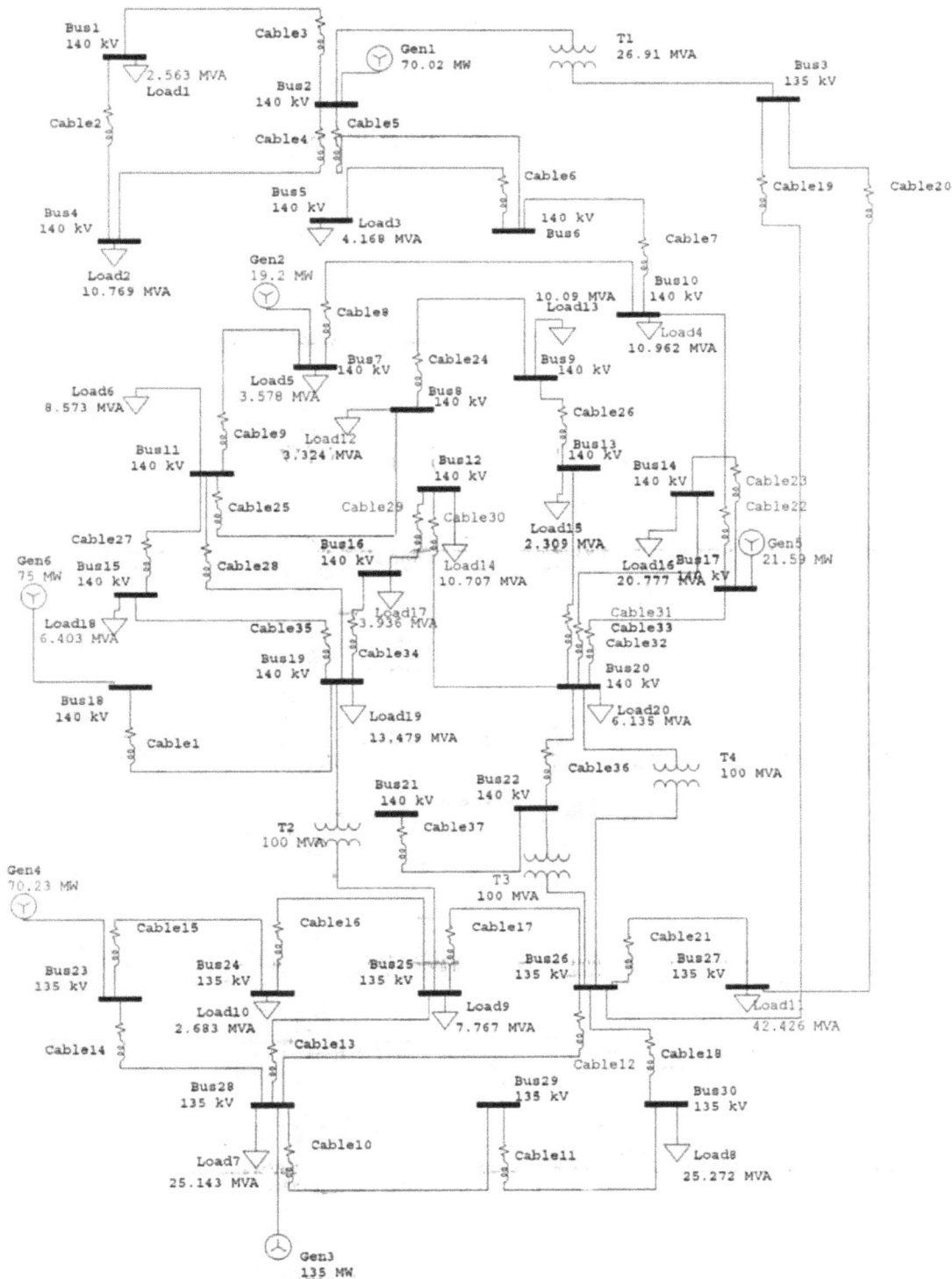

Figure 3. *30 Bus Test System.*

3. Result and Discussion

In this paper we focus on the analysis of transient stability behaviour of a 30-bus multi machine system by implementing PSS and by increasing the inertia of the synchronous machine. For this purpose we have used Accelerated gauss siedel method for initial load flow solution. The maximum number of iteration possible is 2000, the solution precision for the initial load flow is 0.000001, and the acceleration factor for the initial load flow is 1.45. The time increment for integration steps (Δt) is 0.01 and the system frequency is maintained at 50 Hz.

The inertia of the synchronous machine is initially taken

as 3 MW-Sec/MVA and is increased later on to 7 MW-Sec/MVA for the analysis purpose. The inertia of a machine affects the bulkiness of the rotor of the machine; hence inertia cannot be increased much.

The 3- phase fault is created at bus 7 at time 0.04 sec and is cleared after time 0.5 sec. the electromechanical oscillations of electrical power is reduced and field voltage is also kept limited, due to this reason excitation is maintained. The various plots of electrical power, field current, and terminal current individually initial, without PSS and with PSS and inertia combined respectively of generator 6 are as shown in figures.

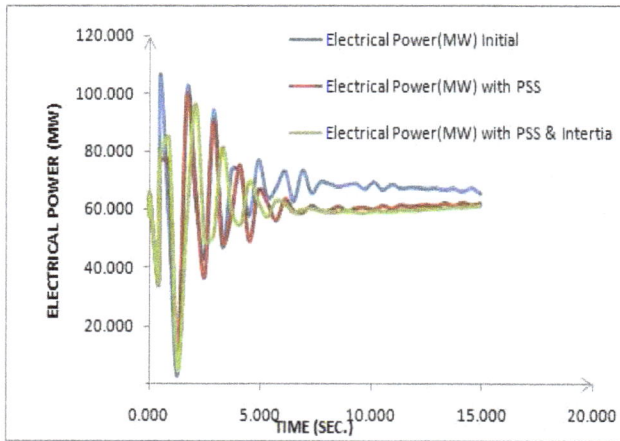

Figure 5. *Terminal current of Generator 6.*

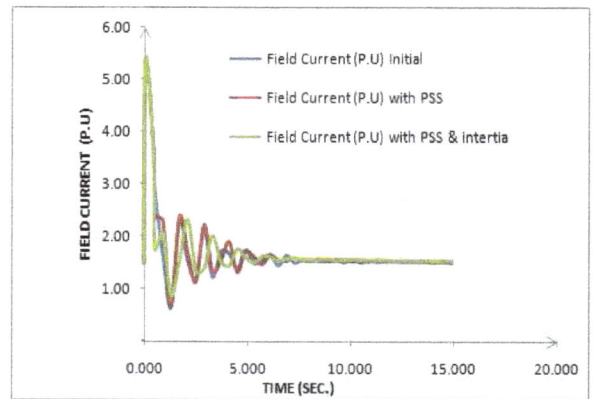

Figure 4. *Electrical Power of Generator.*

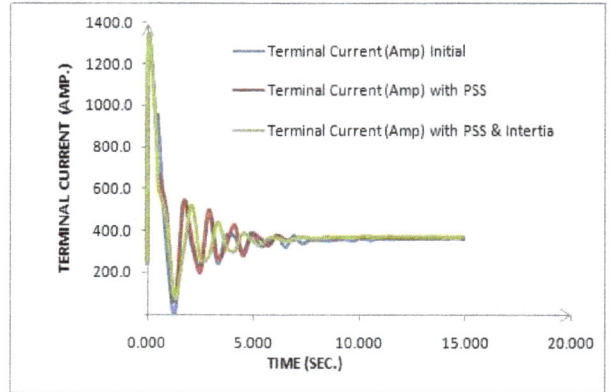

Figure 6. *Field current of Generator 6.*

Table 1. *Synchronous machine parameters.*

Machine			Rating		Positive sequence impedence(%)								Zero seq. Z(%)		
ID	TYPE	MODEL	MVA	KV	R_a	X_d''	X_d'	X_d	X_q''	X_q'	X_q	X_1	X/R	R_0	X_0
Gen1	Generator	Subtransient, Round-Rotor	82.376	140	1	19	28	155	19	65	155	15	7	1	7
Gen2	Generator	Subtransient, Round-Rotor	22.588	140	1	19	28	155	19	65	155	15	7	1	7
Gen3	Generator	Subtransient, Round-Rotor	158.824	135	1	19	28	155	19	65	155	15	7	1	7
Gen4	Generator	Subtransient, Round-Rotor	82.624	135	1	19	28	155	19	65	155	15	7	1	7
Gen5	Generator	Subtransient, Round-Rotor	25.4	140	1	19	28	155	19	65	155	15	7	1	7
Gen6	Generator	Subtransient, Round-Rotor	88.235	140	1	19	28	155	19	65	155	15	7	1	7

Table 2. *Dynamic parameters of Synchronous Machine.*

Machine	Connected bus	Time cons.(sec.)				H(Sec.),,D(MW pu/Hz)&Saturation					Grounding	
ID	ID	T_{do}''	T_{do}'	T_{qo}''	T_{qo}'	H	%D	S100	S120	Sbreak	Conn.	Type
Gen1	Bus2	0.03	6.5	0.03	1.25	12	0	1.7	1.18	0.8	WYE	SOLID
Gen2	Bus7	0.03	6.5	0.03	1.25	12	0	1.7	1.18	0.8	WYE	SOLID
Gen3	Bus28	0.03	6.5	0.03	1.25	12	0	1.7	1.18	0.8	WYE	SOLID
Gen4	Bus23	0.03	6.5	0.03	1.25	12	0	1.7	1.18	0.8	WYE	SOLID
Gen5	Bus17	0.03	6.5	0.03	1.25	12	0	1.7	1.18	0.8	WYE	SOLID
Gen6	Bus18	0.03	6.5	0.03	1.25	12	0	1.7	1.18	0.8	WYE	SOLID

Table 3. Mechanical parameters of synchronous machine.

Machine ID	TYPE	Generator/Motor WR²	RPM	H	Coupling WR²	RPM	H	Prime Mover/Load WR²	RPM	H	Equivalent Total WR²	RPM	H
Gen1	Gen.	475476.97	1500	3	330193	1800	3	330193	1800	3	1426432.75	1500	9
Gen2	Gen.	13379	1500	3	90541	1800	3	90541	1800	3	391137.09	1500	9
Gen3	Gen.	916730	1500	3	636618	1800	3	636618	1800	3	2750189.75	1500	9
Gen4	Gen.	476904	1500	3	331183	1800	3	331183	1800	3	1430711	1500	9
Gen5	Gen.	146609	1500	3	101812	1800	3	101812	1800	3	439827.56	1500	9
Gen6	Gen.	509295	1500	3	353677	1800	3	353677	1800	3	1527884.75	1500	9

WR²: kg-m²   H: MW-Sec/MVA

Table 4. POWER SYSTEM STABILIZER (PSS) INPUT DATA Type: PSS1A.

Generator ID	VSI	KS	VSTMax	VSTMin	VTMin	TDR	A1	A2	T1	T2	T3	T4	T5	T6
Gen 1	SPEED	3.15	0.9	-0.9	0	0.2	0	0	0.76	0.1	0.76	0.1	1	0.1
Gen 2	SPEED	3.15	0.9	-0.9	0	0.2	0	0	0.76	0.1	0.76	0.1	1	0.1
Gen 3	SPEED	3.15	0.9	-0.9	0	0.2	0	0	0.76	0.1	0.76	0.1	1	0.1
Gen 4	SPEED	3.15	0.9	-0.9	0	0.2	0	0	0.76	0.1	0.76	0.1	1	0.1
Gen 5	SPEED	3.15	0.9	-0.9	0	0.2	0	0	0.76	0.1	0.76	0.1	1	0.1
Gen 6	SPEED	3.15	0.9	-0.9	0	0.2	0	0	0.76	0.1	0.76	0.1	1	0.1

Table 5. LOAD RATING.

LOAD	RATING (MVA)
1	02.563
2	10.769
3	04.168
4	10.962
5	03.578
6	08.573
7	25.143
8	25.272
9	07.767
10	02.683
11	42.426
12	03.324
13	10.090
14	10.707
15	02.309
16	20.777
17	03.936
18	06.403
19	13.479
20	06.135

4. Conclusion

The paper presents the improved behaviour of transient stability of a 30-bus multi machine system when implemented with PSS and when inertia is increased, using E-TAP software. The comparison of transient stability performances of the multi machine system initial, with PSS, and with PSS and inertia combined is performed and found that the power swing is damped out, and we get better response in terms of electromechanical oscillations.

References

[1] Farmer, Richard G. "Power System Dynamics and Stability", The Electric Power Engineering Handbook, Ed. L.L. Grigsby,Boca Raton: CRC Press LLC, 2001.

[2] TRANSIENT STABILITY OF POWER SYSTEMS A Unified Approach to Assessment and Control Mania PAVELLA, Damien ERNST, Research Fellow, FNRS Daniel RUIZVEGA University of Li `ege, Belgium Kluwer Academic Publishers Boston /Dordrecht/London

[3] prabha kundur, power system stability and control, tata McGraw Hill publication,edition 2011

[4] Bablesh Kumar Jha, Ramjee Prasad Gupta, Upendra Prasad. Combined Operation of SVC, PSS and Increasing Inertia of Machine for Power System Transient Stability Enhancement. *American Journal of Electrical Power and Energy Systems.*Vol. 3, No. 1, 2014, pp. 7-14. doi: 10.11648/j.epes.20140301.12

[5] W.Watson and G.Manchur, "Experience with supplementary Damping Signals for Generator Static ExicationSystem,"IEEE Trans.,Vol.PAS-92,pp. 199-203,January/February 1973

[6] W.Watson and M.E Coultes ,"Static Exicter Stabilizing Signals on Large Generators-Mechanical Problems,"IEEE Trans.,Vol. PAS-92,pp.204 211,January/February 1973

[7] P.Kundur ,D.C.Lee and H.M. Zein EL-Din,"Power system stabilizer fot thermal units:Analytical Techniques and On-Site Validation,"IEEE Trans.,Vol.PAS-100,pp. 81-95,January 1981

[8] M.L. Shelton, R.F Winklemen, W.A Mittelstandt, and W.L Bellerby,"Bonneville Power Administration 1400 MW Braking Resistor,"IEEE Trans.,VOL.PAS 94,pp.602-611,March/april 1975

[9] P.K. Iyambo, R. Tzonova, "Transient Stability Analysis of the IEEE 14-Bus Electrical Power System", IEEE Conf. 2007

[10] N. Mo, Z.Y. Zou, K.W. Chan, T.Y.G. Pong, "Transient stability constrained optimal power flow using particleswarm optimization", IET Generation, Transmission & Distribution, Vol. 1, pp. 476–483, 2007

[11] D. Chatterjee, A. Ghosh, "Transient Stability Assessment of PowerSystems Containing Series and Shunt Compensators," IEEE Trans. onpower systems, vol. 22, no. 3, August 2007

[12] IEEJ Technical Committee, Standard models of power system, IEEJ Technical Report, Vol. 754, 1999

[13] P. M. Anderson and A. A. Fouad. Power System Control and Stability. The IEEE Press, 1995.

[14] Electric Systems, Dynamics, and Stability with Artificial Intelligence ... By James A. Momoh, Mohamed E. El-Hawary, copyright©2000 by Marcel Dekker Inc.

[15] Kavitha R, "Transient stability of IEEE-30 bus system using E-TAP software," IJSER, vol.3, issue 12, December 2012

21

Transient stability improvement by using PSS and increasing inertia of synchronous machine

Pushpalata Khalkho, Arvind Kumar Singh

Electrical Engg. Deptt. B.I.T Sindri, Dhanbad, Govt. of Jharkhand-828123, India

Email address:

Pushpi215@gmail.com (P. Khalkho), singharvindk67@gmail.com (A. K. Singh)

Abstract: The main objective of this paper is to perform transient stability analysis using the electrical power system design and analysis software namely E-Tap. The purpose of performing transient stability on the power system is to study the stability of a system under various disturbances. The stability of the power system is the ability of generators to remain in synchronization even when subjected to disturbance. In this paper a 9-bus test system is considered fig.(1). Improvement of transient stability by coordination of PSS (Power System Stabilizer) and increasing inertia of synchronous machine has been observed.

Keywords: Power System Stabilizer, ETAP-Software, Transient Stability, Three Phase Fault

1. Introduction

The power system stability is an electromechanical phenomenon, it is thus defined as the ability of designated synchronous machine in the system to remain in synchronism with one another following disturbance such as fault and fault removal at various locations in the system. It also indicates the ability of induction motors in the system to maintain torque to carry load following there disturbances.

Power system stability is the property of a power system that insures the system remains in electromechanical equilibrium throughout any normal and abnormal operating condition.

The term stability can be interpreted as maintenance of synchronism. The transient disturbances are caused by the changes in loads, switching operations, particularly faults and loos of excitation. Thus, maintenance of synchronism during steady state condition and regaining of synchronism or equilibrium after a disturbance are prime importance to the electrical utilities.[1].

The stability of an interconnected power system is its ability to return to normal or stable operation after having been subjected to some form of disturbance. Conversely, instability means a condition denoting loss of synchronism or falling out of step.

1.1. Transient Stability

A synchronous power system has transient stability if after a large sudden disturbance; it can regain and maintain synchronism. A sudden large disturbance includes application of fault, clearing of fault, switching on and off the system element (transmission line, transformers, generator, loads, bus etc.)

1.2. Transient Stability Analysis

Transient stability is the ability of the power system to maintain synchronism when subjected to a severe transient disturbance. The resulting system response involves large excursions of generator rotor angle and is influenced by the nonlinear power-angle relationship. Stability depends on both the initial operation state of the system and the severity of the disturbance. Usually, the system is altered so that the post disturbance steady-state operation differs from

That prior to the disturbance.[1]

A fault in the system will lead to instability and the machine will fall out of synchronism. If the system can't sustain till the fault is cleared, then the whole system will be in stabilized. During the instability not only the oscillation in rotor angle around the final position goes on increasing but also the change in angular speed. In such a situation the system will never come to its final position. The unbalanced condition or transient condition may leads to instability where the machines in the power system fall out of synchronism.

The system is subjected to a large variety of disturbances. The switching on and off of an appliance in the house is also a disturbance depending upon the size and capability of interconnected system. Large disturbances such as lightning strokes, loss of transmission line carrying bulk power do occur in the system. Therefore transient stability is defined as the ability of the power system to survive the transition following the large disturbance and to reach an acceptable operating condition.

The physical phenomenon that occurs during a large disturbance is that there will be an imbalance between the mechanical power input and the electrical power output. This will tend to run the generator at high speed. The result will be the loss of synchronism of the generator and the machine will be disconnected from the system. This phenomenon is referred to as a generator going out of step.[1].

2. Software Used

The simulation software used here is E-Tap or Electrical Transient Stability Analysis Program by Operation Technology. There are different analyses that can be performed on a bus system using this software. Load flow analysis, short Circuit Analysis, Arc Flash analysis, Harmonic Analysis, Transient stability analysis.

E-TAP is a fully graphical enterprise package that runs on Microsoft windows 2003,2008, XP, vista, and 7 operating system. E-TAP is the post comprehensive analysis tool for the design and testing of power system available. Using its standard offline simulation modules, E-TAP com utilize real time operating data for advance monitoring, real time simulation, optimization, energy management system, and high–speed intelligent load shedding.

E-TAP software easily create and edit graphical one-line diagrams (OLD) underground cable raceway system (UGS) ,three-dimensional cable system , advanced time–current coordination and selectivity plots, geographic information system schematics (GIS) , as well as three dimensional ground grid system (GPS) the programmer has been designed to incorporate to three key concepts.

2.1. Power System Stabilizer

The power system stabilizer (PSS) uses auxiliary stabilizing signal to control the excitation system so as to improve power system dynamic performance commonly use input signal to the power system stabilizer are shaft speed, terminal frequency and power. Power system dynamic performance is improved by the damping of system oscillations. This is a very effective method of enhancing small signal stability performance. PSS based on shaft speed signal has been used successfully on hydraulic units since the mid-1960s. A technique developed to derive a stabilizing signal from measurement of shaft speed of a hydraulic unit. The application of shaft speed-based stabilizers to thermal units requires a careful consideration of the effects on torsional oscillations. The stabilizer, while damping the rotor oscillation, can cause instability of the torsional modes.[2-3]

Thermal frequency has been used as the stability signal several (PSS) power system stabilizer application. Normally, the terminal frequency signal is used directly as the stabilizer input signal. In some cases, terminal voltage and current are used to derive the frequency of a voltage behind a simulated machine reactance so as to approximate the machine rotor speed.[3].

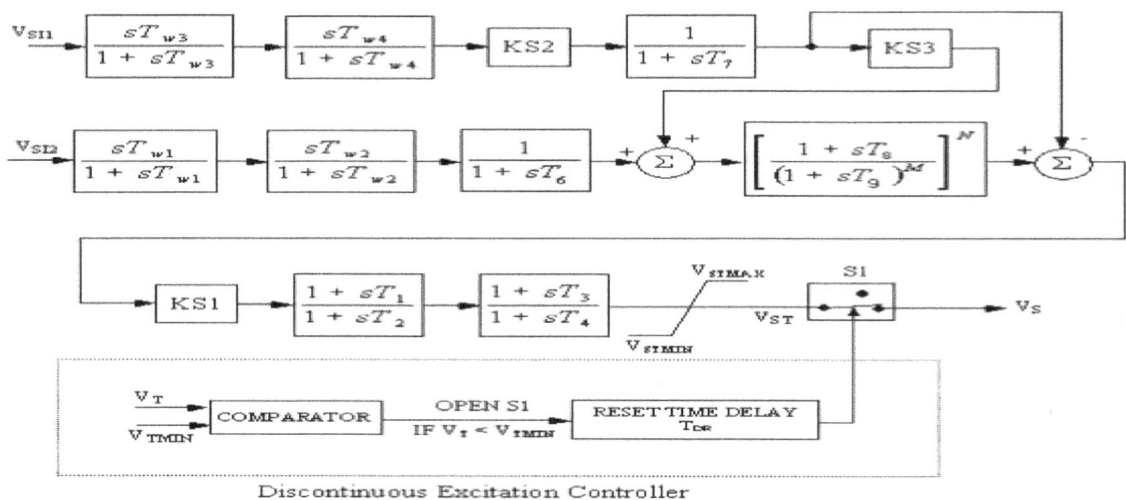

Fig (1) IEEE Type-2 PSS(PSS2A).

2.2. Exciter

The based function of exciter system is to provide direct current to the synchronous machine field winding. The excitation system performs control and protective essential to the satisfactory performance of the power system by controlling the field voltage and the field current. IEEE DC1 type exciter are used in this model.

3. Test System

The test system that has been considered here is the 9-bus multi- machine system. Three generators, three transformers, three statics load and six cable are usedin this system.

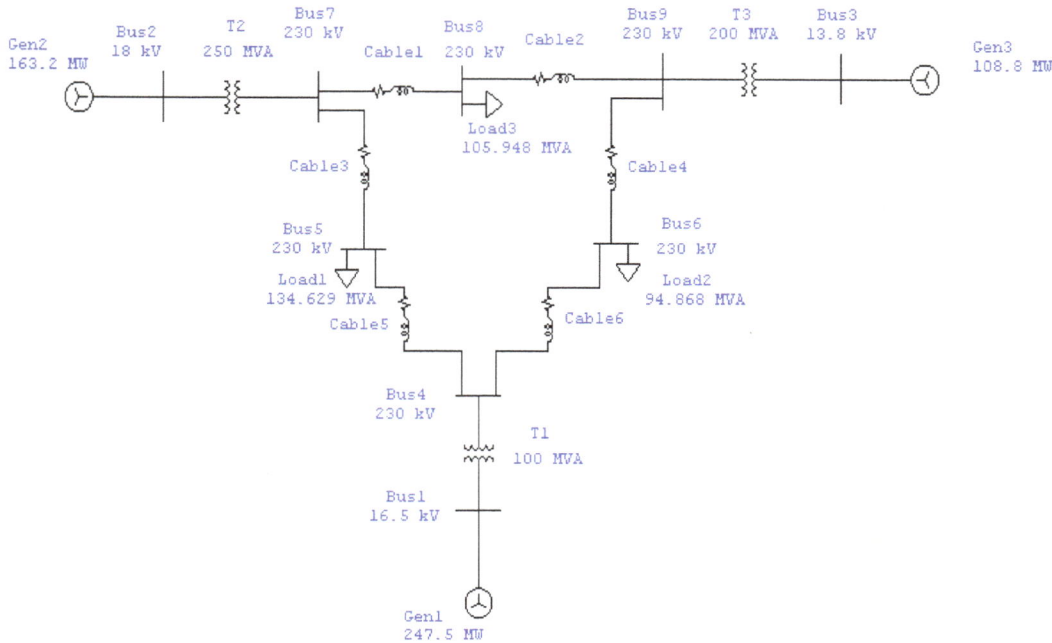

Fig (2)Test system.

3.1. Input Data

Table 1. Synchronous machine parameter input data.

	Machine		Rating			Positive sequence impedance (%)							Zero seq. Z(%)		
ID	Type	Model	MVA	KV	R_a	X_d''	X_d'	X_d	X_q''	X_q'	X_q	X1	X/R	R_0	X_0
GEN1	Generator	Subtransient, Round-Rotor	291.17	16.5	1	19	28	155	19	65	115	15	7	1	7
GEN2	Generator	Subtransient, Round-Rotor	192	18	1	19	28	155	19	65	115	15	7	1	7
GEN3	Generator	Subtransient, Round-Rotor	128	13.8	1	19	28	155	19	65	115	15	7	1	7

Table 2. dynamic parameters of synchronous machine input data.

Machine	Connected Bus	Time constant (sec.)				H(sec.),D(MWpu/Hz) & Saturation					Generator or loading		Grounding	
ID	ID	T_{do}''	T_{do}'	T_{qo}''	T_{qo}'	H	%D	S100	S120	Sbreak	MW	MVar	Conn.	Type
GEN1	BUS1	0.035	6.5	0.035	1.25	3	0	1.07	1.18	0.8	0	0	wye	solid
GEN2	BUS2	0.035	6.5	0.035	1.25	3	0	1.07	1.18	0.8	0	0	wye	solid
GEN3	BUS3	0.035	6.5	0.035	1.25	3	0	1.07	1.18	0.8	0	0	wye	solid

Table 3. Mechanical parameters of synchronous machine input data.

Machine		Generator/Motor			Coupling			Prime mover/load			Equivalent Total		
ID	Type	RPM	WR^2	H	RPM	WR^2	H	RPM	WR^2	H	RPM	WR^2	H
GEN1	GEN	1500	1680671.88	3	1800	0	0	1800	0	0	1500	1680671.88	3
GEN2	GEN	1500	1108225	3	1800	0	0	1800	0	0	1500	1108225	3
GEN3	GEN	1500	738817	3	1800	0	0	1800	0	0	1500	738817	3

WR^2: kg-m^2 H: MW-Sec/MVA

Table 4. Power system stabilizer input data (Type- PSS2A).

Gen. ID	VSI1	VSI2	KS1	KS2	KS3	VSTMax	VSTMin	VTMin	VDR	Tw1	Tw2	Tw3	Tw4
			N	M	T1	T2	T3	T4	T5	T6	T7	T8	
GEN1	Elec.power	Speed	20	0.001	1	0.2	-0.066	0	0.2	10	10	10	10
			4	2	0.16	0.02	0.16	0.02	0	0	0.3	0.15	
GEN2	Elec.power	Speed	20	0.001	1	0.2	-0.066	0	0.2	10	10	10	10
			4	2	0.16	0.02	0.16	0.02	0	0	0.3	0.15	
GEN3	Elec.power	Speed	20	0.001	1	0.2	-0.066	0	0.2	10	10	10	10

Table 5. Exciter input data (Type-DC1).

Machine ID	Control bus ID	KA Efd$_{max}$	KE	KF	TA	TB	TC	TE	TF	TR	VR$_{max}$	VR$_{max}$	SE$_{max}$	SE.75
Gen1	Bus1	46 2.63	0.05	0.1	0.06	0	0	0.46	1	0.005	1	-0.9	0.33	0.1
Gen2	Bus2	46 2.63	0.05	0.1	0.06	0	0	0.46	1	0.005	1	-0.9	0.33	0.1
Gen3	Bus3	46 2.63	0.05	0.1	0.06	0	0	0.46	1	0.005	1	-0.9	0.33	0.1

4. Simulation Result and Discussion

Transient stability study is essentially an action driven time-domain simulation. Actions should be specified at different time instants (events).When to simulate the system response for existing events, such as a recorded fault in the system, we use this type of action, because the recorded fault occurring time and duration are known.

In this paper we discuss the transient stability performance with PSS (Power system stabilizer) and by increasing inertia of synchronous machine. In this system any one method of improving stability may not be adequate. So combinations of two methods are used in this system. Here we use accelerated gauss-seidel for initial load flow calculation. In which maximum number of iteration is 2000 and Solution Precision for the Initial LF is 0.0000010000 And Time Increment for Integration Steps (Δt) is 0.0100and acceleration factor for the initial load flow is 1.45. Initial inertia of the installed machine was 3 MW-Sec/MVA and after increasing its inertia is 7 MW-Sec/MVA. The overall performance of the power system, Solutions to the stability problem of one category should not be affected at the expense of another category. So inertia of the machine is not so much increased because after increasing inertia of machine rotor will be heavier.

The different plot for Gen-1. When a three phase fault on bus-5 at 0.350 sec and cleared at 0.600 sec are shown below in fig. comparisons each-other.

Electrical powers of gen-1, fig.(3) and fig.(4) shows Electrical power (MW) Vs. time(sec.).Electrical power with inertia and Electrical power with PSS2A, increasing inertia are Comparison with initial electrical power.

The electromechanical oscillation for generator electrical power is reduced and steady state power is improved when used PSS2A and increasing inertia of synchronous machine seen in fig.(4). Terminal current of gen-1, fig.(5) and fig.(6)Terminal current (Amp)Vs time(sec.). Terminal current with inertia and terminal current with PSS2A, increasing inertia are comparison with initial terminal

current of synchronous machine.

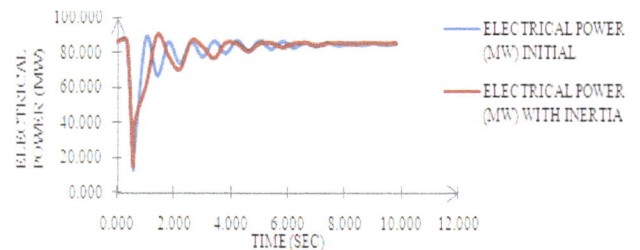

Fig (3) Electrical power (mw) initial & Electrical power (mw) with inertia.

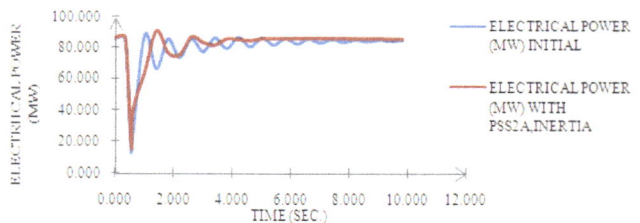

Fig (4) Electrical power (mw) initial & Electrical power (mw) with PSS2A, inertia.

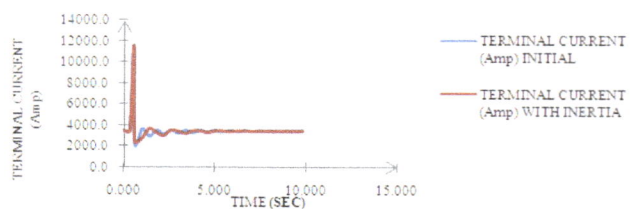

Fig (5) Terminal current (Amp) initial & Terminal current (Amp) with inertia.

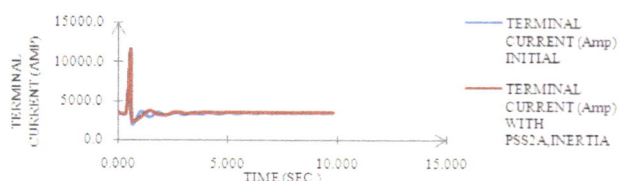

Fig (6)Terminal current (Amp) initial & Terminal current (Amp) with PSS2A, inertia.

The oscillation in terminal current is reduced when used PSS2A and increasing inertia of synchronous machine seen in fig.(6).

Field current of gen-1, fig. (7)and fig.(8) Field current (P.U) Vs time (sec.). Field current with inertia and Field current with PSS2A, increasing inertia are comparison with initial field current of synchronous machine.

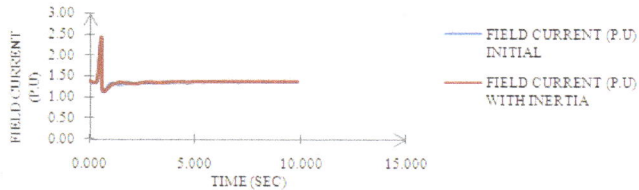

Fig(7) Field current (P.U) initial &Field current (P.U) with inertia.

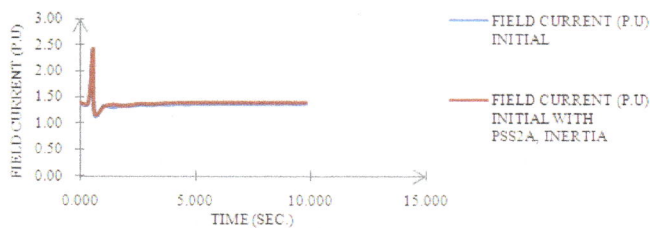

Fig (8) Field current (P.U) initial & Field current (P.U) with PSS2A, inertia.

Oscillation of Field current is also reduced when increasing inertia. Field current are not change when we increasing inertia and PSS2A, increasing inertia in fig.(7) and fig.(8).

Field voltage of gen-1, fig.(9) Field voltage (P.U) Vs time (sec.).Field voltage(P.U) with initial and field voltage with PSS2A, increasing inertia of synchronous machine are Comparison. Field voltage are constant in initial and oscillated initially when used PSS2A, increasing inertia but after some time it is constant and within limit as shown in fig.(9).

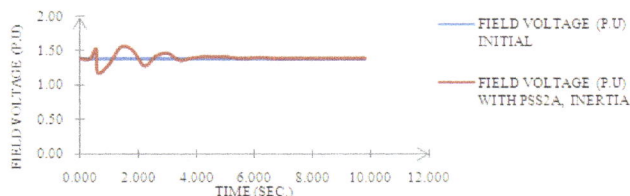

Fig (9) Field voltage (P.U) initial & Field voltage (P.U) with PSS2A, inertia.

5. Conclusion

Transient stability Performances of the multi machine system by using (PSS) power system stabilizer and other method has been compared. When we used (PSS)power

systemstabilizer and increasing inertia then electromechanical oscillation has been achieved better response. The transient stability improvement is not sufficient by using one method. So here we use these two combined method for improving stability. E-Tap provides transient stability study results at all different level of detail, depending on our requirement.

References

[1]   Kundur, P., Power system stability and control, McGraw-Hill, New York, 1994.

[2]   P. Kundur, M. Klein, G.J. Rogers, and M.S. Zywno, " Application of power system stabilizers for Enhancement of overall system stability," IEEE Trans., Vol. PWRS-4, No.2, PP.614- 626,May 1989.

[3]   P. Kundur, D.C. Lee and H.M. Zein EL-Din," Power system stabilizer for thermal units: Analytical Techniques and on-site Validation," IEEE Trans., Vol.PAS-100,PP.81-95,January 1981.

[4]   IEEE Committee Report "Proposed Excitation system Definition for synchronous machine," IEEE Trans., Vol. PAS- 88, PP.1228-1258,August 1969.

[5]   IEEE/CIGRE Joint task force on stability terms and definitions, 'Definition and classification of Power System Stability,' IEEE trans. Power System, Vol. 19,No.2, PP. 1387-1401, May 2004.

[6]   IEEE standard Definitions for excitation system for synchronous machines IEEE standard.

[7]   Kamwa I. ,Grandin R. and Trudal G. , 'IEEE PSS2B versus PSS4B: The limit of performance of modem Power System Stabilizers, IEEE Trans. Power System, Vol., NO.2,PP.903-915,May 2005.

[8]   P. Kundur and D.C lee "Advanced excitation control for power system stability Enhancement," CIGRE Paper 38- 01, paris, France 1986.

[9]   E. W. Kimbark, power system stability, Vol.3: Synchronous machines Jon Wiley &Sons,1956.

[10]   P.K. Iyambo, R. Tzonova, "Transient Stability Analysis of the IEEE 14-Bus Electrical Power System", IEEE Conf. 2007

[11]   K.R. Padiyar, 'Power system Dynamics-Stability and control,' second Edition, (Hyderabad), B.S. Publication, 2002.

[12]   IEEE Task force, "Proposed terms and Definition for power system stability," IEEE trans.,Vol.PAS-101,pp.1894-1898, July 1982.

[13]   F.P. deMello and C. Corcordia, "Concepta of Synchronous Machine Stability as Affected by Excitation control," IEEE Trans., Vol. PAS-88, PP.316-329, April 1969.

22

Fault detection and classification based on DWT and modern approaches for T.L compensated with FACTS

Noha Mahmoud Bastawy, Hossam El-din Talaat, Amr Mohamed Ibrahim

Electrical Power & Machines Dept., Faculty of Engineering, Ain Shams University, Cairo, Egypt

Email address:

nohamahmoudbastawy@hotmail.co.uk (N. M. Bastawy), hossam_talaat@eng.asu.edu.eg (H. El-din Talaat),
amrmohamedhassan@yahoo.com (A. M. Ibrahim)

Abstract: A new approach for detecting and classifying a fault for transmission line compensated with Flexible AC Transmission System (FACTS) is presented in this paper. Unified Power Flow Controller (UPFC) is one of the most advanced FACTS devices that can simultaneously and independently control both the real and reactive power flow in a transmission line. The proposed technique consists of preprocessing module based on Discrete Wavelet Transform (DWT) in combination with Artificial Neural Network (ANN) or Gaussian Process (GP) for detecting and classifying fault events.

Keywords: Neural Network, Gaussian Process, Discrete Wavelet Transform, FACTS

1. Introduction

Electricity market activities and a growing demand for electricity have led to heavily stressed power systems. This requires operation of the networks closer to their stability limits. The flexible alternating current transmission system (FACTS), a new technology based on power electronics, offers an opportunity to enhance controllability, stability, and power transfer capability of ac transmission systems [1-5].

Unified Power Flow Controller (UPFC) is regarded as the most generalized version of FACTS. UPFC consists of a static synchronous series compensator (SSSC) and a static compensator (STATCOM), connected in such a way that they share a common DC capacitor. The UPFC, by means of an angularly unconstrained, series voltage injection, is able to control, concurrently or selectively, the transmission line impedance, the nodal voltage magnitude, and the active and reactive power flow through it. It may also provide independently controllable shunt reactive compensation [6-12].

Although the UPFC improves the power flow in the transmission line, its presence imposes a number of problems including distance protection. The apparent impedance seen by a distance relay is influenced greatly by the location and parameters of UPFC [13], [14]. Thus an adaptive relay setting of the distance protection is required to cope up with the problems of over reach or under reach.

2. Fault Detection and Classification Scheme

"Fig. 1" shows the proposed protection scheme. It consists of two stages, Pre-processing stage based on DWT and fault classification stage based on ANN or GP. The DWT considerably simplifies the input signal of the ANN and GP; it reduces the volume of input data of ANN and GP without loss of information. This dramatically reduces the training stage in and increase the overall performance of the digital relay.

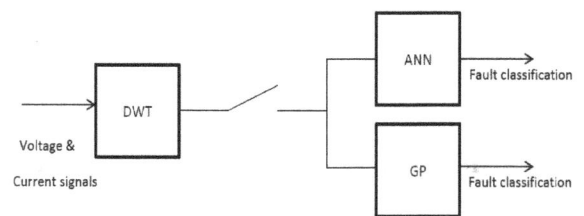

Figure 1. Fault detection and classification scheme

The proposed topology of the protection scheme is composed of two levels as shown in "Fig. 2". Level-1 is used to detect the fault, while level-2 is used to identify faulted phase(s). The output of level-1 activates level-2 if there is a fault. Therefore, the proposed topology determines both the fault type and the faulted phase(s) selection.

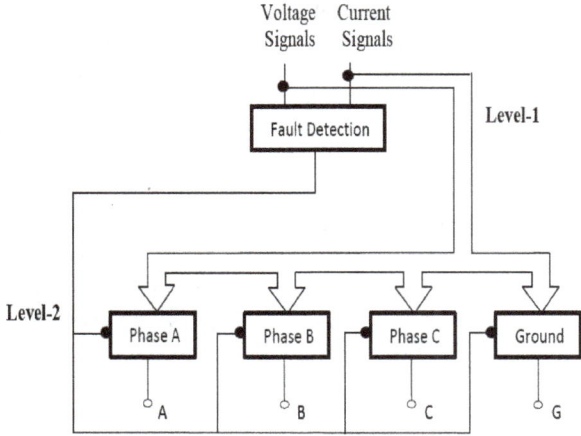

Figure 2. *Proposed protection scheme*

3. Discrete Wavelet Transform

Wavelet analysis is a relatively new signal processing tool and is applied recently by many researchers in power systems due to its strong capability of time and frequency domain analysis. The two areas with most applications are power quality analysis and power system protection [15]. The wavelet transform is the versatile tool with very rich mathematical content and great potential for applications. The wavelet transform decomposes transients into a series of wavelet components i.e. approximation and detail components. The resulting decomposed signals can then be analyzed in both time and frequency domains. Hence, the wavelet transform is feasible and practical for analyzing power system transients [16].

Wavelet transforms are fast and efficient means of analyzing transient voltage and current signals. The wavelet transform not only decomposes a signal into frequency bands, but also, unlike the Fourier transform, provides a non-uniform division of the frequency domain (i.e., the wavelet transform uses short windows at high frequencies and long windows for low frequency components). Wavelet analysis deals with expansion of functions in terms of a set of basic functions (wavelets) which are generated from a mother wavelet by operations of dilatations and translation [17-18]. The discrete wavelet transform is defined by the following equation:

$$DWT(m,n) = \frac{1}{\sqrt{a_0^m}} \sum_k x(k) g\left(a_0^{-m} n - b_0 k\right) \quad (1)$$

Where g(k) is the mother wavelet, x(k) is the signal input and a, b are the scaling and translation parameters. Discrete wavelet transform is implemented by using high-pass filter and low-pass filter respectively, which defined by the following equations:

$$y_{high}[k] = \sum_n x[n] \cdot g[2k - n] \quad (2)$$

$$y_{low}[n] = \sum_k x[n] \cdot h[2k - n] \quad (3)$$

Where $y_{high}(k)$ is the output from the high-pass filter called Detail and $y_{low}(k)$ is the output from the low-pass filter called Approximation, also the output of the low-pass filter down-sampling by a factor of 2 which effectively scales the wavelet by a factor of 2 for next stage. This decomposition has halved the time resolution since only half the number of samples now characterizes the entire signal. However, this operation doubles the frequency resolution, since the frequency band of the signal now spans only half the previous frequency band. The block diagram of filter analysis is shown in "Fig. 3"

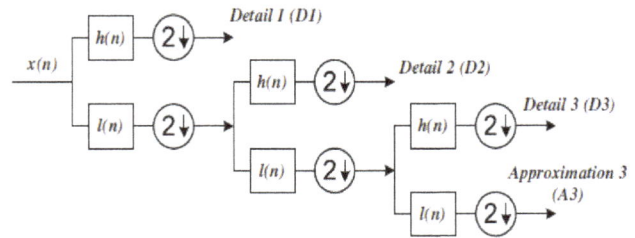

Figure 3. *Wavelet multi-resolution analysis*

4. Artificial Neural Network

Artificial Neural Networks (ANN) are simplified to imitate central nervous system been motivated by the computing performed by human brain. ANN is defined in and as a data processing system consisting of a large number of simple highly interconnected processing elements (artificial neuron) in architecture inspired by the structure of cerebral cortex of the brain [19], [20].

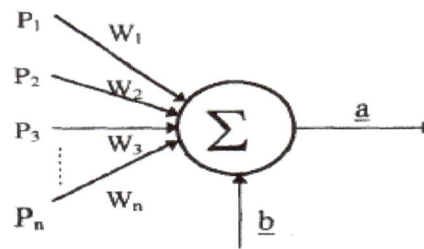

Figure 4. *Perceptron representation*

Once trained, a network response can be, to a degree, insensitive to minor variations in its input. This ability to see through noise and distortion to the pattern that lies within is vital to pattern recognition a real world environment [21], [22]. The neuron is the nervous cell and is represented in the ANN universe as a perceptron. "Fig. 4" shows a simple model of a neuron characterized by a number of inputs P, P2, ..., PN, than weights WI, W2, ...Wn, the bias adjust b and an output a. The neuron uses the input, as well as the information on its current activation state to determine the output a, given as in (4).

$$a = \sum_{k=1}^{n} W_k P_k + b \quad (4)$$

4.1. Feed Forward Neural Network

Feed-forward neural networks can be classified in a single layer or multilayer feed-forward neural networks. Multilayer FNN architecture comprises of input-layer(X); hidden-layer (V); and output-layer (Y); as shown in "Fig. 5" [19], [23].

The algorithm gives a prescription for changing the weights in any feed forward network to learn a training set of input-output pairs. The use of the bias adjust in the ANNs is optional, but the results may be enhanced by it. A multilayer network with one hidden layer is shown in "Fig. 5". This network consists of a set of N input units (Xi, i = 1 ... N), a set of p output units (Yp, p = 1 ... P) and a set of J hidden units (Vj, j = 1 ... J). Thus, the hidden unit Vj receives a net input and produces the output.

$$Vj = F\{\sum_{k=1}^{n} W_{jk} X_k \} \tag{5}$$

Where j= 1....J

The final output is then produced:

$$Yp = F\{\sum_{m=1}^{j} W_{pm} V_{km} \} \tag{6}$$

Where p= 1...P

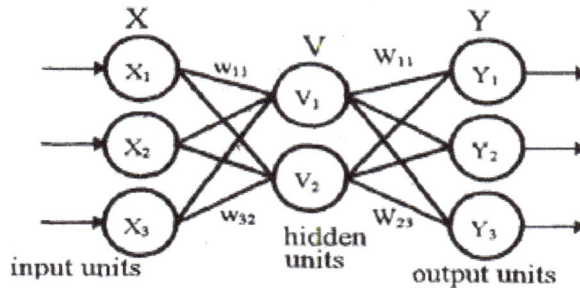

Figure 5. *Multilayer Feed-forward neural network architecture*

5. Gaussian Process

Gaussian process is a supervised learning technique that has been used for regression and classification [24]. The GP models are probabilistic, non-parametric models based on the principles of Bayesian probability [25]. It governs the properties of functions and is fully specified by a mean and a covariance functions. It is based on assigning a prior in the form of a multivariate Gaussian density that imposes a smoothness constraint on the underlying function. For the classification problem this underlying function is the posterior probability [26].

Given a training set S of n observations, S = {(xi, yi) |i = 1...n} where xi denotes an input vector of dimension D and yi denotes the target class of the corresponding input sample i. X refers to the matrix of all the training samples, y denotes the vector of the class labels for all the training samples and f represents the vector of the prior latent functions for all the training samples. One would like to predict the class

membership probability to a test sample x*. This is achieved by obtaining the distribution of the latent function of the test sample f* given the class memberships of the training samples. Since the underlying function corresponds to the posterior probability of class 1, the unrestricted latent function is passed through a squashing function in order to map its value into the unit interval.

The Gaussian process is specified by an a priori multivariate distribution for the latent functions of the training and testing samples. This distribution has a covariance function that ensures that the latent functions of near-by samples are closely correlated. On the other hand, their covariance decreases with increasing the distant between their data samples; this is controlled by hyper-parameters that need to be estimated. During the training phase, the mean and the covariance of the latent function are calculated for each training sample using the algorithms in [26]. The probability that the test sample x* belongs to class 1 is calculated as:

$$P(y *= 1|X,y,x *) = \int_{-\infty}^{\infty} p(y *= 1|f *)p(f * |X,y,x *)df * \tag{7}$$

P (y* = 1|f*) is evaluated using sigmoid activation function:

$$P(y *= 1|f *) = \sigma(y * f *) \tag{8}$$

Substituting in (7), we get

$$P(y *= 1|X,y,x *) = \int_{-\infty}^{\infty} \sigma(y * f *)p(f * |X,y,x *)df * \tag{9}$$

Where

$$P(f *|X,y,x *) = \int_{-\infty}^{\infty} p(f * |f,X,x *)p(f|X,y)df \tag{10}$$

6. Simulations and Results

The proposed classification scheme is implemented on MATLAB software. It is trained and tested using measurements of three-phase voltage and current samples obtained from the PSCAD/EMTDC. "Fig. 6" shows the 220 kV, 50 Hz simulated system one-line diagram. "Fig. 7" shows 3ph voltages and currents waveforms for single phase to ground fault at 50% of first line section at fault resistance 10 ohm and fault inception angle 45.

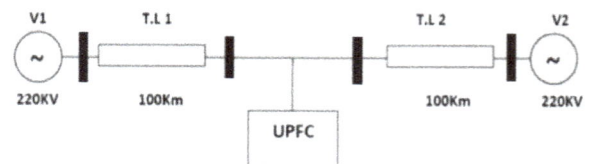

Figure 6. *Transmission system with UPFC*

The samples are analysed using DWT, the mother wavelet considered is Daubechies (db)4. The behaviour of the DWT for actual fault current and voltage waveforms is illustrated in "Fig. 8" for single phase to ground fault at 50% of first line length at fault resistance 10 ohm and fault inception angle 45.

All possible fault types are simulated at fault resistance 0, 10, 20 ohm and fault inception angle 0, 45, 90. Fault locations at 10%, 20%, 40%, 50%, 70%, 80%, and 100% from the length of each line are taken for the training process of ANN & GP. The test will begin with fault occurrence simulation at distance 30%, 60 and 90% from length of each line section.

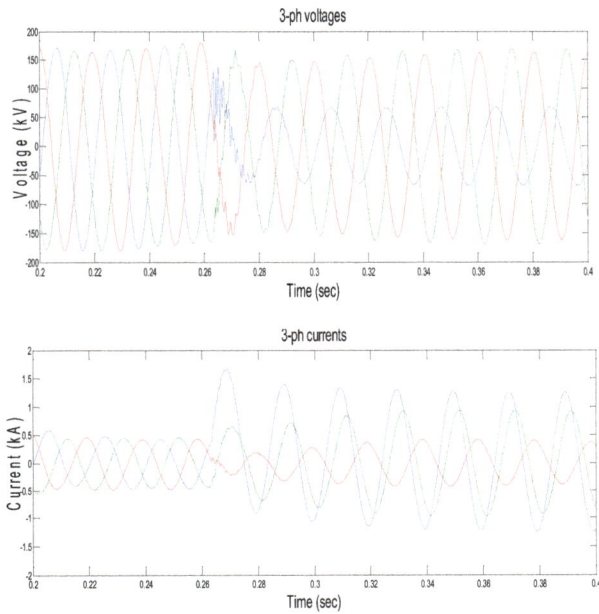

Figure 7. *3ph voltages and currents waveforms for 1ph to ground fault at 50% of first line section*

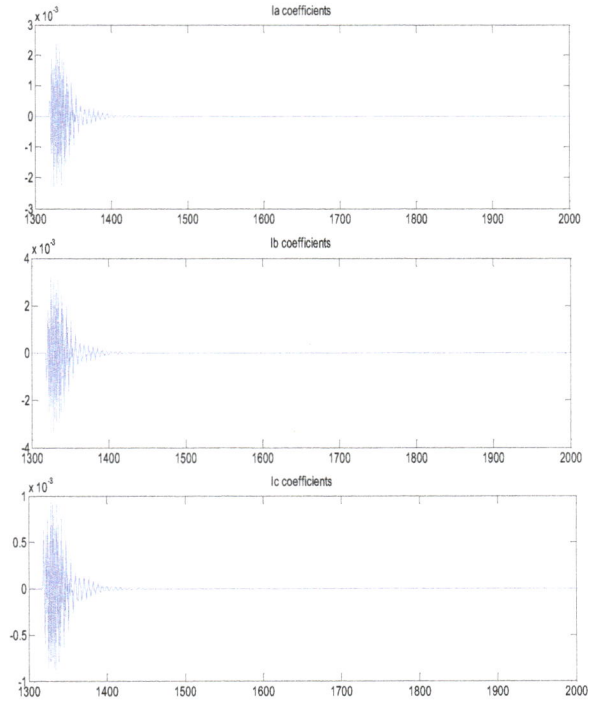

Figure 8. *Detailed coefficients of 3ph voltages and currents at time of fault*

Table 1. *Estimated output of ANN for detecting fault (Level-1)*

Fault Type	Fault Location		Desired Value	Actual Value
A-G	30%	1st line section	1	1
A-G	60%	1st line section	1	0.99
A-G	90%	1st line section	1	1
A-G	30%	2nd line section	1	1
A-G	60%	2nd line section	1	0.99
A-G	90%	2nd line section	1	1
A-B	30%	1st line section	1	1
A-B	60%	1st line section	1	1
A-B	90%	1st line section	1	1
A-B	30%	2nd line section	1	0.98
A-B	60%	2nd line section	1	0.95
A-B	90%	2nd line section	1	0.96
AB-G	30%	1st line section	1	1
AB-G	60%	1st line section	1	1
AB-G	90%	1st line section	1	1
AB-G	30%	2nd line section	1	0.99
AB-G	60%	2nd line section	1	0.98
AB-G	90%	2nd line section	1	1
ABC-G	30%	1st line section	1	1
ABC-G	60%	1st line section	1	0.99
ABC-G	90%	1st line section	1	1
ABC-G	30%	2nd line section	1	1
ABC-G	60%	2nd line section	1	1
ABC-G	90%	2nd line section	1	1

Table 2. Estimated output of GP for detecting fault (Level-1)

Fault Type	Fault Location		Desired Value	Actual Value
A-G	30%	1st line section	1	1
A-G	60%	1st line section	1	1
A-G	90%	1st line section	1	1
A-G	30%	2nd line section	1	1
A-G	60%	2nd line section	1	1
A-G	90%	2nd line section	1	1
A-B	30%	1st line section	1	1
A-B	60%	1st line section	1	1
A-B	90%	1st line section	1	1
A-B	30%	2nd line section	1	1
A-B	60%	2nd line section	1	1
A-B	90%	2nd line section	1	1

Fault Type	Fault Location		Desired Value	Actual Value
AB-G	30%	1st line section	1	1
AB-G	60%	1st line section	1	1
AB-G	90%	1st line section	1	1
AB-G	30%	2nd line section	1	1
AB-G	60%	2nd line section	1	1
AB-G	90%	2nd line section	1	1
ABC-G	30%	1st line section	1	1
ABC-G	60%	1st line section	1	1
ABC-G	90%	1st line section	1	1
ABC-G	30%	2nd line section	1	1
ABC-G	60%	2nd line section	1	1
ABC-G	90%	2nd line section	1	1

Table 3. Estimated output of ANN for classifying fault (Level-2)

	Phase A		Phase B		Phase C		Ground	
	Desired	Actual	Desired	Actual	Desired	Actual	Desired	Actual
A-G at 30%1st line	1	1	0	0	0	0	1	1
A-G at 60%1st line	1	1	0	0	0	0	1	1
A-G at 90%1st line	1	1	0	0	0	0.03	1	0.99
A-G at 30%2nd line	1	0.89	0	0	0	0	1	1
A-G at 60%2nd line	1	1	0	0	0	0	1	1
A-G at 90%2nd line	1	1	0	0	0	0	1	1
A-B at 30%1st line	1	0.99	1	0.99	0	0.001	0	0
A-B at 60%1st line	1	1	1	1	0	0	0	0
A-B at 90%1st line	1	1	1	1	0	0	0	0
A-B at 30%2nd line	1	0.97	1	1	0	0.16	0	1
A-B at 60%2nd line	1	0.96	1	1	0	0	0	0
A-B at 90%2nd line	1	1	1	1	0	0	0	0
AB-G at 30%1st line	1	1	1	1	0	0.002	1	1
AB-G at 60%1st line	1	1	1	0.99	0	0	1	0.99
AB-G at 90%1st line	1	1	1	1	0	0	1	0.99
AB-G at 30%2nd line	1	1	1	1	0	0.09	1	1
AB-G at 60%2nd line	1	1	1	1	0	0	1	1
AB-G at 90%2nd line	1	0.97	1	1	0	0.01	1	0.92
ABC-G at 30%1st line	1	1	1	1	1	1	1	0.99
ABC-G at 60%1st line	1	1	1	0.99	1	1	1	0
ABC-G at 90%1st line	1	0.98	1	0.99	1	1	1	0.88
ABC-G at 30%2nd line	1	1	1	1	1	1	1	1
ABC-G at 60%2nd line	1	1	1	1	1	1	1	1
ABC-G at 90%2nd line	1	0.96	1	1	1	0	1	1

Table 4. Estimated output of GP for classifying fault (Level-2)

Fault type & location	Phase A		Phase B		Phase C		Ground	
	Desired	Actual	Desired	Actual	Desired	Actual	Desired	Actual
A-G at 30%1st line	1	1	-1	-1	-1	-1	1	1
A-G at 60%1st line	1	1	-1	-1	-1	-1	1	1
A-G at 90%1st line	1	1	-1	-1	-1	-1	1	1
A-G at 30%2nd line	1	1	-1	1	-1	1	1	1
A-G at 60%2nd line	1	1	-1	-1	-1	1	1	1
A-G at 90%2nd line	1	1	-1	-1	-1	-1	1	1
A-B at 30%1st line	1	1	1	1	-1	-1	-1	-1
A-B at 60%1st line	1	1	1	1	-1	-1	-1	-1
A-B at 90%1st line	1	1	1	1	-1	-1	-1	-1
A-B at 30%2nd line	1	1	1	1	-1	1	-1	-1
A-B at 60%2nd line	1	1	1	1	-1	-1	-1	-1
A-B at 90%2nd line	1	1	1	1	-1	-1	-1	-1
AB-G at 30%1st line	1	1	1	1	-1	-1	1	1
AB-G at 60%1st line	1	1	1	1	-1	-1	1	1
AB-G at 90%1st line	1	1	1	1	-1	-1	1	1
AB-G at 30%2nd line	1	1	1	1	-1	1	1	1
AB-G at 60%2nd line	1	1	1	1	-1	1	1	1
AB-G at 90%2nd line	1	1	1	1	-1	-1	1	-1
ABC-G at 30%1st line	1	1	1	1	1	1	1	1
ABC-G at 60%1st line	1	1	1	1	1	1	1	1
ABC-G at 90%1st line	1	1	1	1	1	1	1	1
ABC-G at 30%2nd line	1	1	1	1	1	-1	1	-1
ABC-G at 60%2nd line	1	1	1	1	1	1	1	1
ABC-G at 90%2nd line	1	1	1	1	1	1	1	-1

Table 1 and Table 2 show some parts of calculations of test results for artificial neural network and Gaussian process at fault resistance 10 ohm and fault inception angle 45 for detecting faults (Level-1). The output of ANN for level-1 is either 0 or 1 indicating that there is a fault or not and for GP is either 1 indicating a fault or -1 indicating no fault condition.

Table 3 and Table 4 show some parts of calculations of test results for ANN and GP at fault resistance 10 ohm and fault inception angle 45 for classifying faults (Level-2).

The neural network achieves higher test accuracy than the Gaussian process method. The test accuracy in detection level reached 100% for ANN & 100% for GP. The overall test accuracy in classification level reached 96.2% for ANN & 90% for GP.

7. Conclusion

The use of an ANN and GP as a pattern classifier to improve the performance of distance relay in transmission system compensated with FACTS is discussed in this paper. The neural network achieves higher test accuracy than the Gaussian process method. The proposed approach is designed to detect the faults, to classify the fault type, and to identify the faulted phase. The wavelet transform provides an efficient way to extract signal components at different frequency bands.

References

[1] John J. Paserba, "How FACTS Controllers Benefit AC Transmission Systems", Transmission and Distribution Conference and Exposition, 2003 IEEE PES, Vol. 3, pp. 949-956, September 2003.

[2] John Wiley, "FACTS Modeling and Simulation in Power Networks", 2004.

[3] T. Manokaran, V.Karpagam, "Performance of Distance Relay Estimation in Transmission Line with UPFC", International Journal of Computer and Electrical Engineering, Vol. 2, No. 1, pp. 158-161, February 2010.

[4] Pavlos S. Georgilakis, Peter G. Vernados, "Flexible AC Transmission System Controllers", Materials Science Forum Vol. 670, pp 399-406, 2011.

[5] Lijun Cai, "Robust Coordinated Control of FACTS Devices in Large Power Systems", published by Logos Verlag Berlin 2004.

[6] Tomasz Okon and Kazimierz Wilkosz, "Influence of UPFC Device on Power System State Estimation", PowerTech IEEE, pp. 1-8, June 2011.

[7] Tomasz Okon, Kazimierz Wilkosz," Consideration of Different Operation Modes of UPFC in Power System State Estimation", Environment and Electrical Engineering (EEEIC) 10th International Conference, pp. 1-4, May 2011.

[8] Bo Hu, Kaigui Xie, Rajesh Karki, "Reliability Evaluation of Bulk Power Systems Incorporating UPFC", IEEE 2010.

[9] C. D. Schauder, L. Gyugyi, M. R. Lund, D. M. Hamai, T. R. Rietman, D. R. Torgerson, and A. Edris, "Operation of The Unified Power Flow Controller (UPFC) Under Practical Constrains", IEEE Transactions on Power Delivery, Vol. 13, No. 2, pp. 630-639, April 1998.

[10] Nampetch P., S.N. Singh, and Surapong C., "Modeling of UPFC and Its Parameters Selection", Power Electronics and Drive Systems 4th IEEE International Conference, Vol. 1, pp. 77-83, October 2011.

[11] Ali Ajami, S.H. Hosseini, and G.B. Gharehpetian, "Modelling and Controlling of UPFC for Power System Transient Studies", ECTI Transactions on Electrical Eng., Electronics, and Communications, Vol. 5, No. 2, pp. 29-35, August 2007.

[12] Mehrdad Ahmadi Kamarposhti, Mostafa Alinezhad, Hamid Lesani, Nemat Talebi, "Comparison of SVC, STATCOM, TCSC, and UPFC Controllers for Static Voltage Stability Evaluated by Continuation Power Flow Method", IEEE Electrical Power & Energy Conference, pp. 1-8, October 2008.

[13] P. K. Dash, A. K. Pradhan, G. Panda, A. C. Liew, "Digital Protection of Power Transmission Lines in The Presence of Series Connected FACTS Devices", IEEE, Vol. 3, pp. 1967-1972, January 2000.

[14] P.K. Dash, A.K. Pradhan, G. Panda, "Distance protection in the presence of unified power flow controller", Electric Power Systems Research 54, pp. 189–198, 2000.

[15] N.Zhang, M.Kezunovic, "Transmission Line Boundary Protection Using Wavelet Transform and Neural Network", IEEE Transactions on Power Delivery, Vol. 22, Issue 2, pp. 859-869, April 2007.

[16] Sriya Chakraborty, Shalini Singh, Anu Bhalla, Pallavi Saxena, Ramesh Padarla, "Wavelet Transform Based Fault Detection and Classification in Transmission Line", International Journal of Research in Engineering & Applied Sciences, ISSN: 2249-3905, Vol. 2, Issue 5, May 2012.

[17] Abdulhamid A. Abohagar, M.W.Mustafa, "Back Propagation Neural Network Aided Wavelet Transform for High Impedance Fault Detection and Faulty Phase Selection", IEEE International Conference on Power and Energy (PECon), pp. 790-795, December 2012.

[18] Francisco Martín, José A. Aguado, "Wavelet-Based ANN Approach for Transmission Line Protection", IEEE Transactions on Power Delivery, Vol. 18, No. 4, pp. 1572-1574, October 2003.

[19] Anant Oonsivilai, Sanom Saichoomdee, "Appliance of Recurrent Neural Network toward Distance Transmission Lines Protection", IEEE, pp. 1-4, January 2009.

[20] Janison R. de Carvalho, Denis V. Coury, Carlos A. Duque, David C. Jorge, "Development of Detection and Classification Stages for a New Distance Protection Approach Based on Cumulants and Neural Networks", Power and Energy Society General Meeting IEEE, pp. 1-7, July 2011.

[21] A.P.Vaidy, Prasad A. Venikar, "ANN Based Distance Protection of Long Transmission Lines by Considering the Effect of Fault Resistance", IEEE - International Conference On Advances In Engineering, Science And Management, pp. 590-594, March 2012.

[22] D.V Coury, D.C Jorge, "Artificial Neural Network Approach to Distance Protection of Transmission Lines", IEEE Transactions on Power Delivery, Vol. 13, No. 1, pp. 102-108, January 1998.

[23] E.A. Feilat and K. AI-Tallaq, "A New Approach For Distance Protection Using Artificial Neural Network", Universities Power Engineering Conference UPEC, Vol. 1, pp. 473-477, September 2004.

[24] Hannes Nickisch, Carl Edward Rasmussen, "Approximations for binary Gaussian process classification", Journal of Machine Learning Research 9, pp. 2035–2078, 2008.

[25] Pavle Boškoski, Matej Gašperin, Dejan Petelin, "Bearing fault prognostics based on signal complexity and Gaussian process models", IEEE conference, 2012.

[26] C. E. Rasmussen, C. K. I. Williams, "Gaussian Processes for Machine Learning", MIT Press, Cambridge 2006.

23

Assessment of a viability of wind power in Iraq

Osama Tarek Al-Taai[1], Qassim Mahdi Wadi[2], Amani Ibraheem Al-Tmimi[1]

[1]Department of Atmospheric Sciences, College of Science, Al-Mustansiriyah University, Baghdad, Iraq
[2]Al-MamonUniversity College, Department of Electrical Power Technical Engineering, Baghdad, Iraq

Email address:

Aus_tar77@yahoo.com (O. T. Al-Taai), qassiimwadi2014@gmail.com (Q. M. Wadi), d.amani_altmimi@yahoo.com (A. I. Al-Tmimi)

Abstract: Wind energy is now being used in almost every country of the world as an important and pollution free renewable source of energy. This study deals with the feasibility of utilizing winds in generating electricity in Iraq by using the daily average wind speed data since (1/4/2011 to 1/4/2012) for fifteen stations from different regions have been selected for this purpose and for three height (12,50,100 m)and applied it with wind turbine (3Kw). The wind power and the wind power density for the selected stations have been calculated and the maximum values at Basrah (Albrjsuh), the medium values at Baghdad (Abughraib) and the lowest values in northern region Mosul (Bashiqah). Byusing the graphical method to calculate the Weibull distribution .the highest values foe Weibull parameters at Basrah (Albrjsuh), the medium values at Baghdad (Abughraib) and the lowest values in Mosul (Bashiqah). The regression statistical analysis correlation was found for all stations and the most effective values was in at Basrah (Albrjsuh), the medium values at Baghdad (Abughraib) and the lowest values was Mosul (Bashiqah).

Keywords: Wind Power, Average Daily Wind Speed, Graphical Method, Regression Statistical Analysis, Iraq

1. Introduction

Renewable energy refers to energy resources that occur naturally and repeatedly in the environmental where it can be harnessed for human benefit that poses no major environ mental problems. Wind power is one of leading ones among the new, renewable, clean and cheap resource. It describes the processes by which the wind is used to generate mechanical power or electricity by wind turbines which convert the kinetic energy in wind in to mechanical power. This mechanical can be used for specific tasks such as grinding grain, pumping water, or can convert this mechanical power by generator in to electricity. The total quality of this source is extremely large and various with time at any given location [1]. Iraq is an oil country, but this does not prevent the possession of other resources are inexhaustible renewable energies such as solar, hydro and wind power. This research includes wind characteristics and assesses the wind power in Iraq, using the distribution of Weibull and the possibility of using it for generating electricity.

1.1. The Wind Power

Calculations of wind power are derived from the equation for kinetic energy (*KE*), which is:

$$KE = \frac{1}{2}mv^2 \qquad (1)$$

Where *m* is mass, *v* is acceleration, Air mass is equal to the product of its density and volume. Volume is dependent on the area through which the air is passing, the speed with which it is moving, and the amount of time it travels. Air mass can therefore be calculated as:

$$m = \rho.A.v.t \qquad (2)$$

Where ρ is the air density (1.225 kg/m3 dry air at 1 atm. and 15[0] C), *A* is the area through which the air passes, *v*is wind speed and *t* is time. Since power is energy divided by time, the equation for wind power (*WP*) can be written as [2]:

$$WP = \frac{1}{2}\rho.A.v^3 \qquad (3)$$

1.2. Wind Power Density

Wind power density is the amount of wind power available per unit of area perpendicular to the wind flow. In practice, wind power density is used to estimate the potential electrical output of a wind farm once the area swept by wind turbine rotors and the power system efficiency are known. The equation for wind power density (*WPD*) is simply the wind power (equation3) divided by the area, or [2]:

$$WPD = \frac{1}{2}\rho v^3 \qquad (4)$$

The best wind capacity is the density of wind capacity (Wind Power Density), because it will gives a clear picture of how to distribute wind speed on average (mean) and this quantity can be estimated in practice by using (Weibull Distribution) which it depends on the two parameter (*c*) scale and (*k*) shape parameters, where ΓGamma function [3,4]:

$$WPD = \frac{1}{2}\rho c^3 \Gamma\left(\frac{1+3}{k}\right) \qquad (5)$$

1.3. Vertical Estimation of Wind Speed Using the Power-Law Model

As described above, winds are slowed by friction at the earth's surface, so that wind speeds tend to be greater at higher elevations. For regions with relatively level terrain and little vegetation, the method most commonly used to obtain this extrapolation is the 1/7 power-law model. The equation of the 1/7 power-law model is:

$$\frac{v(Z)}{v(Z_0)} = \left(\frac{Z}{Z_0}\right)^{\frac{1}{7}} \qquad (6)$$

Where Z is the height at which the wind speed is to be estimated, $v(Z)$ is the wind speed to be estimated, and Z_0 and are $v(Z_0)$ the reference height and wind speed, respectively [5].

1.4. Frequency Distribution of Wind Speed

The wind speed probability density distributions and their functional forms represent the major aspects in wind related literature. The probability distributions most commonly used are those of Weibull and Rayleigh [6]. The Weibull distribution has been found to fit a wide collection of recorded wind data. The variations in wind velocity are characterized by the two functions; the probability density function and the cumulative distribution function. Obtaining the Weibull density distribution is necessary to determine the shape (*k*) and scale(*c*) parameters, the common methods for determining k and care: graphical, standard deviation, moment, maximum likelihood and energy pattern factor methods.

In this research we applied the two methods, Graphical which is one of the most reliable ways to get real results close to the observed, and a new method called energy pattern factor method (*EPF*).The concept of this method is useful in calculating the available energy in the wind along with the knowledge of the annual or monthly wind speed. It is also useful in choosing a location with limited wind data,

because long-term data from neighboring sites can be correlated with one-site short-term measurements.

Graphical method is another way to determine the k and c from Weibull distribution [7]. We transform the cumulative distribution function into a linear form, adopting logarithmic scales. The expression for the cumulative distribution of wind velocity can be rewritten as:

$$1 - F(v) = e^{-(v/c)^k} \qquad (7)$$

Considering the logarithm twice, we get

$$ln\{-ln[1 - F(v)]\} = kln(v) - klnc \qquad (8)$$

Where $F(v)$ the probability density function, k the shape parameter, c the scale parameter, By Plotting different values of ln[-ln(1-F(v)] vs. ln (v) ,a straight line is fitted to the points . The slope of line is, k and the intercept on the ln[-ln(1-F(v)] axis is $-klnc$. The scale parameter:

$$c = e^{\left(-\frac{y-intercept}{k}\right)} \qquad (9)$$

2. Materials and Methodology

2.1. Study Area and Data Collection

The data used in this study are the monthly average of daily values of wind speed at 3m was obtained from ministry of agriculture for locations distributed in different regions in Iraq [8], as shown in figure (1), the stations are [Mosul (Bashiqah), Kirkuk (Daquq), Salahaldeen (Tikrit), Dealaa (Khalis), Baghdad (Abughraib), Anbar (Aldawar), Wasat (Essaouira), Karbla - Lake Razzaza, Babel (Kifli), Qadisiyah (Dewaneia), Najaf (Mashkhab), Mayssan (Ekhala), The-Qar (Shatrah), Basrah (Albrjsuh)] as shown in table(1).

Table 1. Stations Sequences from North to South.

Stations Sequences from N to S	Latitude (N)	Longitude (E)	Sea Level Altitude in Meters
Mosul (Bashiqah)	36 45 09	43 33 88	228.0
Kirkuk (Daquq)	35 16 93	44 42 97	227.9
Salahaldeen (Tikrit)	34 65 28	43 63 61	116.0
Dealaa (Khalis)	33 75 17	44 62 22	40.9
Baghdad (Abughraib)	33 32 21	44 23 93	30.4
Anbar (Aldawar)	33 27 57	43 02 08	45.6
Wasat (Essaouira)	33 00 45	44 49 27	27.2
Karbla (Lake Razzaza)	32 33 20	43 58 37	49.0
Babel (Kifli)	32 30 63	44 39 16	21.1
Qadisiyah (Dewaneia)	32 01 93	44 89 85	24.0
Najaf (Mashkhab)	31 53 27	44 30 11	19.0
Mayssan (Ekhala)	31 48 04	47 11 25	9.0
The-Qar (Shatrah)	31 45 57	46 19 52	7.0
Muthna (Khader)	30 17 32	45 37 51	7.0
Basrah (Albrjsuh)	30 17 32	47 04 04	7.0

Figure 1. *Map of Selected Stations in Iraq.*

By using eq. (6) we calculated the wind speed in three high (12, 50, 100 m) as shown in figures (2, 3, 4) and table (2).

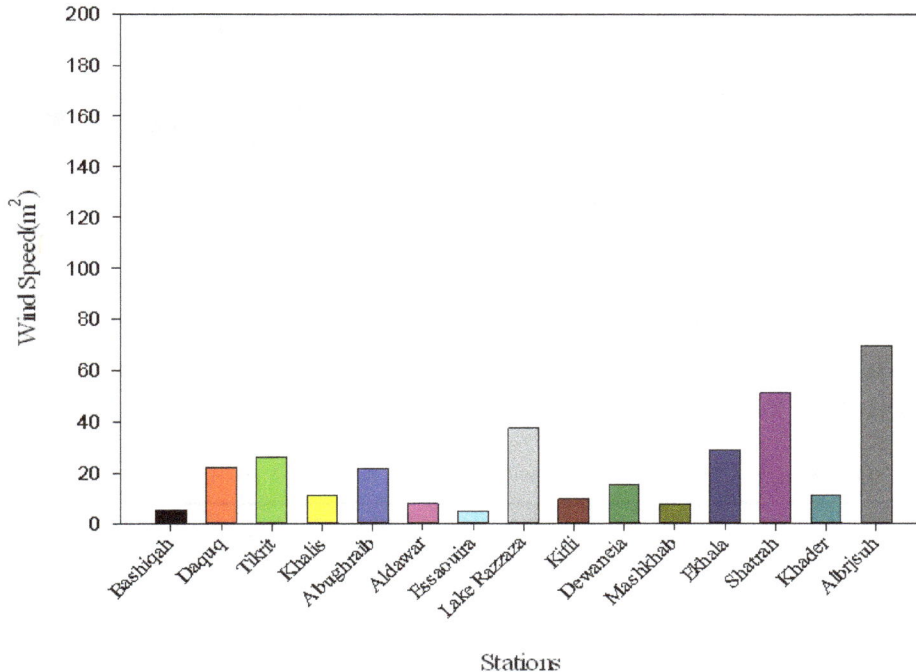

Figure 2. *Wind Speed for Selected stations at (12 m).*

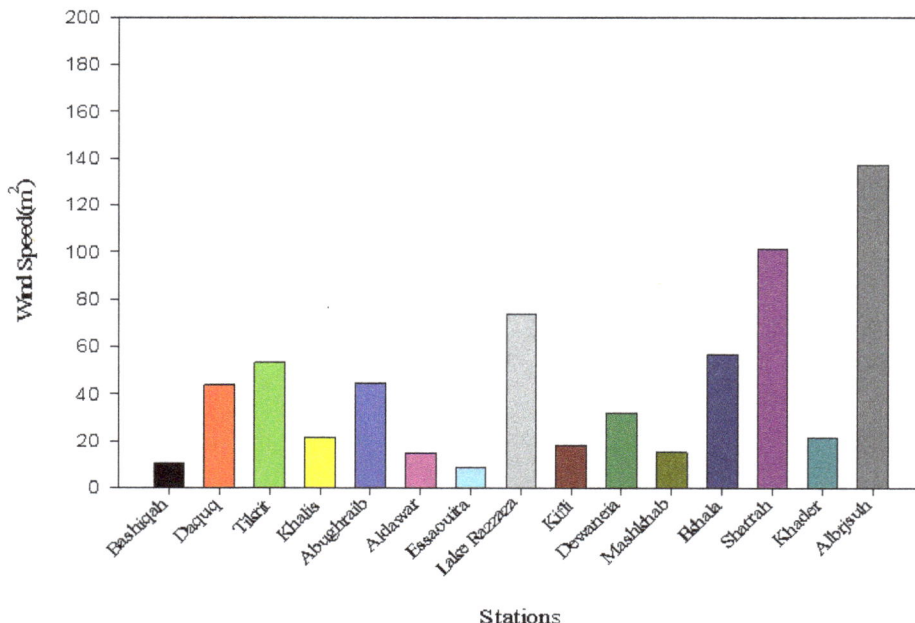

Figure 3. *Wind Speed for Selected stations at (50 m).*

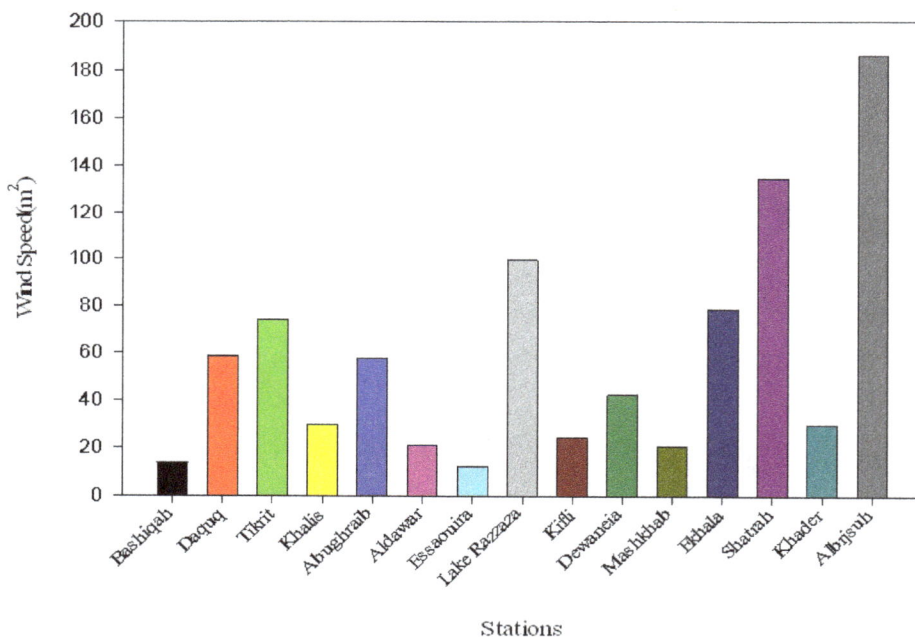

Figure 4. *Wind Speed for Selected stations at (100 m).*

By using equations (8 and 9) we calculate the Weibull distribution parameters k- Shape parameter and c (m/sec)-scale parameter in three high (12, 50, 100 m) as shown in table (3).

Table 2. *Mean Wind Speed $\bar{x}$ (m/s) and Stander Deviation (S.D.) at high (12, 50, 100 m) for all Selected Stations.*

Stations	Mean and Stander deviation	High (12 m)	High (50 m)	High (100 m)
Basrah (Albrjsuh)	$\bar{x}$	4.18	5.298	5.825
	S.D.	1.75	2.12	2.39
Karbla (Lake Razzaza)	$\bar{x}$	1.59	2.014	2.199
	S.D.	0.85	0.98	1.12
Baghdad (Abughraib)	$\bar{x}$	2.049	2.541	2.845
	S.D.	0.80	0.99	1.1
Kirkuk (Daquq)	$\bar{x}$	2.917	3.705	3.978

Stations	Mean and Stander deviation	High (12 m)	High (50 m)	High (100 m)
	S.D.	1.11	1.34	1.62
Dealaa (Khalis)	$\bar{x}$	2.224	2.736	3.037
	S.D.	0.989	1.283	1.437
Mayssan (Ekhala)	$\bar{x}$	2.922	3.691	4.107
	S.D.	1.511	1.85	2.079
Muthna (Khader)	$\bar{x}$	3.873	4.88	5.374
	S.D.	1.44	1.79	1.96
Babel (Kifli)	$\bar{x}$	2.108	2.629	2.903
	S.D.	0.93	1.164	1.29
The-Qar (Shatrah)	$\bar{x}$	2.239	2.783	3.11
	S.D.	1.01	1.24	1.36
Anbar (Aldawar)	$\bar{x}$	2.837	3.601	3.977
	S.D.	1.18	1.51	1.57
Wasat (Essaouira)	$\bar{x}$	3.401	4.270	4.726
	S.D.	1.422	1.770	1.940
Salahaldeen (Tikrit)	$\bar{x}$	3.047	3.821	4.268
	S.D.	1.23	1.600	1.700
Qadisiyah (Dewaneia)	$\bar{x}$	2.437	3.08	3.371
	S.D.	1.13	1.49	1.66
Mosul (Bashiqah)	$\bar{x}$	1.77	2.243	2.244
	S.D.	0.72	0.90	0.99
Najaf (Mashkhab)	$\bar{x}$	1.82	2.299	2.537
	S.D.	1.00	1.27	1.40

By using equations (3 and 5) we calculated the wind power and wind power density by Weibull distribution at high (12, 50,100 m) for all selected stations as shown in table (4)and figure (5), we also chosen the wind turbine 3kw [9]see table(5).

Table 3. Weibull Parameters at high (12, 50, 100 m) for all Selected Stations.

Stations	Weibull Parameters	High (12 m)	High (50 m)	High (100 m)
Basrah (Albrjsuh)	k	2.561	2.688	2.609
	c	4.709	5.958	6.557
Karbla (Lake Razzaza)	k	1.947	2.156	2.054
	c	1.795	2.274	2.482
Baghdad (Abughraib)	k	2.751	2.763	2.786
	c	2.303	2.855	3.196
Kirkuk (Daquq)	k	2.84	3.007	2.639
	c	3.27	4.149	4.477
Dealaa (Khalis)	k	2.395	2.258	2.235
	c	2.509	3.089	3.429
Mayssan (Ekhala)	k	2.024	2.084	2.072
	c	3.298	4.167	4.636
Muthna (Khader)	k	2.908	2.96	2.98
	c	4.343	5.476	6.019
Babel (Kifli)	k	2.416	2.407	2.397
	c	2.378	2.965	3.275
The-Qar (Shatrah)	k	2.357	2.388	2.426
	c	2.526	3.14	3.507
Anbar (Aldawar)	k	2.576	2.545	2.728
	c	3.195	4.057	4.47
Wasat (Essaouira)	k	2.566	2.578	2.614
	c	3.831	4.809	5.320
Salahaldeen (Tikrit)	k	2.649	2.561	2.548
	c	3.428	4.304	4.802
Qadisiyah (Dewaneia)	k	2.267	2.172	2.134
	c	2.751	3.478	3.806
Mosul (Bashiqah)	k	2.607	2.684	2.651
	c	1.993	2.523	2.651
Najaf (Mashkhab)	k	1.875	1.878	1.882
	c	2.05	2.59	2.651

Table 4. *Wind Power Density by Weibull Distribution (w/m²) and Wind Power at high (12, 50, 100 m) for all Selected Stations.*

Stations	WPD, WP in (w/m²)	High (12 m)	High (50 m)	High (100 m)
Basrah (Albrjsuh)	WPD by Weibull dis.	69.7	137.2	186
	WP	1181.50	2355.01	3169.61
Karbla (Lake Razzaza)	WPD by Weibull dis.	4.9	8.9	12.2
	WP	82.48	164.41	221.29
Baghdad (Abughraib)	WPD by Weibull dis.	7.8	14.9	20.8
	WP	136.95	272.99	367.41
Kirkuk (Daquq)	WPD by Weibull dis.	22	43.9	58.5
	WP	392.69	782.72	1053.47
Dealaa (Khalis)	WPD by Weibull dis.	11	21.5	29.7
	WP	196.52	391.72	527.22
Mayssan (Ekhala)	WPD by Weibull dis.	29	56.7	78.6
	WP	503.51	1003.61	1350.77
Muthna (Khader)	WPD by Weibull dis.	51.1	101.6	134.5
	WP	836.52	1667.39	2244.15
Babel (Kifli)	WPD by Weibull dis.	9.7	18.1	24.5
	WP	154.19	307.34	413.65
The-Qar (Shatrah)	WPD by Weibull dis.	11.4	21.7	29.8
	WP	199.58	397.82	535.43
Anbar (Aldawar)	WPD by Weibull dis.	21.7	44.7	57.5
	WP	375.94	749.35	1008.55
Wasat (Essaouira)	WPD by Weibull dis.	37.5	73.9	99.2
	WP	623.8156	1243.4113	1673.50
Salahaldeen (Tikrit)	WPD by Weibull dis.	26.3	53.2	73.9
	WP	437.5615	872.1629	1173.84
Qadisiyah (Dewaneia)	WPD by Weibull dis.	15.2	31.8	42.3
	WP)	274.30	546.75	735.88
Mosul (Bashiqah)	WPD by Weibull dis.	5.2	10.4	13.7
	WP	84.95	169.33	227.91
Najaf (Mashkhab)	WPD by Weibull dis.	7.6	15.2	20.4
	WP	127.66	254.46	342.48

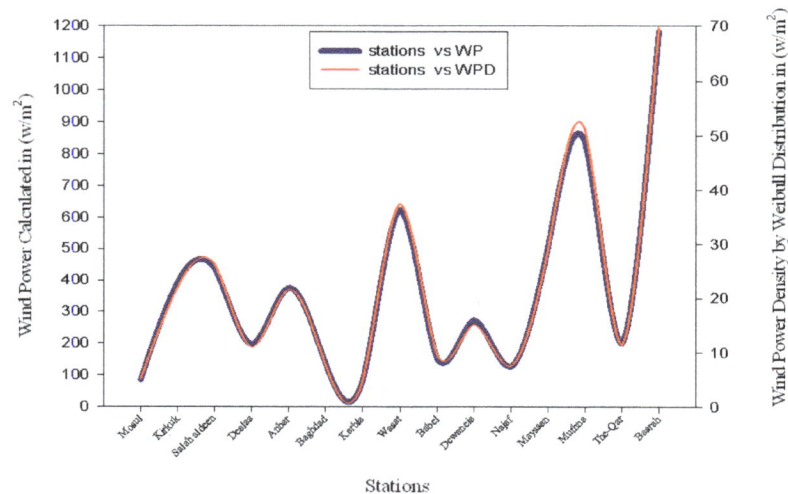

Figure 5. *Compare between Wind Power calculated and Wind Power Density by Weibull Distribution.*

Table 5. *Some Specifications of Wind Turbine (3kw).*

Nominal Power	3 kw
Rotor diameter	4.5 m
Area swept	15.89 m^2
Maximum power	4000w
Rated output voltage	240v
Rated wind speed	10 m/s
Start wind speed	2 m/s
Maximum wind speed	45 m/s
Tower height	12 m
Number of blades	3
Rated rooter speed	220 r/m

Also in this paper we have been employed the relation wind speed and wind power with the time by illustrating a Parameters Liner Regression statistical analysis formula with first order type as show below [10].

$$f = y_o + ax \qquad (10)$$

Where f is the function for wind speed or wind power calculated, a is the regression slope, y_o is the regression constant (equation 11), x is the number of day. As shown in figures (6, 7 and 8)

$$y_o = \frac{\sum_{i=1}^{n}(xi-x)-(fi-f)}{\sum_{i=1}^{n}(xi-x)^2} \qquad (11)$$

And for this reason we selected three stations represent north middle and south [Mosul (Bashiqah), Baghdad (Abughraib) and Basrah (Albrjsuh)] in Iraq as shown in tables (6 and 7).

Table 6. *Parameters Liner Regression for Wind Speed (WS) at high (12, 50, 100 m).*

Stations	Parameters Liner regression	Pearson correlation	High (12 m)	High (50 m)	High (100 m)
		R	0.298	0.298	0.298
Basrah (Albrjsuh)	y_o		5.156	6.489	7.164
	a		-0.05	-0.006	-0.007
		R	0.224	0.224	0.224
Baghdad (Abughraib)	y_o		2.379	2.994	3.306
	a		-0.0018	-0.0023	-0.0025
		R	0.322	0.322	0.322
Mosul (Bashiqah)	y_o		2.185	2.750	3.037
	a		-0.002	-0.002	-0.003

Table 7. *Parameters Liner Regression for Wind Power (WP) at high (12, 50,100 m).*

Stations	Parameters Liner regression	Pearson correlation	High (12 m)	High (50 m)	High (100 m)
		R	0.255	0.255	0.255
Basrah (Albrjsuh)	y_o		1938.35	3863.59	5200.00
	a		-4.124	-8.221	-11.064
		R	0.157	0.157	0.157
Baghdad (Abughraib)	y_o		217.25	433.03	582.81
	a		-0.47	-0.872	-1.173
		R	0.241	0.241	0.241
Mosul (Bashiqah)	y_o		130.47	260.06	350.02
	a		-0.248	-0.494	-0.665

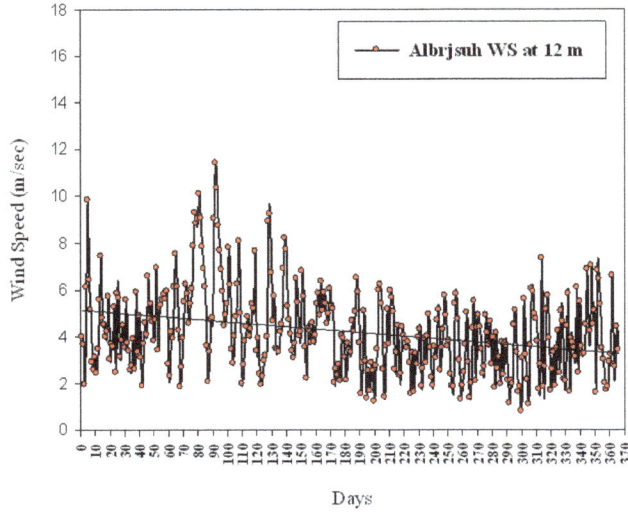

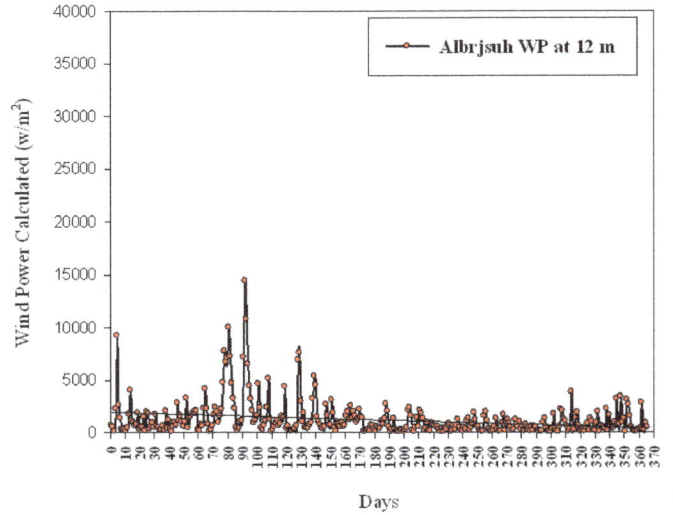

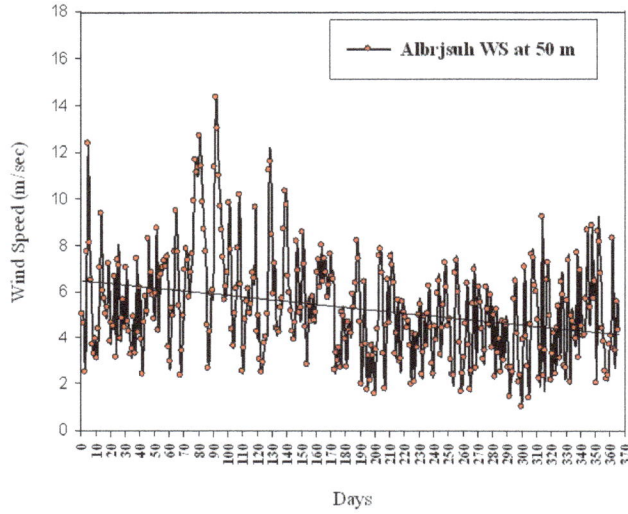

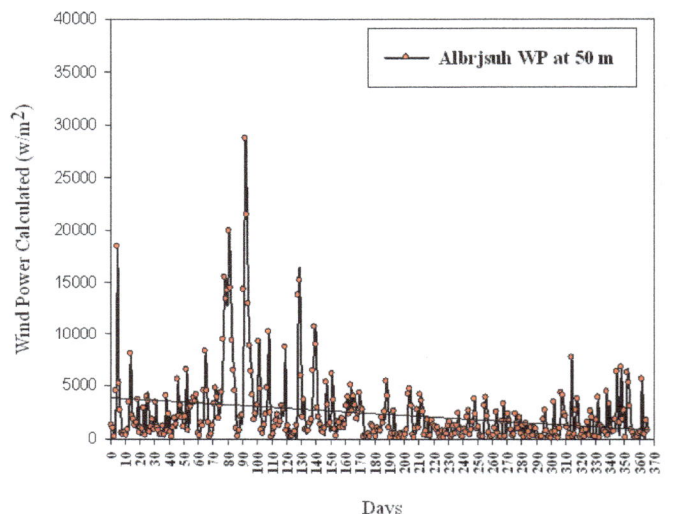

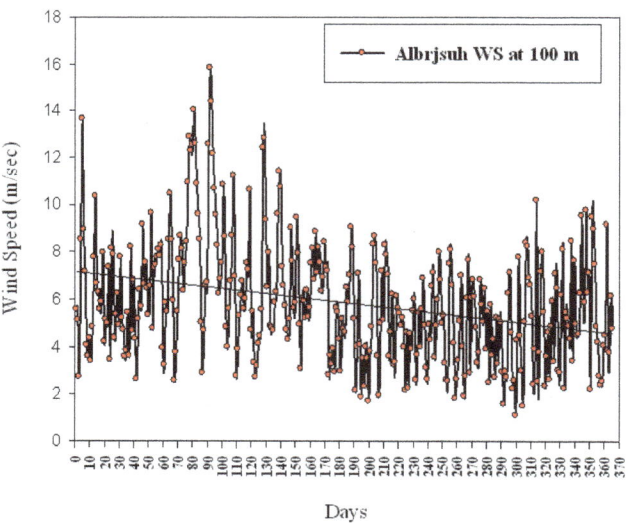

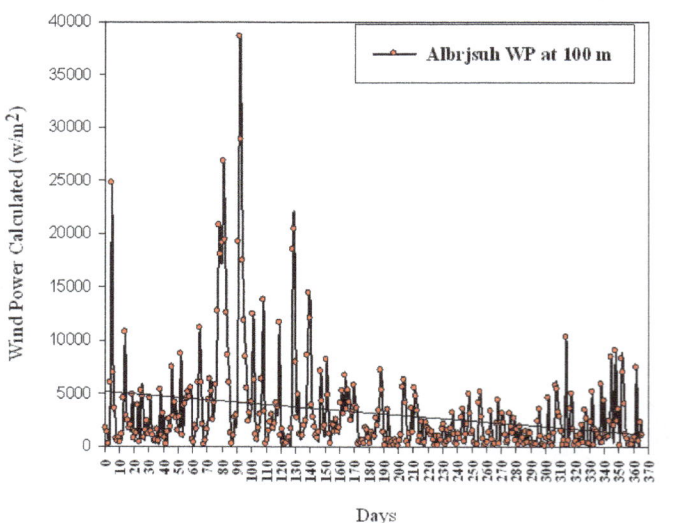

Figure 6. *The Day Average of Basrah (Albrjsuh) Station Wind Speed and Wind Power calculated at (12, 50, 100 m).*

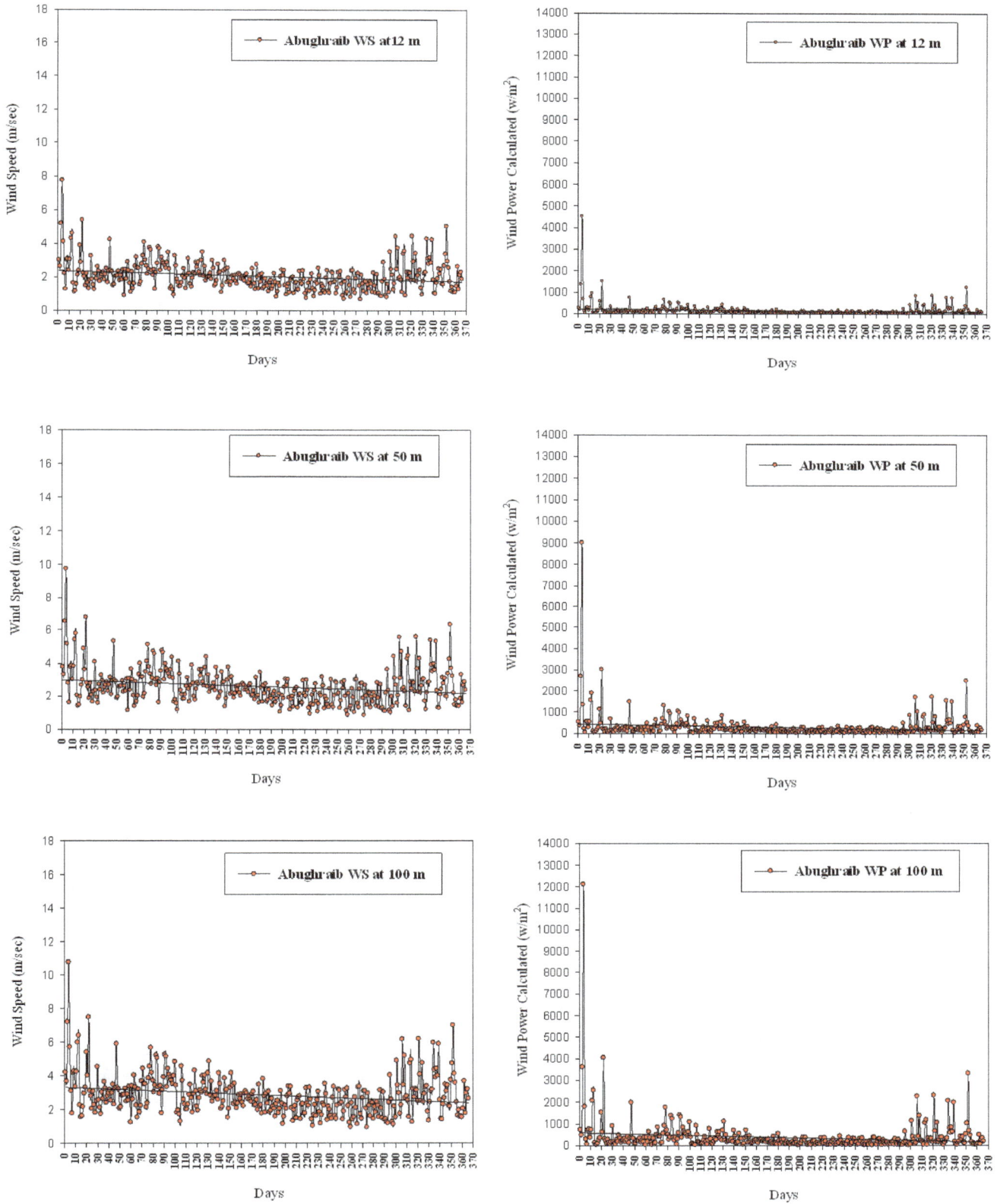

Figure 7. *The Day Average of Baghdad (Abughraib) Station Wind Speed and Wind Power calculated at (12, 50, 100 m).*

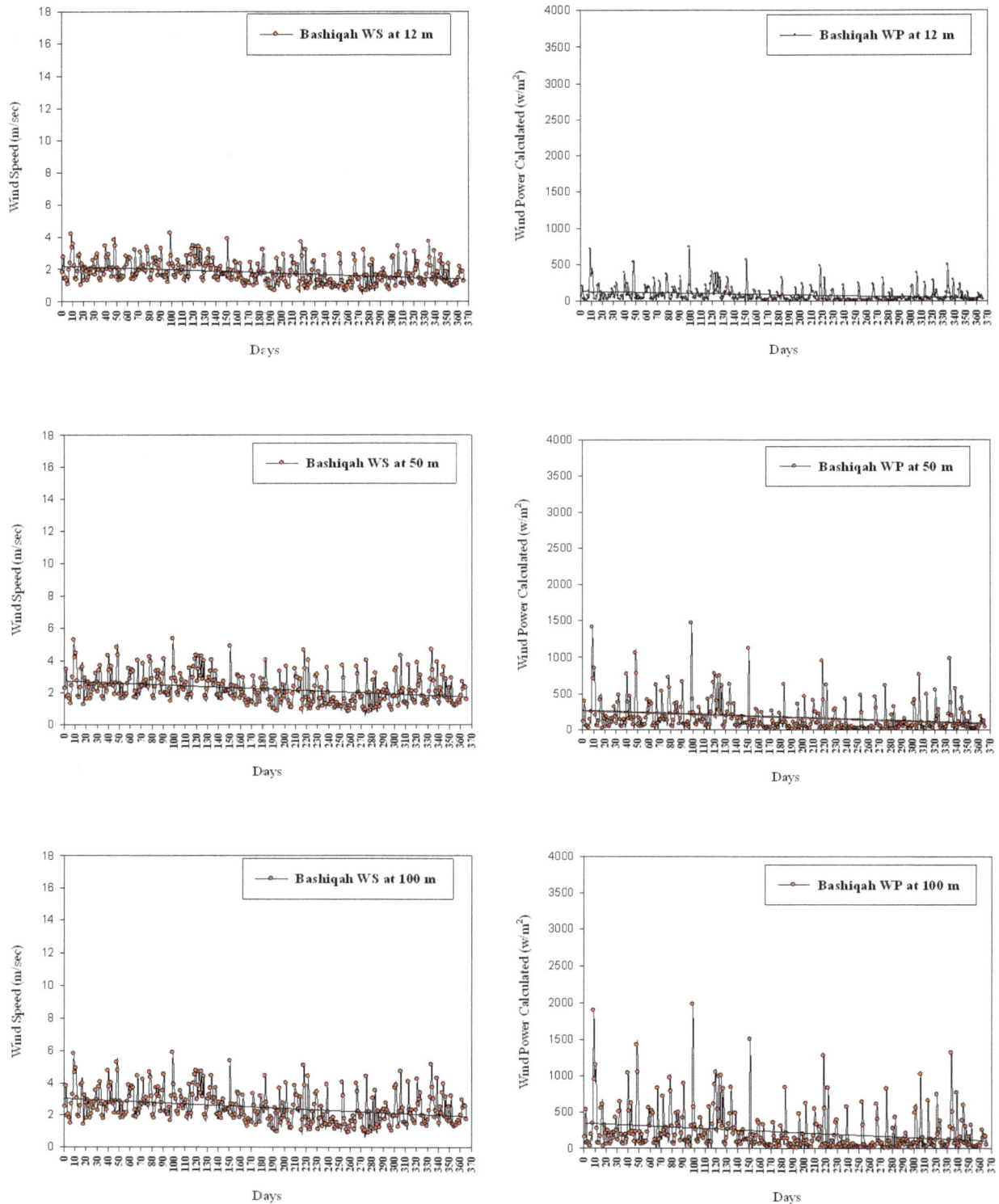

Figure 8. *The Day Average of Mosul (Bashiqah) Station Wind Speed and Wind Power calculated at (12, 50, 100 m).*

3. Result and Discussion

At all the stations under study, the wind power generation and the wind power density can be higher in the south region as the daily average wind speed are strongest during this period and for all heights. The maximum values at Basrah (Albrjsuh), the medium values at Baghdad (Abughraib) and the lowest values in northern region Mosul (Bashiqah).Medium size or small wind turbines can be used to generate electricity in those regions .The true estimation of Weibull parameters is important for the manufacturer of the wind turbines who need to know the performance of their turbines "The graphical method", using the real life wind speed data .The highest values foe Weibull parameters at Basrah (Albrjsuh), the medium values at Baghdad (Abughraib) and the lowest values in Mosul (Bashiqah). The regression statistical analysis correlation was found for all stations and the most effective values was in at Basrah (Albrjsuh), the medium values at Baghdad (Abughraib) and the lowest values was Mosul (Bashiqah).

References

[1] Mix Ma. (1997)." Adaptive Extremism Control and wind turbine control."Ph. P. thesis Technical university of Denmark.

[2] Physics of Wind Power.

[3] C. Dunder and D. Inam,) 1996. ''The analysis of Wind Data and Wind Energy Potential in Bardirwa.'' Turkey.

[4] K. Rosen, (1998). '' An Assessment of the Potential for utility- scale wind power Generation in Eritrea.''

[5] H. Panofsky and J. Dutton, (1984).''Atmospheric Turbulence: models and methods for engineering applications.'' Pennsylvania State University: John Wiley and Sons.

[6] Sanusi, Y.K. and Abisoye, S.G., (2011). "Estimation of Wind Energy Potential in Southwestern Nigeria." Pacific Journal of Science and Technology; 12(2):160-166. http://www.akamaiuniversity.us/PJST.htm

[7] Alghou, M. A., Sulaiman, M.Y., B.Z.Azmi and Abd. Wahab, M., (2007). "Wind Energy Potential of Jordan." International Energy Journal; 8:pp.71-78.

[8] Ministry of Agriculture, unpublished data.

[9] Congress of the United States, Congressional Budget Office, (2008). "The Economic Effects of Recent Increases in Energy Prices." Available [Online] at http://www.cbo.gov/ftpdocs/74xx/doc7420/07-21-Energy%20DIST.

[10] Gieleck, Mark i, F. Mayes, and L. Prete., (2008). "Incentives, Mandates, and Government Programs for Promoting Renewable Energy." Available [Online] at http://www.eia.doe.gov/cneaf/solar.renewables/rea_issues/in cent.html.

24

Evaluation of electricity consumption of a residential flat in Egypt

Mofreh M. Nassief

Faculty of Engineering Zagazig University Egypt

Email address:

Mofreh_melad@yahoo.com

Abstract: Energy saving is one of the most important demands in our world .The aim of this paper is to evaluate the annual electrical energy consumption of a residential flat in EGYPT and how to reduce it. This evaluation includes the compatibility of a flat for the Egyptian code requirements of residential buildings. This flat is of a total area of 160 m^2 with one façade opening. A software package Visual-Doe 4.0 was used for this analysis. Three different climatic regions were used for this analysis: Cairo, Alexandria and Aswan cities with different weather conditions (due to their geographical location on latitude 30.1, 31.2 and 24 respectively). The results give the minimum requirements of walls insulation (2.5cm) for different orientation and different climatic location. Also an average maximum reduction of about 17.11 % of the total electrical energy consumption with insulation 5cm compared to the actual consumption is fulfilled.

Keywords: HVAC, Systems, Electricity, Consumption, Thermal, Insulation

1. Introduction

Buildings sector today represents 40% of world's energy demand, 33% in commercial buildings and 67% in residential ones according to the Economic Co-operation and Development Organization (ECDO). The effort announced today for transforming the way buildings are conceived, constructed, operated and dismantled has ambitious targets. By 2050 new buildings will consume zero net energy from external power supplies and produce zero net carbon dioxide emissions while being economically viable to construct and operate Guirguis [5].

K.F.Fong et al.[11] proposed a hybrid renewable cooling system (HRCS) for office building application by utilizing both the solar energy and the ground source, in the HRCS appropriate design and operation between the ground-source radiant cooling and the solar absorption cooling was worked out. It was found that the HRCS could have 43.8%, 53.3% and 68% primary energy saving when compared to the sole ground –source heat pump system, the sole solar absorption cooling system and the conventional vapor compression air –conditioning system respectively

Burton et al. discussed an experimental technique to investigate the optimum speed (for thermal comfort) of a ceiling mounted fan at various temperatures. The experiments took place in an office without windows and known width and heights. Results indicate that the effect of temperature on optimum for speed was highly significant.

Baker et al. summarized the results drawn from a series of field studies in predominantly overheated circumstances in order to develop the application of passive cooling techniques in southern Europe and also to develop suitable criteria for assessing proposed buildings. The results indicated that field studies failed to show directly the difference in predicated and observed room conditions.

Green building is the practice of increasing the efficiency of new buildings, and reducing their impact on human health and the environment through better site location, design, construction, operation and maintenance, Building_Codes.pdf, 2008.

Hanna et al. [6] studied the effect of building envelope to save energy associated with the total electricity consumed for residential buildings in Egypt (Cairo and Alexandria cities). The analysis shows that the over-all thermal transfer value (OTTV) for the exterior walls should not exceed 30 W/m^2 for Cairo but Alexandria does not need any insulation. The roof needs 50 mm insulation to reach 25W/m^2.

Hanna et al. [6] summarized the results of energy simulation analysis to determine the effectiveness of building characteristics in reducing electrical energy consumption and saving for office building in Egypt. The simulation includes different variables such as window to wall area ratio, shading, light power density, different HVAC systems and window type. The main conclusion of these results is that a significant energy saving can be achieved by selecting materials with appropriate design techniques.

Sheble [13] evaluates the effect of window to wall area ratio (WWR) for different buildings types to save energy associated with the total electricity consumed. The analysis was agreed for different outdoor climate conditions in Egypt. The results show that decreasing the WWR generally saves more energy. For the very hot dry region it is recommended to reduce the WWR and prevention of natural ventilation during the day. The WWR up to 20% is preferred for commercial buildings to have energy efficiency.

Karlsson et al. [12] agreed a comprehensive investigation of a low-energy building. The ventilation rates, energy performance, occupants, perception of the indoor climate, and finally the environmental performance, with respect to CO2-emissions were investigated to study the buildings at different system levels. The results show that the heating and ventilation system can be improved to prevent draughts and ventilation short-cuts, and the heating system could be changed to district heating to achieve lower CO2 emissions and minimize the environmental impact. The pay-back period, with today's Swedish electricity prices for a building like this is about 19.5 years compared with a comparable Swedish building, within reasonable limits. If the electricity price is assumed to increase with 5%/yearly in the future, as the case has been during the last 5 years in Sweden, the pay-off time will decrease to about 14 years.

Wang et al.[16] discussed possible solutions for zero energy building design in UK. Simulation software Energy Plus and TRNSYS 16 are used in this study, where Energy Plus simulations are applied to enable facade design studies considering building materials, window sizes and orientations. TRNSYS is used for the investigation of the feasibility of zero energy houses with renewable electricity, solar hot water system and energy efficiency heating systems under Cardiff weather conditions. Various design methods are compared and optimal design for typical homes and energy systems are provided.

Virtal et al. [14] presented energy simulation of office buildings located in France. It takes into account the energy consumption of different HVAC systems. Two different kinds of buildings have been used. Reference building has external structures of a typical new building in Paris today. In advanced building external walls and windows have lower U-value and improved tightness. Windows have also better shading coefficient and external overhang of 500 mm for solar shading. Result shows that with right HVAC system and building design can reduce energy consumption nearly 75 %.

Guirguis [5] shows a remarkable effect of the building envelope construction on the electricity consumption for different weather cities in Egypt. The prefabricated panel with 10 cm insulation gives lower electricity consumption and higher energy saving reaches up to 40% in comparison with other higher U-value brick walls. The wall insulation gives energy saving nearly 40%, 33% and 41% for Cairo, Alexandria and Aswan cities respectively.

2. Theoretical Work

In this present analysis the Visual-Doe program,2007 is one of the Green Design tools software programs that enable the architects, engineers and energy analysts to quickly evaluate the energy savings of building design options. The program uses the hourly simulation tools as the calculation engine so that energy use and peak demand are accurately evaluated on an hourly basis and annual electricity calculations. Different building shapes can be created. This program includes library for data entry such as building constructions, HVAC systems, lighting and equipment. All can be controlled using operating schedules during the year. Here after, the equations used in this analysis, ASHRAE Handbook Code of thermal insulation, 2007,Ventilating Engineering Guide,1970, ECP306/1, 2005.

a) *Total thermal resistance* (R_t) which is the sum of all thermal resistances for a composed wall beside and outside surface thermal resistances, So

$$R_t = R_{so} + R_1 + R_2 + \ldots R_n + R_{si}, (m^{2\circ}C/W) \qquad (1)$$

Where:

$$Rso = 1/(\text{€ hr} + ho) \quad \& \quad Rsi = 1/(1.2 \text{ € hr} + hi)$$

b) *Thermal transmittance* for a wall (U) which is the quantity of heat passing through a unit area of wall per unit time normally when there is a unit difference in temperature of inside & outside, So

$$U = 1/R_t \qquad (2)$$

Also, the total transmittance (U_o) for walls contain openings like Windows, Doors, Balconies, etc, can be calculated from.

$$U_o = \Sigma(A_w.U_w + A_d.U_d + A_g.U_g + \ldots)/(A_w + A_d + A_g + \ldots), (W/m^{2\circ}C) \qquad (3)$$

c) *Energy Savings* "Q_{saving}", (kW hour/m^2/year)

The main goal for thermal insulation in buildings is the reduction of heat transfer which results in improving the inside environment of a building also in saving of electrical and thermal energies, so the quantity of energy per year can be calculated from

$$Q_{saving} = (U_{before insulation} - U_{after insulation}) \times DD \times 24 \times 10^{-3} \qquad (4)$$

Where:

$$DD = \Sigma_{(24hours)} |T_b - T_o|, (°C/day)$$

The simulated flat is of total area about 160 m² with ceiling height 3 m. It consists of a hall, three rooms, a kitchen and a bath room as shown in Fig. 1. (The main façade containing four windows). The two sides and back walls are surrounded with adjacent flat walls which can be classified as adiabatic walls.

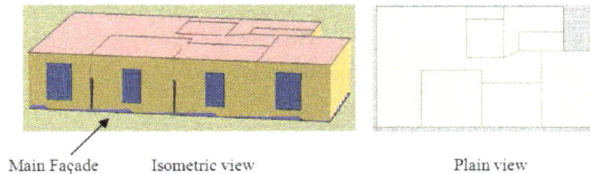

Figure 1. *Sketch diagram of a single flat model with total area 160 m² and height 3 m.*

In this work we simulate different walls and insulations (as shown in Fig. 2). The simulation includes also roof insulation, the roof construction is shown in Fig. 3. The annual total electricity was evaluated. The simulations were applied for different climate regions, Cairo, Alexandria and Aswan using climatic data. Different flat orientations (East, East-North, North, North-West, West, South, South-West and South-East) take place to evaluate the suitable and less wall insulation thicknesses for each. Different HVAC systems, Visul DOE 4.0 User Manual, [15] were used in the simulation for comparison of energy consumption.

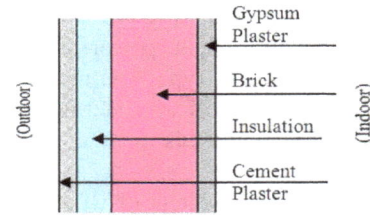

Figure 2. *Wall construction.*

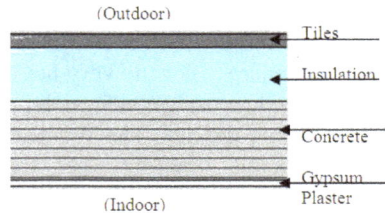

Figure 3. *Roof construction.*

Table 1 shows some information of basic design data about the components of HVAC systems taking in account heating max., supply air temp.is 46.1°C and the system of heating is "Furnace High". Its thermal efficiency is 0.8.The auxiliary power is zero kW, its pilot light power is 146.5 W and the baseboard source is hot water from plant. Cooling: min. supply temp is 12.8°C and the system of cooling is DX High Eff. Which has coil by pass factor of 0.19 Supply fan of mechanical, drive and motor efficiency of 0.55, 0.95 and 0.9 respectively and its static pressure is 497.6 pa

Table 1. *Different HVAC systems used in the analysis.*

No.	System	Description of the main components	Diagram
1	Residential Variable Volume, Variable Temp	Economizer - heating and cooling units - supply fan - closed and open loops	
2	Single Zone Variable Temp	Economizer - preheating - heating and cooling units - supply fan - closed and open loops	
3	Two Pipe Fan Coil	Economizer-heating and cooling units - supply fan- closed and open loops	
4	Unit Heater	heating unit - supply fan - closed loop	
5	Residential System	heating and cooling units -supply fan - closed loop	
6	Unit Ventilator	heating unit - supply fan - closed and open loops	

3. Results

The HVAC system types and their options affect energy saving. Here in this research we use 6 different HAVC systems. The pre-heater components and open loops increase the energy consumption on the other side the closed loops reduce energy consumption. Also the flat orientation affects the energy consumption. To evaluate the effect of flat orientation with the energy consumption through the HVAC cooling system, the sides and rear walls are kept adiabatically (2.5 to25 cm insulation thickness) with varying only the main façade wall constructions and insulation. Fig. 4 shows annual total electricity consumption of different HAVC systems for different climate regions in EGYPT (North direction) .

The figure shows that the HAVC system number 2 (table 1) gives the maximum energy consumption than other systems due to the presence of the preheating unit and open loop. On the other hand, the HAVC system number 6 gives the lower energy consumption. Aswan city has maximum energy consumption because of its geographical location on latitude 24 and Alexandria has low energy consumption than other cities due to its geographical location on latitude 31.2. From the comparison of Figs. 4 to 11 for different flat orientation it is clear that the annual total electricity consumption for system 6 (unit ventilator HVAC system) is the least one of the 3 cities Aswan, Cairo and Alexandria respectively when the main façade is oriented to the North direction. Also the annual total electricity consumption for system 2 (Single zone variable temperature system) is the highest one of the 3 cities Alexandria, Cairo, Aswan respectively when the main façade is oriented to both East and West directions . The moderate annual total electricity consumption is for system number 1 (Residential variable volume, variable temperature),the other 3 systems numbers 3, 4 and 5 (two pipe fan, coil unit heater and residential system respectively) for the 3 cities and for all different climate regions in EGYPT are almost equal in the consumption of the annual total electricity . In addition, in all figures, it is clear that Alexandria is considered the least city in energy consumption (for different climate regions and different HVAC systems) this because of its geographical location, on adverse Aswan is considered the highest city in energy consumption (for different climate regions and different HVAC systems) this is because of its hot climate. Between the six different HVAC systems used in analysis, the unit ventilator HVAC system is chosen with different Air Change/Hour, ACH, because it satisfies the main goal of the research, that is saving energy.

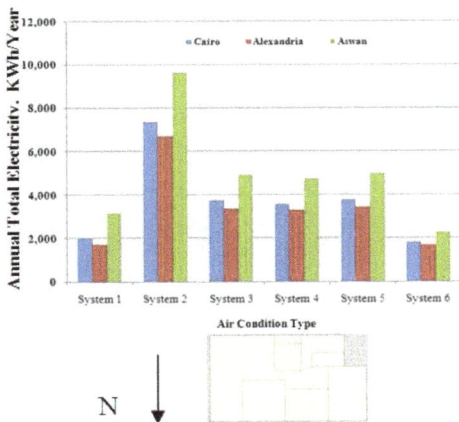

Figure 5. Annual total electricity consumption of different HAVC systems for different climate regions in EGYPT (North WEST direction).

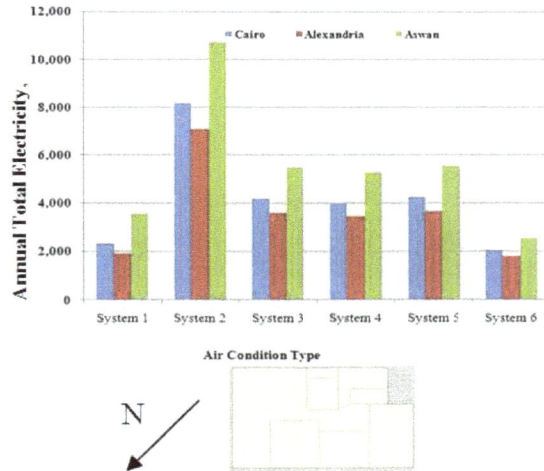

Figure 6. Annual total electricity consumption of different HAVC systems for different climate regions in EGYPT (West direction).

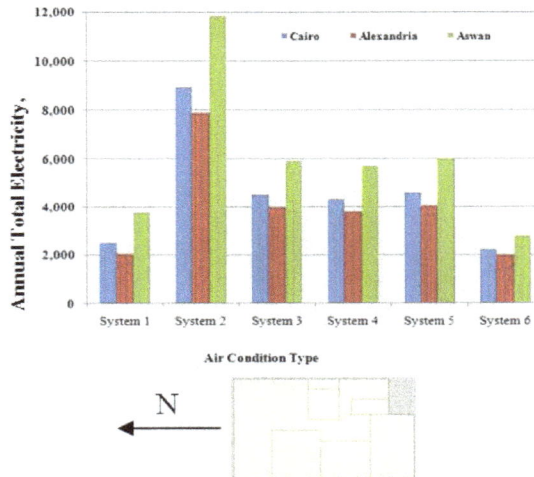

Figure 4. Annual total electricity consumption of different HAVC systems for different climate regions in EGYPT (North direction).

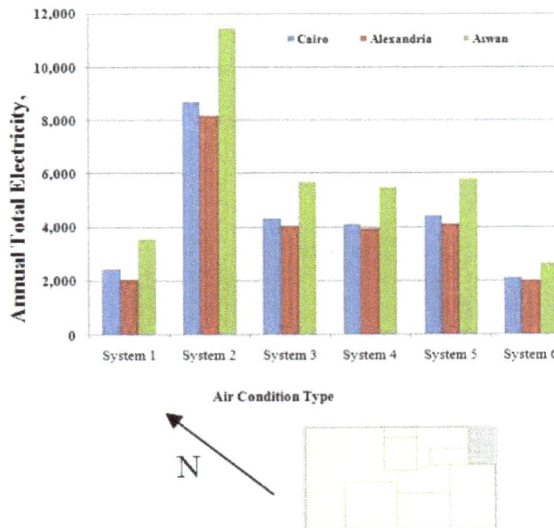

Figure 7. Annual total electricity consumption of different HAVC systems for different climate regions in EGYPT (South West direction)

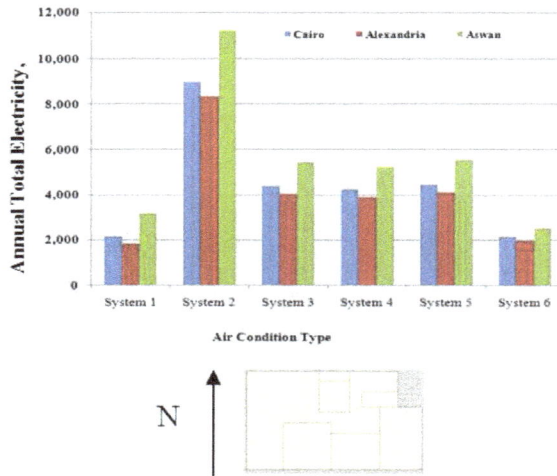

Figure 8. *Annual total electricity consumption of different HAVC systems for different climate regions in EGYPT (South direction).*

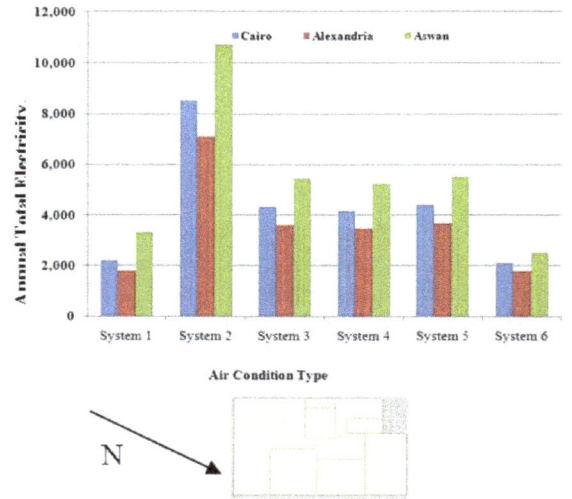

Figure 9. *Annual total electricity consumption of different HAVC systems for different climate regions in EGYPT (South East direction)*

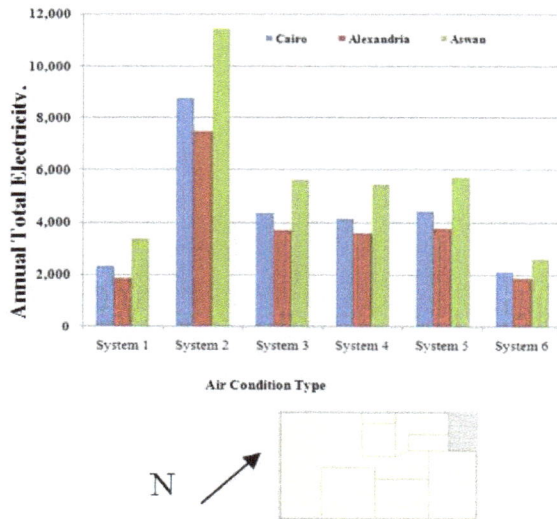

Figure 10. *Annual total electricity consumption of different HAVC systems for different climate regions in EGYPT (East direction)*

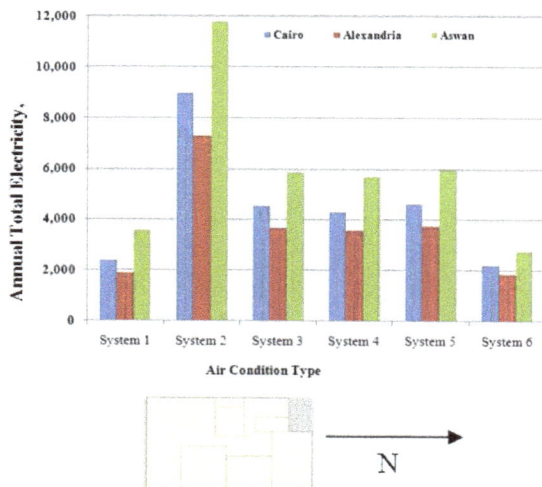

Figure 11. *Annual total electricity consumption of different HAVC systems for different climate regions in EGYPT (North East direction).*

Figs 12 to 19 show the annual total electricity energy consumption for both different ACH and climate regions in EGYPT (Cairo, Alexandria and Aswan) using the more economizer system for energy which is number 6 (the unit ventilator HAVC system) and the main façade of the study flat is oriented to different directions, North, North West, West, South West, South, South East, East and North East respectively. It is clear from the figures that as ACH increases the energy consumption increases. Also at Zero ACH for different directions, the annual total energy consumption for both Alexandria and Cairo approximately differs a little because of the difference in temperature during the year is approximately in the range of 3°C to 5 °C and due to the average rate of humidity during the year in Cairo is more than Alexandria & Aswan, so with the increase of ACH the energy consumption increases, we find that the annual total energy consumption is approximately for Cairo 2.5 times for Aswan and 1.25 for Alexandria.

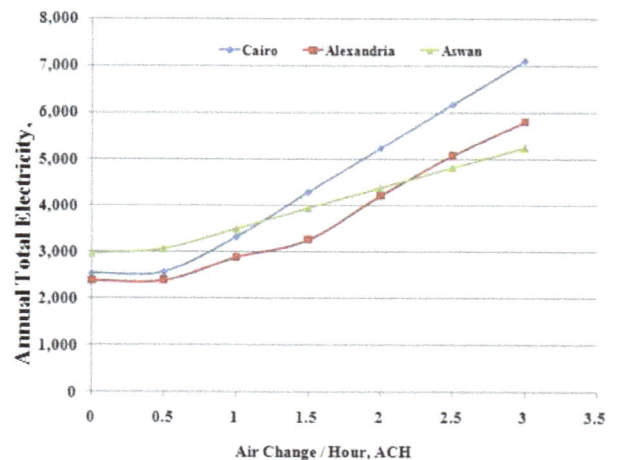

Figure 12. *Annual total electricity consumption of ACH for different climate regions in EGYPT (North direction)*

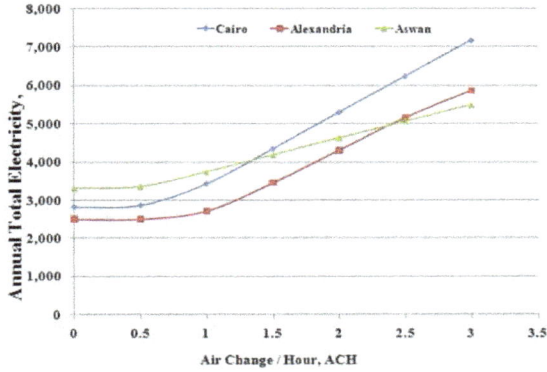

Figure 13. *Annual total electricity consumption of ACH for different climate regions in EGYPT (North West direction).*

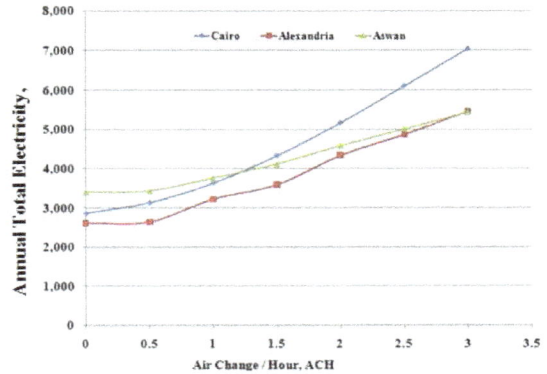

Figure 14. *Annual total electricity consumption of ACH for different climate regions in EGYPT (West direction)*

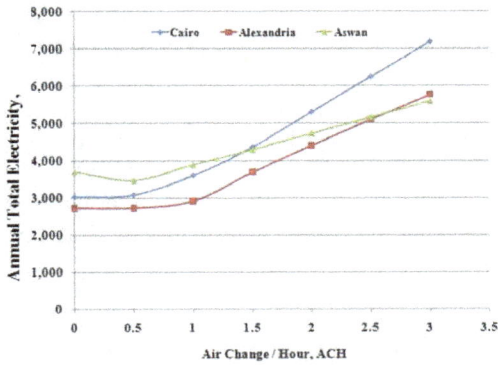

Figure 15. *Annual total electricity consumption of ACH for different climate regions in EGYPT (South West direction)*

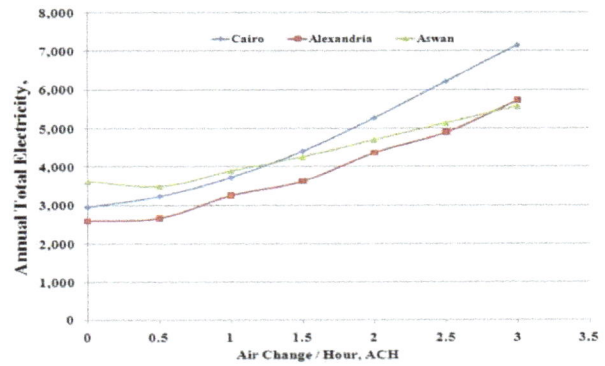

Figure 16. *Annual total electricity consumption of ACH for different climate regions in EGYPT (South direction)*

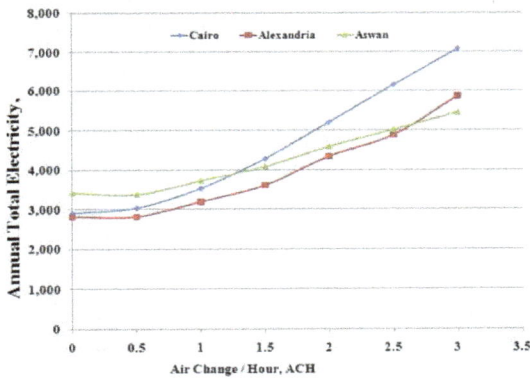

Figure 17. *Annual total electricity consumption of ACH for different climate regions in EGYPT (South East direction)*

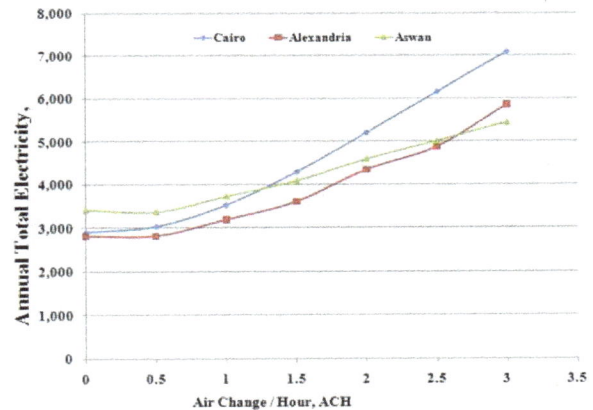

Figure 18. *Annual total electricity consumption of ACH for different climate regions in EGYPT (East direction).*

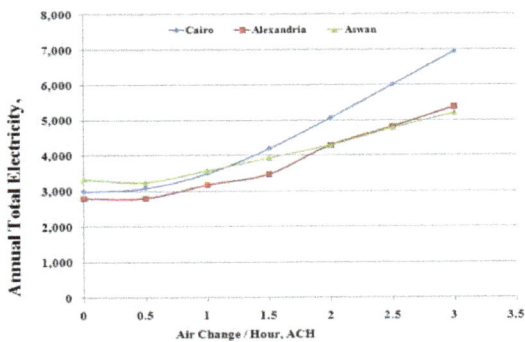

Figure 19. *Annual total electricity consumption of ACH for different climate regions in EGYPT (North East direction).*

Fig. 20 shows the annual total energy consumption for different wall insulation in Cairo for both different oriented directions (North, South, East and West) and R-value (thermal resistance). It is noted from figure that at zero insulation the energy consumption is relatively high for all directions but it is the least one for the North direction. Also as the insulation thickness increases the energy consumption decreases and the R-value increases. Finally the best oriented direction for façade is the North direction for both energy consumption and R-value.

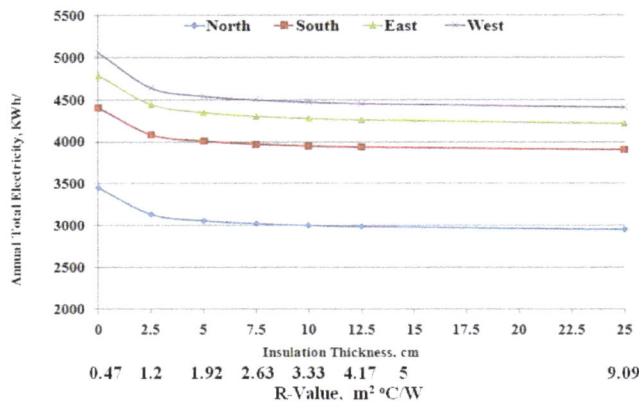

Figure 20. Annual total electricity consumption for different wall insulation thickness Cairo-EGYPT

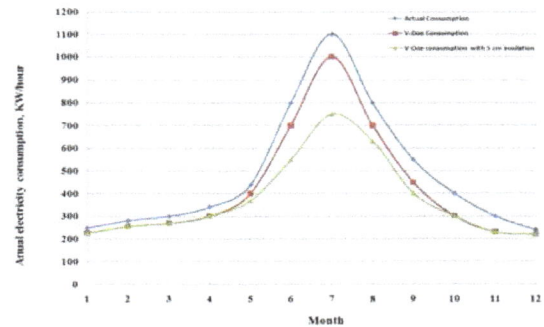

Figure 21. Monthly comparison between actual electricity consumption and the results obtained from V –DOE with and without insulation for Cairo -EGYPT

Fig 21 shows a comparison between the actual case and the case study with and without insulation of the annual total electric energy consumption. It is clear that the maximum consumption of energy is at July (because of the average higher temperature in this month) and the consumption is reduced with 5cm insulation by17.11 % compared to the actual case.

The wall construction affects the thermal performance of buildings in which low thermal resistance(R-value) yields to more heat flow by conduction inside the building through walls. Thick construction or insulation (higher R-value) reduces the heat gain which reduces the indoor air temperature inside the buildings, as illustrated in table (2) . In this table there is a compression between the values of R for two different thickness of insulation (2.5 and 5cm) for the flat in Cairo city and the values that satisfy the requirement of Egyptian energy code ECP, 2005.

Table 2. Requirements of satisfaction of the energy efficiency of the external walls according to Egyptian energy code (ECP), Absorptive =0.38, West direction, WWR>30%

Zone	R- value according to ECP	R- value without insulation	R- value with insulation	
			2.5 cm	5cm
Cairo	1.07	0.47	1.2	1.92

4. Conclusions

1-The research shows a noticeable effect of the building construction on the energy consumption for the different three cities of different weather.

2-Between the six HAVC systems used in analysis, the more economizer system for energy consumption is the unit ventilator HAVC system.

3-The North and North-East façade need less wall insulation than other façade.

4-The walls need more thick insulation especially in Aswan city.

5-As the thickness of insulation increases, the R-value increases and the annual energy consumption decreases especially for the North direction.

6-For zero insulation the R-value and the annual total electricity are maximum for the four directions, North, South, East, and West, but the West direction is the most consumption for energy and the North direction is the least one.

7-Due to the average rate of humidity for Cairo during the year (more than Aswan and Alexandria) the annual total energy consumption is approximately for Cairo 2.5 and 1.25 respectively times for Aswan and Alexandria.

8-The exposed East, West and South-West walls need more thick insulation especially for Cairo and Aswan cities.

9-By using 5cm insulation thickness for the walls ,the annual total energy consumption is reduced approximately by 17.11 %.

Footnotes

T_o: the daily average temperature,(°C) .

T_b: base temperature, (18.3 °C for warming &25°C for cooling).

DD:degree-day (the sum of temperature difference), (°C/day).

A_w: net area of solid walls, (m^2).

A_g: net area of glass openings, (m^2) .

A_d: net area of doors, (m^2).

h_i , h_o: inside and outside surface thermal conductance, (W/m^2°C).

h_r: heat transfer coefficient by radiation, (W/m^2°C).

R_t: Total thermal resistance, (m^2°C/W).

R-value: Total thermal resistance of the composed wall (m^2 °C/W) .

U_w: total thermal transmittance for solid walls, (W /m^2°C)

U_g: total thermal transmittance for glass openings, (W/m^2 °C).

U_d: total thermal transmittance for doors, (W /m^2 °C).

€: coefficient of emissivity,0.9 for building materials, (non).

Acknowledgement

The author like to thank the team of aerodynamic lab. Housing & Building National Research Center-Egypt and Mr Badr for their valuable help in this work

References

[1] ASHRAE Handbook Fundamentals,. American Society of Heating, Refrigerating and air-conditioning Engineering, 1997.

[2] Baker, N., Standeven, M.,"A Behavioral approach to thermal comfort assessment in naturally building", CIBSE National conference, 76-84, 1995.

[3] Burton, D.R., Robeson, K.A. and Nevius,R.G., "The effect of temperature on the preferred air velocity for sedentary subjects dressed in shorts", ASHRAE ,157-168,1975.

[4] Egyptian Energy Code Program (ECP) 306/1-2005,edition 2005.

[5] Guirguis, N.M., "Energy Efficiency of Flat for Different Egyptian Regions", World Renewable Energy Conference, WREC, Abu Dhabi,2010.

[6] Hanna, G.B., Guirguis, N.M., Osman, H.S., Hussein, M.A., " Energy analysis for new residential buildings in Egypt", International Conference : Future Vision and Challenges for Urban Development, Cairo , Egypt, 20-22,2004

[7] Hanna, G.B., Guirguis, N.M.,Sheble,S.S., " ENERGY CODE FOR NEW OFFICE BUILDING", IIR Middle East Electricity Conference, Dubai,2005.

[8] Housing & Building Research Center (HBRC), Code of Thermal Insulation,3rd edition Cairo-Egypt,2007.

[9] http://www.iea.org/g8/2008/Building_Codes.pdf, 2008.

[10] Institution of Heating & ventilating Engineers Guide, Book A, 1970.Cairo-Egypt.

[11] K. F. Fong, C.K. Lee," Investigation On Hybrid System Design Of Renewable Cooling For Office Building In Hot And Humid Climate". Vol.75, 2014,PP 1-9.

[12] Karlsson, J.F., Moshfegh, B., "A comprehensive investigation of a low-energy building in Sweden", Renewable Energy 32, 1830–1841, 2007.

[13] Sheble, S.S., "Effect of window to wall ratio and different climate conditions on energy consumption for residential and commercial buildings in Egypt", Journal of Housing and Building National Research Centre (HBRC), Cairo – Egypt, Vol 3 No. 2,2007.

[14] Virta1, M., Itkonen1, H.,Mustakallio1,P., Kosonen1,R., " Energy efficient HVAC-system and building design", Paper No. R5-TS29-OP05.CLIMA, Antalya, TURKEY,2010.

[15] Visual DOE 4.0 User Manual, August, 2004. http://www.archenergy.com

[16] Wang,L., Gwilliam, J., Jones, P., " Case study of zero energy house design in UK", Energy and Buildings, 41,1215-1222,2009.

25

The linear induction motor (LIM) & single linear induction motor (SLIM)

Nahid Ahmadinia

M. A. Electrical Power Engineering, Science and Research Branch, Islamic Azad University, Broujerd, Iran

Email address:

Nahmadinia@yahoo.com

Abstract: First of all mentions linear induction machines in 1890, only two years after the discovery of the rotary induction principle. Basically the concept of the linear device consists in imagining a rotary machine to be cut along a radial plane and 'unrolled' so that the primary member then consists of a single row of coils in slots in a laminated steel core. The differences between rotary and linear motors are outlined and reasons for the slow application of linear motors are explained. Principal developments in linear machines since the 1950s are described. Induction motor which can be used to power capsules in an xv capsules in a pneumatic capsule pipeline system. Several optimal design schemes of a single sided linear induction motor (SLIM) adopted in linear metro are presented in this paper Firstly the equivalent circuit of SLIM fully considering the end effects, half-filled slots, back iron saturation and skin effect is proposed ,based on one dimensional air gap magnetic equations In the circuit, several coefficients including longitudinal end effect coefficients Kr(s) and Kx(s), transversal end edge effect coefficients Cr(s) and Cx (s), and skin effect coefficient K fare achieved by using the dummy electric potential method and complex power equivalence between primary and secondary sides Furthermore, several optimal design restraint equations of SLIM are provided in order to improve the operational efficiency and reduce the primary weight. The result tries to establish a new concept for elevators through a new construction technique and assembly of the system with counterweight, which increases the reliability and comfort with cost reduction.

Keywords: Electric Motors, linear Induction Motor (LIM), Single-Sided linear (SLIM)

1. Introduction

The SLIM has the following merits comparing with the rotary induction motor (RIM) : greater ability to exert thrust on the secondary without mechanical contacts, higher acceleration or deceleration, less wear of the wheels, smaller turn circle radius, and more flexible road line. Because of its cut-open magnetic circuit, the linear induction motor (LIM) possesses the inherent characteristics such as longitudinal end-effect, transversal edge-effect and normal force. In addition, it also has half-filled slots in the primary ends. Therefore, an accurate equivalent circuit model of LIM is difficult to be obtained compared to that of RIM. Many analysis techniques of SLIM have been studied and developed in the past years. However, effective methods on the design scheme of SLIM have not been obtained due to the following reasons. The selection of electric loading and magnetic loading by loading distribution is so difficult that one cannot calculate the apparent power (kVA) easily. The

Power factor and efficiency are affected by the end effect which is again affected by the design technique of SLIM. It uses a rotary electrical motor as source of motion in order to convert the rotary mot ion into a linear motion. Often, it is necessary to use a complex mechanical system of gears, axles and screws jacks. When used directly, these transmission systems for movement have great losses. Among the reasons is the increased an abrasive wear due to the friction of the mechanical parts, even when using low viscosity fluids for the lubrication. This results into higher operational and maintenance costs Therefore, for transport applications, the use of an electrical machine that produces directly the linear motion would result in lowers operational and maintenance costs as well as higher reliability and efficiency. The purpose of this paper is to propose mover position control of linear induction motor (LIM) using an adaptive backstopping approach based on field orientation.

2. History

The history of linear electric motors can be traced back at least as far as the 1840s, to the work of Charles Wheatstone at King's in London, but Wheatstone's model was too inefficient to be practical. In 1824, the French physicist François Arago formulated the existence of rotating magnetic fields, termed Arago's rotations, which, by manually turning switches on and off, Walter Baily demonstrated in 1879 as in effect the first primitive induction motor. Practical alternating current induction motors seem to have been independently invented by Galileo Ferraris and Nikola Tesla, a working motor model having been demonstrated by the former in 1885 and by the latter in 1887. In April 1888, the *Royal Academy of Science of Turin* published Ferraris's research on his AC polyphone motor detailing the foundations of motor operation. George Westinghouse, who was developing an alternating current power system at that time, licensed Tesla's patents in 1888 and purchased a US patent option on Ferraris' induction motor concept. Tesla was also employed for one year as a consultant. Westinghouse employee C. F. Scott was assigned to assist Tesla and later took over development of the induction motor at Westinghouse. Steadfast in his promotion of three-phase development, Mikhail Dolivo-Dobrovolsky's invented the cage-rotor induction motor in 1889 and the three-limb transformer in 1890. The General Electric Company (GE) began developing three-phase induction motors in 1891. By 1896, General Electric and Westinghouse signed a cross-licensing agreement for the bar-winding-rotor design, later called the squirrel-cage rotor.

Figure 1. Three-phase totally enclosed fan-cooled (TEFC) induction motor, with and, at right, without end cover to show cooling fan. In TEFC motors, interior losses are dissipated indirectly through enclosure fins mostly by forced air convection

3. Linear Electric Motor & Linear Induction Motor

A linear electric motor's primary typically consists of a flat magnetic core (generally laminated) with transverse slots which are often straight cut with coils laid into the slots, with each phase giving an alternating polarity and so that the different phases physically overlap. The secondary is frequently a sheet of aluminum, often with an iron backing plate. Some LIMs are double sided, with one primary either

side of the secondary, and in this case no iron backing is needed. Two sorts of linear motor exist, *short primary*, where the coils are truncated shorter than the secondary, and a *short secondary* where the conductive plate is smaller. Short secondary LIMs are often wound as parallel connections between coils of the same phase, whereas short primaries are usually wound in series

The primaries of transverse flux LIMs have a series of twin poles lying transversely side-by-side, with opposite winding directions. These are typically made either with a suitably cut laminated backing plate or a series of transverse U-cores .Induction motors use electromagnetic induction to rotate a shaft or rotor and create mechanical energy from electric energy. With minimal design alterations an electric motor like this also has the ability to create electricity from mechanical energy, although these motors are referred to as generators. Electromagnetic induction is the production of an electrical potential difference (or Voltage) across a conductor situated in a changing magnetic field. A Faradays Law state that electromotive force (EMF) produced along a closed path is proportional to the rate of change of the magnetic flux through any surface bounded by that path. In practice, this means that an electrical current will flow in any closed conductor, when the magnetic flux through a surface bounded by the conductor changes. This applies whether the field itself changes in strength or the conductor is moved through it.

The most basic Induction motor is the Squirrel cage motor. Squirrel cage motors are the most common industrial AC motor, when running off a constant AC supply they are simply constant speed devices. They acquired there name from the rotors used which resemble a squirrel running cage. This type of motor has revolutionized the way factories operate, improving efficiency and safety compared to earlier mechanical methods.

When a squirrel cage motor is used with a three-phase power source, it forms the simplest electric motor. This motor is made up of three fixed coils around a rotating shaft which consists of a core filled with a series of conductors set up in a circle pattern around the shaft. With the three-phase current flowing through the coils a rotating magnetic field is produced, the field then induces a current in the conductors on the shaft. This is where three-phase power outperforms single phase. For the shaft to rotate smoothly without stopping the three coils situated at 120° from each other are supplied with power at precise intervals. This can be better described by the graphics below.

Usually, transportation of laminated products is done through rolls, put in rotation by electric motors. The acting moving force depends solely on the adherence of the laminated product to the moving rolls, which reduces the speed of the latter. Since a large acceleration is not possible due to the resulting high inertial forces, a sliding of the bars on the rolls will occur. A solution to this problem may be obtained by using a linear asynchronous motor whose moving body would be the laminated product.

The experiments were performed under various practical

conditions and changes were made in the system in order to practically obtain an optimal design of the model over which the experiments were carried on. The experimental data were obtained by changing the various parameters of the LIM such as air gap, magnetic gap, changing the materials of the secondary of LIM, by creating breaks in the electrical and magnetic circuits of the LIM, by de-aligning the primary with respect to secondary etc.by using different supply and controls. The tests were successfully conducted for direct voltage control, frequency control and variable voltage variable frequency control i.e. V/f control by using VVVF Drive and three phase auto Transformer. The data taken after experiments are plotted individually and relatively in order to compare the results under different situations finally, the observations made and the graphs plotted have been explained and results successfully approved with the practical requirements for Traction Applications of Linear Induction Motor.

A linear electric motor's primary typically consists of a flat magnetic core (generally laminated) with transverse slots which are often straight cut with coils laid into the slots, with each phase giving an alternating polarity and so that the different phases physically overlap. The secondary is frequently a sheet of aluminum, often with an iron backing plate. Some LIMs are double sided, with one primary either side of the secondary, and in this case no iron backing needed.

Two sorts of linear motor exist, short primary, where the coils are truncated shorter than the secondary, and a short secondary where the conductive plate is smaller. Short secondary LIMs are often wound as parallel connections between coils of the same phase, whereas short primaries are usually wound in series. The primaries of transverse flux LIMs have a series of twin poles lying transversely side-by-side, with opposite winding directions. These are typically made either with a suitably cut laminated backing plate or a series of transverse U-cores. For this reason, induction motors are sometimes referred to as asynchronous motors. An induction motor can be used as an induction generator, or it can be unrolled to form a linear induction motor which can directly generate linear motion.

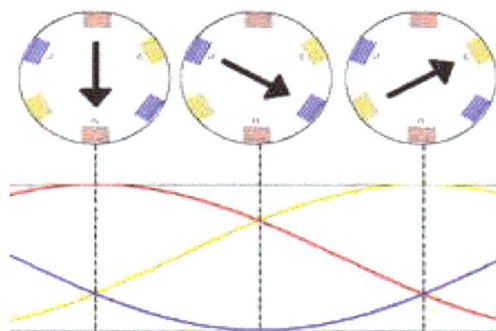

Figure 2. A three-phase power supply provides a rotating magnetic field in an induction motor.

In both induction and synchronous motors, the AC power supplied to the motor's stator creates a magnetic field that rotates in time with the AC oscillations. Whereas a synchronous motor's rotor turns at the same rate as the stator field, an induction motor's rotor rotates at a slower speed than the stator field. The induction motor stator's magnetic field is therefore changing or rotating relative to the rotor. This induces an opposing current in the induction motor's rotor, in effect the motor's secondary winding, when the latter is short-circuited or closed through external impedance. The rotating magnetic flux induces currents in the windings of the rotor in a manner similar to currents induced in a transformer's secondary winding(s). The currents in the rotor windings in turn create magnetic fields in the rotor that react against the stator field.

4. Synchronous Speed

An AC motor's synchronous speed, n_s, is the rotation rate of the stator's magnetic field, which is expressed in revolutions per minute as

$$n_s = \frac{120 \times f}{p}$$

where f is the motor supply's frequency in Hertz and p is the number of magnetic poles. That is, for a six-pole three-phase motor with three pole-pairs set 120° apart, p equals 6 and n_s equals 1,000 RPM and 1,200 RPM respectively for 50 Hz and 60 Hz supply systems.

4.1. Slip

Slip, s, is defined as the difference between synchronous speed and operating speed, at the same frequency, expressed in rpm or in percent or ratio of synchronous speed. Thus

$$s = \frac{n_s - n_r}{n_s}$$

where n_s is stator electrical speed, n_r is rotor mechanical speed. Slip, which varies from zero at synchronous speed and 1 when the rotor is at rest, determines the motor's torque. Since the short-circuited rotor windings have small resistance, a small slip induces a large current in the rotor and produces large torque. At full rated load, slip varies from more than 5% for small or special purpose motors to less than 1% for large motors. These speed variations can cause load-sharing problems when differently sized motors are mechanically connected. Various methods are available to reduce slip, VFDs often offering the best solution.

5. Principles

In this model of electric motor, the force is produced by linearly moving magnetic field acting on conductors in the field. Any conductor is it a loop, a coil or simply a piece of plate metal that is placed in this field will have eddy currents induced in it thus creating an opposing magnetic field, in accordance with Lenz's law. The two opposing fields will repel each other, thus creating motion as the magnetic field sweeps through the metal:

$$n_s = 2f_s / p$$

Where, f_s is supply frequency in Hz, p is the number of poles, and n_s is the synchronous speed of the magnetic field in revolutions per second.

The travelling field pattern has a velocity of:

$$v_s = 2tf_s$$

v_s Is velocity of the linear travelling field in m/s, t is the pole pitch.

For a slip of s, the speed of the secondary in a linear motor is given by

$$v_r = (1 - s)v_s$$

6. Thrust

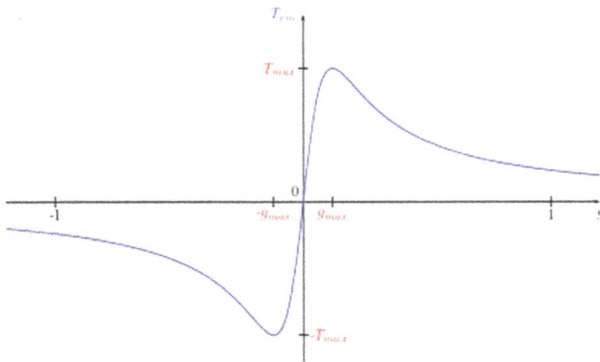

Figure 3. *Thrust generated as a function of slip*

As explained earlier, the input power to the stator windings is utilized in producing useful mechanical power which is exerted on the rotor and to account for the rotor copper losses. In terms of the equivalent circuit com ponents, the mechanical power developed by the rotor is the power transferred across the air-gap from the stator to rotor minus the rotor copper loss. Under normal operations, the LIM develops a thrust proportional to the square of the applied voltage, and this reduces as slip is reduced similarly to that of an induction motor with a high rotor resistance.

6.1. Levitation

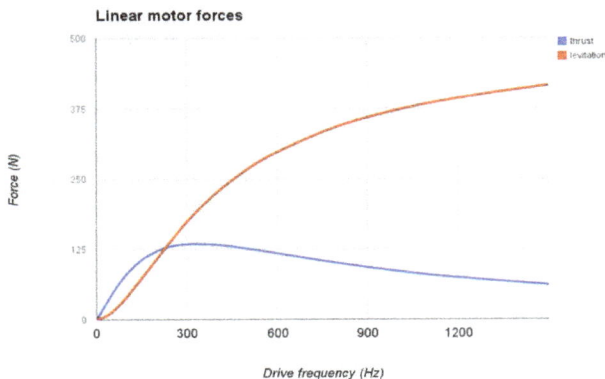

Figure 4. *Levitation and thrust force curves of a linear motor*

In addition, unlike a rotary motor, an electro dynamic levitation force is shown, this is zero at zero slip, and gives a roughly constant amount of force/gap as slip increases in either direction. This occurs in single sided motors, and levitation will not usually occur when an iron backing plate is used on the secondary, since this causes an attraction that overwhelms the lifting force. A linear induction motor (LIM) is an AC asynchronous linear motor that works by the same general principles as other induction motors but is very typically designed to directly produce motion in a straight line.

linear induction motors have a finite length primary, which generates end-effects, whereas with a conventional induction motor the primary is arranged in an endless loop. Despite their name, not all linear induction motors produce linear motion; some linear induction motors are employed for generating rotations of large diameters where the use of a continuous primary would be very expensive. As with rotary motors, linear motors frequently run on a 3 phase power supply and can support very high speeds. However, there are end-effects which reduce the force, and it's often not possible to fit a gearbox to trade off force and speed. Linear induction motors are thus frequently less energy efficient than normal rotary motors for any given required force output. LIMs are often used where contactless force is required, where low maintenance is desirable, or where the duty cycle is low. Their practical uses include magnetic levitation, linear propulsion, and linear actuators. They have also been used for pumping liquid metals.

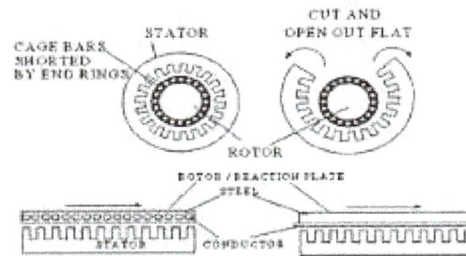

Figure 5. *Cutting and Unrolling a RIM to a LIM*

This paper is interested in thrust and its relation to other variable parameters. The normal force is perpendicular to the stator in the z-direction. Lateral forces are undesirable forces which are developed in a SLIM because of the orientation of the stator. These forces are a matter of concern in high frequency operation (60Hz) where they increase in magnitude. A set of guided mechanical wheel tracks is sufficient to eliminate a small lateral force. For the analysis and design of a SLIM having negligible end-effects per-phase conventional equivalent circuit shown in Fig 1 may be used. The circuit components are determined from the SLIM parameters. The SLIM performances to be determined are thrust and efficiency. The approximate equivalent circuit of a LIM is presented as shown in Fig 1. This circuit is on a per phase basis. The core losses are neglected because a realistic air gap flux density leads to mode rate flux densities in the core and hence, rather low core losses. Skin effect is small at rated frequency for a flat linear induction motor with a thin conductive sheet on the secondary. Therefore, equivalent rotor inductance is negligible. The remaining non-negligible parameters are shown in Fig 1 and are discussed blew. The drive generated by linear induction motors is somewhat similar to conventional induction motors; the drive forces show a roughly similar characteristic shape relative to slip, albeit modulated by end effects. Unlike a circular induction motor, a linear induction motor shows 'end effects'. These end effects include losses in performance and efficiency that are believed to be caused by magnetic energy being carried away and lost at the end of the primary by the relative movement of the primary and secondary. With a short secondary, the behavior is almost identical to a rotary machine, provided it is at least two poles long, but with a short primary reduction in thrust occurs at low slip. However, because of end effect, linear motors cannot 'run light'- normal induction motors are able to run the motor with a near synchronous field under low load conditions. Due to end effect this creates much more significant losses with linear motors.

7. The Basic Principle of LIM Operation

The principle of operation of a LIM is the same as that of a rotary induction motor. A linear Induction motor is basically obtained by opening the rotating squirrel cage induction motor and laying it flat. This flat structure produces a linear force instead of producing rotary torque from a cylindrical machine. LIMs can be designed to produce thrust up to several thousands of Newton's. The winding design and supply frequency determine the speed of a LIM. The basic principle of LIM operation is similar to that of a conventional rotating squirrel-cage induction motor. Stator and rotor are the two main parts of the conventional three phase rotary induction motor. The stator consists of a balanced poly phase winding which is uniformly placed in the stator slots along its periphery. The stator produces a sinusoid ally distributed magnetic field in the air-gap rotating at the uniform speed $2\omega/p$, with ω representing the network pulsation (related to the frequency f by $\omega = 2\pi f$) and (p) the number of poles. The relative motion between the rotor conductors and the magnetic field induces a voltage in the rotor. This induced voltage will cause a current to flow in the rotor and will generate a magnetic field. The interaction of these two magnetic fields will produce a torque that drags the rotor in the direction of the field. This principle would not be modified if the squirrel cage were re placed by a continuous sheet of conducting.

8. Conclusion

In this Paper, the best way to cover a vast range of the rotor speed was found by defining an efficiency criterion at which the system would reach the steady state before certain time, which in this case was 2s. This was developed as a first stage in the control of the LIM, where setting a desired speed of the rotor and reaching it rapidly were the main concerns. In order to achieve this compromise of reaching the reference speed in a small time was found that the best way was not only by changing the frequency of the source but also the voltage amplitude. For that purpose two multiplexer blocks were implemented in the simulation model. The first one addressed the look up table corresponding to the range under which the speed reference of the rotor was covered outputting the optimum value of frequency. The second multiplexer addressed the value of voltage amplitude corresponding to the speed reference of the rotor (upon defined ranges) necessary for the system to reach the desired conditions.

References

[1] Chapman, S. (1999). Electric Machinery Fundamentals, New York: McGraw-Hill.

[2] Control Techniques. (2007). User Guide Unidrive Models size 0 to 6. Emerson Industrial Automation

[3] Doyle, M. Electromagnetic Aircraft Launch System - EMALS. Naval Air Warfare Center, Aircraft Division Lakehurst, NJ 08733

[4] Meeker D. Indirect Vector Control of a Redundant Linear Induction Motor for Aircraft Launch. NAVAIR Public Release 08-642.

[5] Nowak, L. (2001). Movement simulation in 3D eddy current transient problem COMPEL: The International Journal for Computation and Mathematics in Electrical and Electronic Engineering, Vol. 20, No. 1, 2001, pp. 293 – 302, MCB University Press.

[6] Ogata, K. (2002). Modern Control Engineering, New Jersey: Prentice Hall.

[7] Roa, C. (2010). Electric Motor Control System with Application to Marine Propulsion, MSc Thesis Dissertation, Department of Ocean and Mechanical Engineering, Florida Atlantic University, Florida.

[8] Upadhyay, J. Mahendra S.N, Electric Traction, Allied publishers Ltd, New Delhi, 2000.

[9] Xiros, N. (2010). Nonlinear Modeling of Linear Induction Machines for Analysis and Control of Novel Electric Warship Subsystems

26

Combined operation of SVC, PSS and increasing inertia of machine for power system transient stability enhancement

Bablesh Kumar Jha, Ramjee Prasad Gupta, Upendra Prasad

Electrical Engg. Deptt., B.I.T Sindri, Dhanbad ,Jharkhand, India

Email address:
bableshjha@gmail.com(B. Kumar Jha), Ramjee_gupta@yahoo.com(R. P. Gupta), Upendra_bit@yahoo.co.in (U. Prasad)

Abstract: In this paper improvement of transient stability by coordination of PSS (Power System Stabilizer) and SVC (Static var Compensator) and increasing inertia of synchronous machine has been observed. Because single method is not sufficient for improving stability. For this purpose a 9 bus multi machine system has been considered. Transient stability improvement has been tested subjected to three phase fault at bus 3 after 0.5 second and fault has been cleared after 1 second. By the use of PSS, SVC and by increasing inertia method for the test system the electromechanical oscillation for generator electrical power has been reduced and the steady state power transfer has been enhanced. In this paper the Inertia of the machine is not so much increased. Because after increasing inertia of the machine rotor will be havier.so that it is kept always within limit as considering its reliability and economy. And field voltage is also kept limited.

Keyword: Transient Stability, ETAP, PSS, Exciter, SVC

1. Introduction

Power system stability has been recognized as an important problem for secure system operation. Transient instability has been the dominant stability problem on most systems, and has been the focus of much of the industry's attention concerning system stability. As power systems have evolved through continuing growth in interconnections, use of new technologies and controls, and the increased operation in highly stressed conditions, different forms of system instability have emerged. For example, voltage stability, frequency stability and interarea oscillations have become greater concerns than in the past.

Classification of power system stability

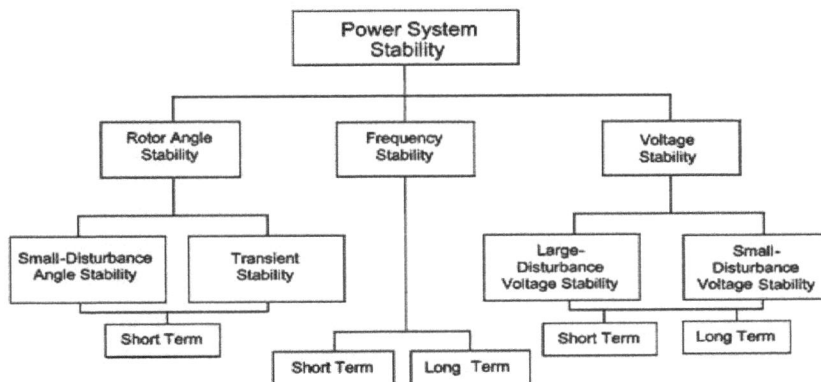

Fig-(1) classification of power system stability[1]

For convenience of analysis, stability problems are generally divided into two major categories:

· Steady-state stability
· Transient stability

The steady state stability is the stability of the system under conditions of gradual or relatively slow change in load. The load is assumed to be applied at a rate which is slow when compared either with the natural frequency of oscillation of the major parts of the system or with the rate of change of field flux in the rotating machine in response to change in loading.

The transient state stability refers to the maximum flow of power possible through a point without losing the stability with sudden and large changes in the network conditions such as brought about by faults, by sudden large increment of loads.

1.1. Transient Stability Analysis

A fault in the system will lead to instability and the machine will fall out of synchronism. If the system can't sustain till the fault is cleared, then the whole system will be in stabilized. During the instability not only the oscillation in rotor angle around the final position goes on increasing but also the change in angular speed. In such a situation the system will never come to its final position. The unbalanced condition or transient condition may leads to instability where the machines in the power system fall out of synchronism.

The system is subjected to a large variety of disturbances. The switching on and off of an appliance in the house is also a disturbance depending upon the size and capability of the interconnected system. Large disturbances such as lightning strokes, loss of transmission line carrying bulk power do occur in the system. Therefore transient stability is defined as the ability of the power system to survive the transition following the large disturbance and to reach an acceptable operating condition.

The physical phenomenon that occurs during a large disturbance is that there will be an imbalance between the mechanical power input and the electrical power output. This will tend to run the generator at high speed. The result will be the loss of synchronism of the generator and the machine will be disconnected from the system. This phenomenon is referred to as a generator going out of step.

The Etap Transient Stability Analysis is designed to investigate the system dynamic response disturbance. The program models dynamic characteristics of a power system, implements the user-defined events and action, solves the system network equation and machine differential equation interactively to find out system and machine response in time domain.

In this paper improvement of transient stability analysis of 9-bus multi machine system by using the coordinated effect of power system stabilizer (PSS), static var compensator (or SVC). In this analysis we create a three phase fault on specified bus and then investigation is to analyze the behavior of the synchronous machine. For this work we used the licensed packaged of ETAP software.

The paper is organized as follows: section 2 gives a brief introduction of power system stabilizer (or PSS) and static var compensator (or SVC) and exciter which has been used. A 9-bus multi machine system or test system is described In section 3. The computer simulation results for system under study are presented and discussed in Section 4 and in Section 5 conclusions are given.

2. Implementation of SVC and PSS

Static var compensators are shunt-connected static generator and/or absorbers whose outputs are varied so as to control specific parameters of the electric power system. The term 'static' is used to indicate that SVCs, unlike synchronous compensators, have no moving or rotating main components. By rapidly controlling the voltage and reactive power, an SVC can contribute to the enhancement of the power system dynamic performanance.Normallly, voltage regulation is the primary mode of control, and this improves voltage stability and transient stability. However, the contribution of an SVC to the damping of the system oscillation resulting from voltage regulation alone is usually small; supplementary control is necessary to achieve significant damping.[2]The effectiveness of an SVC in enhancing system stability depends on location of the SVC.).

A commonly used topology of a svc shown in fig.(2).Comprises a parallel combination of TCR and fixed capacitor. It is basically a shunt connected static var generator/absorber. Whose output is adjusted to exchange capacitive or inductive current so as to maintain or control specific parameters of electrical power system, typically bus voltage.

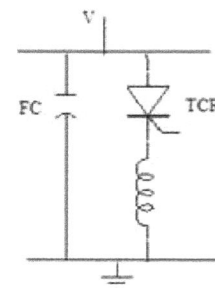

Fig(2) SVC

The reactive power injection of a SVC connected to bus k is given by

$$Q_k = V_k^2 B_{SVC}$$

$B_{svc} = B_c - B_L$; the symbol B_c and B_L are the respective susceptance of the fixed capacitor and TCR.it is also important to note that a svc does not exchange real power with the system.

The small signal dynamic model of a SVC is shown in fig.(3).ΔB_{svc} is defined as ΔB_c-ΔB_L.the differential equation from this block diagram can easily be defined as

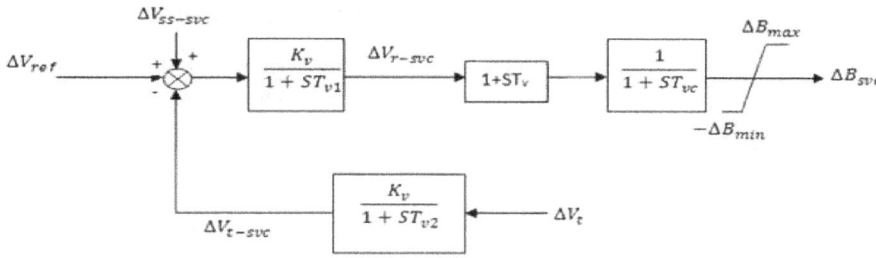

Fig(3) SVC Dynamic model

$$\frac{d}{dt}\Delta B_{svc} = \frac{1}{T_{svc}}\left\{-\Delta B_{svc} + \left(1 - \frac{T_{v1}}{T_{v2}}\right)\Delta V_{r-svc} - \frac{K_v T_{v1}}{T_{v2}}\Delta V_{t-svc}\right\} + \frac{K_v T_{v1}}{T_{v2}T_{svc}}\left\{\Delta V_{ss-svc} + \Delta V_{ref.}\right\}$$

$$\frac{d}{dt}\Delta V_{r-sv} = \frac{1}{T_{v2}}\left\{-\Delta V_{r-svc} - K_v\Delta V_{t-svc} + K_v V_{ref.} + K_v V_{ss-svc}\right\}$$

$$\frac{d}{dt}\Delta V_{t-svc} = \frac{1}{T_m}\left\{\Delta V_t - \Delta V_{t-svc}\right\}$$

K_v, T_{v1}, T_{v2} are the gain and time constant of voltage controller respectively.T_{svc} is the time constant associated with SVC response.T_m is the voltage sensing circuit time constant.

Svc control model which has been used is shown above in fig.(3).An SVC comprising a fixed capacitor and a thyristor-controlled reactor is considered for enhancement of the system stability.

2.1. Power System Stabilizer

The basic of a power system stabilizer (PSS) is to add damping to the generator oscillation by using auxiliary stabilizing signal(s).To provide damping, the stabilizer must produce a component of electrical torque in phase with the rotor speed variation. This is achieved by modulating the generator excitation so as to develop a component of electrical torque in phase with rotor speed deviation. Shaft speed, integral of power and terminal frequency are among the commonly used input signals to PSS.[3].PSS based on shaft speed signal has been used

successfully since the mid-1960s.a technique developed to derive a stabilizing signal from measurement of shaft speed of a system. Among the important consideration in the design of equipment for the measurement of speed deviation is the minimization of noise caused by shaft run out and other causes.[3-4]the allowable level of noise is dependent on its frequency. For noise frequency below 5Hz, the level must be less than 0.02%, since significant changes in terminal voltage can be produced by low-frequency changes in the field voltage. The application of shaft speed stabilizer to thermal unit requires a careful consideration of the effects on torsional oscillation. The stabilizer, while damping the rotor oscillation, can cause instability of the torsional modes. One approach successfully used to circumvent the problem is to sense the speed at a location on the shaft near the nodes of the critical torsional modes [5-6].In addition ,an electronic filter is used in stabilizing path to attenuate the torsional components.

Power system stabilizer which has been used in this research is shown below in fig.(4).

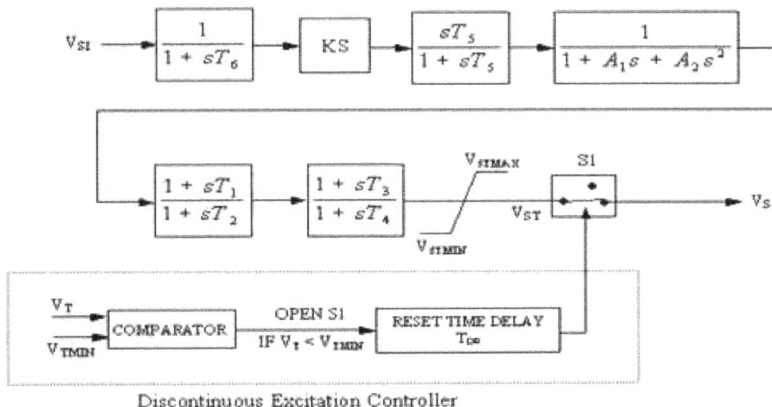

Discontinuous Excitation Controller

Fig.-(4) IEEE Type-1 PSS(PSS1A)

2.3. Exciter

The IEEE type of DC1 exciter is field-controlled dc commutator exciters, with continuously acting voltage regulators. The exciter may be separately excited or self exicted, the latter type being more common. When self-excited, Ke is selected so that initially Vr =0, representing operator action of tracking the voltage regulator by periodically trimming the shunt field rheostat set point.[7]

3. Model System

The test system that has been considered here is the 9-Bus Multi-Machine System as shown below in Fig.(5).

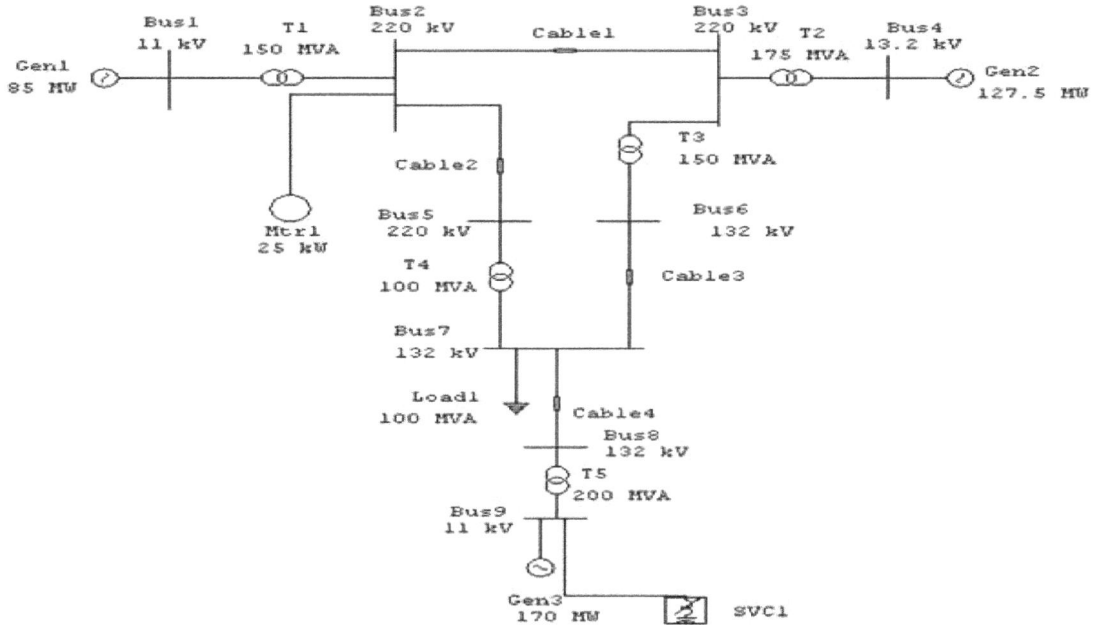

Fig.(5) test system

Which consisted 9-bus, three generators, four cables, five transformers and two loads one is static load of 100 MVA and another is an induction motor of 25 MW. Gen-1, Gen-2 and Gen-3 rated of 85 MW, 127.5 MW and 170 MW respectively. All other input parameters of generators are shown below in Table-1,2 and 3.

Table-1 Synchronous machine parameters

Machine			Rating		Positive sequence impedence(%)								Zero seq. Z(%)		
ID	TYPE	MODEL	MVA	KV	R_a	X_d''	X_d'	X_d	X_q''	X_q'	X_q	X_1	X/R	R_0	X_0
Gen1	Generator	Subtransient, Round-Rotor	100	11	1	19	28	155	19	65	155	15	7	1	7
Gen2	Generator	Subtransient, Round-Rotor	150	13.2	1	19	28	155	19	65	155	15	7	1	7
Gen3	Generator	Subtransient, Round-Rotor	200	11	1	19	28	155	19	65	155	15	7	1	7

Table-2 Dynamic Pparameters Of Synchronous Machine

Machine	Connected bus	Time cons.(sec.)					H(Sec.),,D(MW pu/Hz)&Saturation				Grounding	
ID	ID	T_{d0}''	T_{d0}'	T_{q0}''	T_{q0}'	H	%D	S100	S120	Sbreak	Conn.	Type
Gen1	Bus1	0.03	6.5	0.03	1.25	12	0	1.7	1.18	0.8	WYE	SOLID
Gen2	Bus4	0.03	6.5	0.03	1.25	12	0	1.7	1.18	0.8	WYE	SOLID
Gen3	Bus9	0.03	6.5	0.03	1.25	12	0	1.7	1.18	0.8	WYE	SOLID

Table-3 Mechanical parameters of synchronous machine

Machine		Generator/Motor			Coupling			Prime Mover/Load			Equivalent Total		
ID	TYPE	WR²	RPM	H	WR²	RPM	H	WR²	RPM	H	WR²	RPM	H
Gen1	Gen.	32406	1500	4	32406	1500	4	32406	1500	4	97217.99	1500	12
Gen2	Gen.	48609	1500	4	48609	1500	4	48609	1500	4	145826.98	1500	12
Gen3	Gen.	64811	1500	4	64811	1500	4	64811	1500	4	194432.98	1500	12

WR²: kg-m²   H: MW-Sec/MVA

The IEEE type of DC1 exciter, with continuously acting voltage regulators is installed with all generators. The exciter is self-excited. When self-excited, Ke is selected so that initially Vr =0, representing operator action of tracking the voltage regulator by periodically trimming the shunt field rheostat set point. Input data of exciter is shown below in Table-4.

Table-4 Power system stabilizer (pss) input data   type: pss1a

Generator ID	VSI	KS	VSTMax	VSTMin	VTMin	TDR	A1	A2	T1	T2	T3	T4	T5	T6
Gen1	SPEED	3.15	0.9	-0.9	0	0.2	0	0	0.76	0.1	0.76	0.1	1	0.1
Gen2	SPEED	3.15	0.9	-0.9	0	0.2	0	0	0.76	0.1	0.76	0.1	1	0.1
Gen3	SPEED	3.15	0.9	-0.9	0	0.2	0	0	0.76	0.1	0.76	0.1	1	0.1

An IEEE type of PSS1A is connected with all generators. The parameter of power system stabilizer is shown in Table-5.

Table-5 Exciter input data   type: dc1

Machine ID	Control Bus ID	KA Efd_max	KE	KF	TA	TB	TC	TE	TF	TR	VR_max	VR_min	SE_max	SE.75
Gen1	Bus1	46 2.63	0.05	0.1	0.06	0	0	046	1	0.005	1	-0.9	0.33	0.1
Gen2	Bus4	46 2.63	0.05	0.1	0.06	0	0	0.46	1	0.005	1	-0.9	0.33	0.1
Gen3	Bus9	46 2.63	0.05	0.1	0.06	0	0	0.46	1	0.005	1	-0.9	0.33	0.1

The rating of the SVC is assumed to be 200 Mvar capacitive and 200Mvar inductive .The voltage regulator gain is set at 10 to provide a 10% slope in the control range.

4. Simulation Result and Discussion

The Etap Transient Stability Analysis is designed to investigate the system dynamic response disturbance. The program models dynamic characteristics of a power system, implements the user-defined events and action, solves the system network equation and machine differential equation interactively to find out system and machine response in time domain.

In this paper we discuss the transient stability performance with PSS, SVC and by increasing inertia of synchronous machine. The transient stability improvement is not only sufficient by using one method. So here we use these three combined method for improving stability. Here we use ACCELERATED GAUSS-SEIDEL for initial load flow calculation. In which maximum number of iteration is

2000 and Solution Precision for the Initial LF is 0.000001 And Time Increment for Integration Steps (Δt) is 0.0100 and acceleration factor for the initial load flow is 1.45.Intial inertia of the installed machine was 4 MW-Sec/MVA and after increasing its inertia is 7 MW-Sec/MVA.Inertia of the machine is not so much increased. Because after increasing inertia of the machine rotor will be havier.so that it is kept always within limit as considering its reliability and economy.. The electromechanical oscillation for generator electrical power is reduced as well as the steady state power is also enhanced as seen in fig-(6).oscillation in terminal current and field current is also reduced and the magnitude of field current is also reduced as seen in Fig(7) &Fig.(9).Field voltage of Gen-1 is initially oscillated but after some time it is constant and within limit as shown in fig.(8).if only inertia of generator is increased then field voltage was does not change.

The different plot for Gen-1. When a three phase fault on bus-3 at 0.5 sec and cleared at 1 sec are shown below in fig.

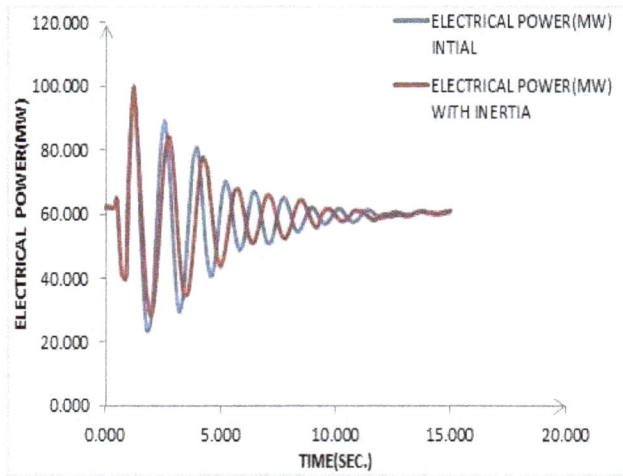

(1)

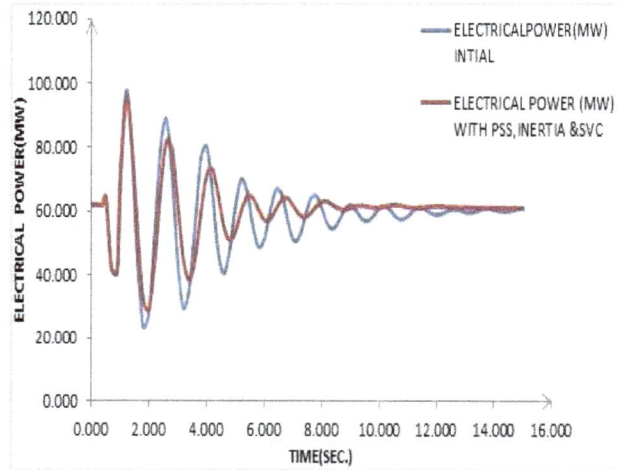

(2)

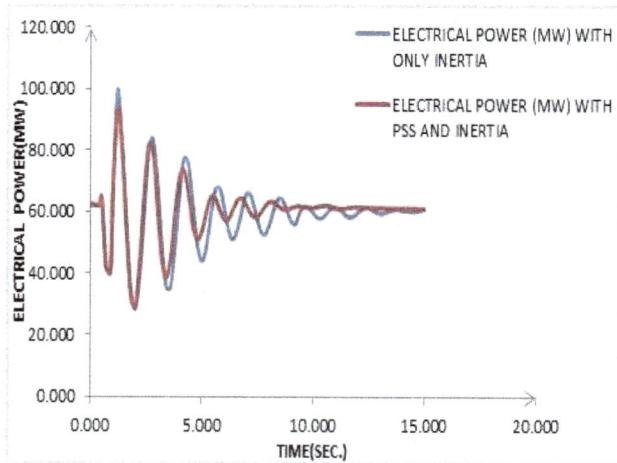

(3)

Fig(6) *Electrical power of Gen-1 (1)only inertia is increased (2) implementation of pss and inertia (3) implementation of inertia ,pss&svc.*

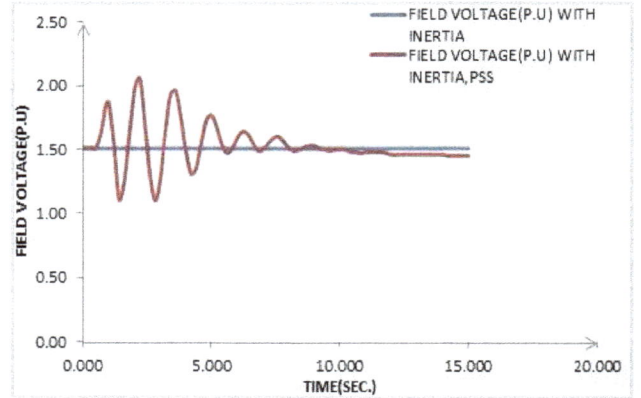

(1)

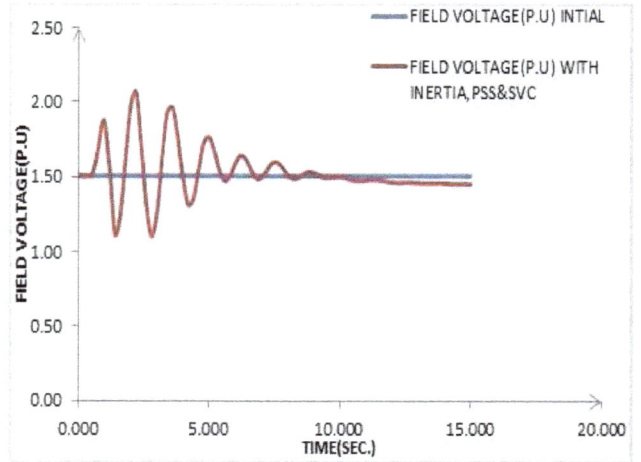

(2)

Fig(7) *Field voltage of Gen-1 (1)implementation of inertia&pss (2)implementation of inertia,pss and svc.*

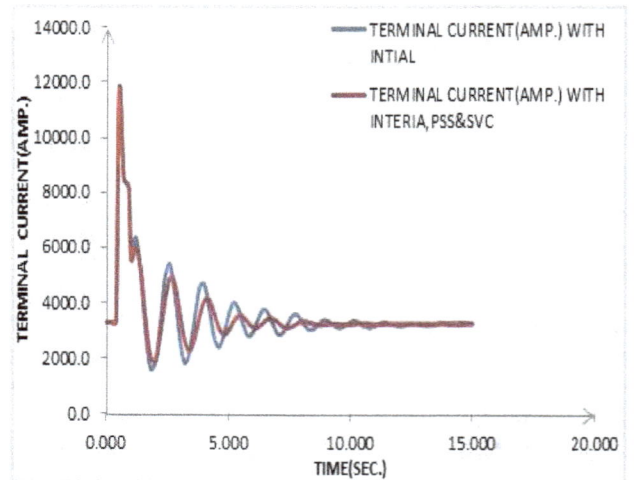

(1)

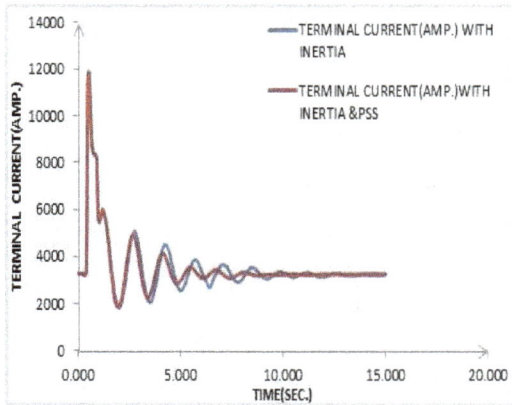

(2)

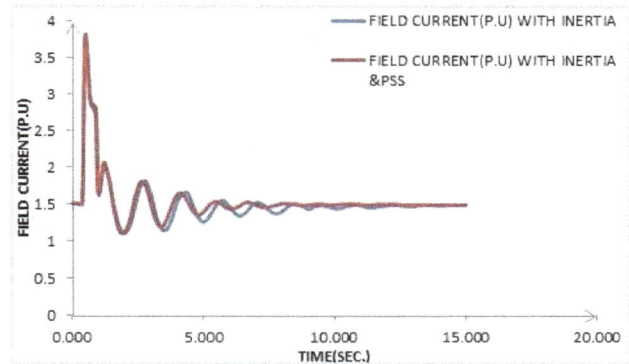

(3)

Fig(8) *terminal current of Gen-1 (1)only inertia is increased (2) implementation of inertia &svc (3) implementation of inertia,pss&svc.*

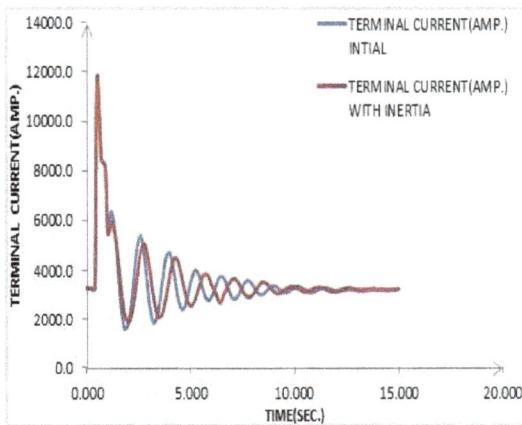

(1)

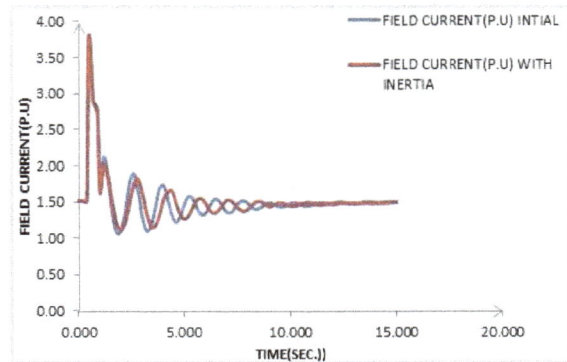

(2)

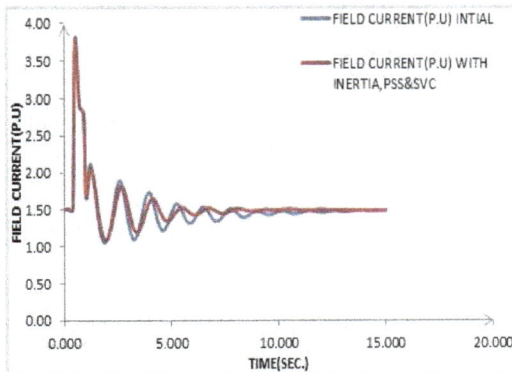

(3)

Fig(9) *Field current of Gen-1 (1)only inertia is increased (2) implementation of inertia and pss (3) implementation of inertia ,pss&svc.*

5. Conclusion

In this paper a new optimal control approaches for improvement of transient stability. Here Transient stability Performances of the multi machine system by using coordinated effect of PSS, SVC and by increasing inertia of machine and conventional method has been compared. And we see that better response in terms of electromechanical oscillation has been achieved in case of with PSS and SVC.The proposed method also has the advantage of considering the permissible system conditions. In general, analytical analysis and simulation results using E-TAP software show that the proposed and good flexibility for transient stability improvement.

Nomenclature

Parameter	Defintion	Parameter	Definition
Ra	armature resistance in ohm	Xd", Xq"	direct-axis,quadrature-axis synchronous subtransient reactance in percent
Xd',Xq'	direct-axis,quadrature-axis synchronous transient reactance in percent	Xd,Xq	direct-axis,quadrature-axis synchronous reactance in percent
X1	positive sequence reactance	R0,X0	zero sequence resistance.reactance
X/R	armature X/R ratio	Td0" ,Tq"	direct-axis,quadrature-axis subtransient open circuit time constant in seconds

Td0',Tq'	direct-axis,quqdrature-axis transient open-circuit time constant in seconds	H	intertia of synchronous machine
D	shaft mechanical damping term in percent	S100,S120	saturation factor at 100%,120% terminal voltage
Sbreak	per unit of terminal voltage at which the generator saturation curve skews from the air-gap line.	VSI	PSS input (speed, power or frequency) in pu
KS	PSS gain(p.u)	VSTmax, VSTmin	Maximum, Minimum PSS output(p.u)
TDR	Reset time delay for discontinuous controller(sec.)	A1,A2	PSS signal conditioning frequency filter constant(p.u)
T1,T3	PSS lead compensation time constant(sec.)	T2,T4	PSS leg compensation time constant(sec.)
T5,T6	PSS washout time constant(sec	KA	Regulator gain(p.u)
Efdmax	Maximum exciter output voltage(p.u)	KE	Exciter constant for self-excited field(p.u)
Kf	Regulator stabilizing circuit gain(p.u)	TA	Regulator amplifier time constant(sec.)
TB,TC	Voltage regulator time constant(sec.)	TE	Exciter time constant(sec.)
TF,	Regulator stabilizing circuit,Input filter time constant(sec.)	TR	Regulator Input filter time constant(sec.)
VRmax,	Maximum value of the regulator output voltage(p.u)	VRmin	Minimum value of the regulator output voltage(p.u)
SEmax	The value of excitation function at Efdmax	SE.75	The value of excitation function at 0.75 Efdmax
K	Voltage regulator gain(p.u)	a1,a2	Additional control signal gain
T	Voltage regulator time constant(sec.)	tm	Measurement time constant(sec.)
Tb	Thyristor phase control time constant(sec.)	td	Thyristor phase control delay(sec.)
t1,t2	Voltage regulator time constant(sec.)	tbmax,tbmin	Maximum,minimumsusceptance limit(p.u)

References

[1] P.L. Dandeno, A.N Karas, K.R. McClymont, and W.Watson,"Effect of High-Speed Rectifier Exication System on Generator Stability Limits," IEEE Trans.,vol. PAS-87,pp.190-201,January 1968

[2] W.Watson and G.Manchur, "Experience with supplementary Damping Signals for Generator Static ExicationSystem,"IEEE Trans.,Vol.PAS-92,pp. 199-203,January/February 1973

[3] W.Watson and M.E Coultes ,"Static Exicter Stabilizing Signals on Large Generators-Mechanical Problems,"IEEE Trans.,Vol. PAS-92,pp.204-211,January/February 1973

[4] P.Kundur ,D.C.Lee and H.M. Zein EL-Din,"Power system stabilizer fot thermal units:Analytical Techniques and On-Site Validation,"IEEE Trans.,Vol.PAS-100,pp. 81-95,January 1981

[5] M.L. Shelton, R.F Winklemen, W.A Mittelstandt, and W.L Bellerby,"Bonneville Power Administration 1400 MW Braking Resistor,"IEEE Trans.,VOL.PAS-94,pp.602-611,March/april 1975

[6] P.K. Iyambo, R. Tzonova, "Transient Stability Analysis of the IEEE 14-Bus Electrical Power System", IEEE Conf. 2007

[7] N. Mo, Z.Y. Zou, K.W. Chan, T.Y.G. Pong, "Transient stability constrained optimal power flow using particleswarm optimization", IET Generation, Transmission & Distribution, Vol. 1, pp. 476–483, 2007

[8] D. Chatterjee, A. Ghosh, "Transient Stability Assessment of PowerSystems Containing Series and Shunt Compensators," IEEE Trans. onpower systems, vol. 22, no. 3, August 2007

[9] IEEJ Technical Committee, Standard models of power system, IEEJ Technical Report, Vol. 754, 1999

[10] P. M. Anderson and A. A. Fouad. Power System Control and Stability.The IEEE Press, 1995.

[11] M. H. Haque, "Evaluation of First Swing Stabilityof a Large Power System With Various FACTS Devices",IEEE Trans. Power Systems, vol.23, no.3,pp.1144-1151, August 2008

[12] P.Kundur,"effective use of power system stabilizer for enhancement of power system stability."in proc. 1999.IEEE PES Power engg. Society Summer meeting pp. 96-103

[13] G.K morison ,B.Gao and P.kundur,"volatage stability analysis using static and dynamic approaches,"IEEE Trans. Power system ,vol.8,pp. 1159-1171,aug.1993

[14] LerchE,PovhDAdvancedSVCcontrolfordampingpower system oscillations. IEEE Trans Power Syst 16(2)pp.:524–535, (1991)

27

Disruptions and malfunction control in ORC using spiral predictive model

Fareed ud Din, Abdul Rehman Raza, Muhammad Azam

Faculty of Engineering and Technology, Lahore, Pakistan

Email address:

fareeduddin@superior.edu.pk(F. U Din), abdurraza@superior.edu.pk(A. R. Raza)

Abstract: This paper provides a critical and analytical assay in the process vicinity of an Organic Rankine Cycle (ORC) resulting in a representation of a controlling model named as Spiral Model as the best approach to implement for an efficient Plant Management (PM) and Risk Mitigation Planning (RMP), focusing on the robust and elegant energy production. There have been so many predictive and sensing process models presented for a gist and substantial control of the ORC plant in recent years but the proposed Spiral Predictive Model (SPM), eliminating all the limitation of all previously implemented models, provides the robustness by performing all the roles in increments; e.g. in the changing controllers, complex time-frequency characteristics, fault detectors for turbines against disruptions and the multi-switching techniques needs to be cascaded ahead of time with predictive and detective techniques. The proposed model optimizes the performance of ORC by response tracking and recursive correction which relegates the errors and sudden disturbance in the process flow. Fast response and recursive correction nicely handles Demand Response (DR) and parameters variations at different working modules which ultimately provide the dynamic performance capability. This study will be elaborating efficient model design and implementation to conjure up a well-designed working flow in an ORC plant.

Keywords: Spiral Predictive Model (SPM), Organic Rankine Cycle (ORC), Demand Response (DR), Plant Management (PM), Risk Mitigation Planning (RMP)

1. Introduction

Organic Rankine Cycle (ORC) is a process for elegant energy production by using an organic, high molecular mass fluid with low boiling point than the water-steam phase change. The organics fluid used allows Rankine cycle to use and recover heat from temperature sources such as biomass combustion, industrial waste heat, geothermal heat, solar ponds etc. The heat is converted into useful work that can itself be converted into electricity. To convert excess heat of the system into electrical power using efficient generator and robust turbines; controlled by sensitive controllers and the process model which takes control the overall system. The ORC units and accompanying control system with associated equipment upgrade and present data, quantifying the energy saving benefit; which is also the main focus of the paper. The Rankine Cycle is a well known and understood thermodynamic cycle used to convert heat into work, most commonly applied in power generation. In the conventional Rankine Cycle, the working fluid (usually water) is heated to saturate in a boiler, traverse through a turbine while producing work, returns to the liquid state in a condenser, and is pumped back into the boiler to repeat the cycle [1]. The ORC differs from the traditional Rankine Cycle because instead of water, a high molecular mass organic fluid is used as the working fluid. This organic fluid (normally selected organic fluids are R134a, R113, R425ca, R245fa, R123) is typically characterized by a lower boiling point than that of water, enabling the ORC to operate at lower temperatures and take advantage of waste heat generated at lower temperatures than other recovery methods [2]. The simple structure of an ORC is shown in the figure-1.

As shown by different studies it has been proven that there is an unexpected increase in electrical consumption and load but the intensification in generating electric resources is less [2],[3]. The main objective of this paper is to optimize the control process of the plant by taking a deeper look onto the sudden disturbances which causes the problem and irregularity in energy production [7]. The proposed system to optimize the functionality of ORC with specification and customization provide the chance to take maximum benefit of the working fluids to the peak extent of theirs via heat recovery system of ORC.

2. Selection of Process Model

The very first question arises in designing a novel process model is that are the previously designed models are not giving the outputs as per the expectations; as the most widely used strategies in thermal power plants use its simple structure with no precise modeling due to the uncertainty, nonlinearity, long delay and time-varying dynamics of the boiler-turbine systems and it cannot provide satisfactory performance with its monotonous control mechanism for various changing load demands and parameters variations of the complex process of thermal power plants [1, 8, 9].

Many developed countries like China and Japan are applying ORC to generate power due to the great advantages of improving efficiency, efficiently saving energy and less generated pollution [1]. So it is essential to cope up the continuously varying Demand Response of electrical consumption while maintaining the temperature within a designed range.

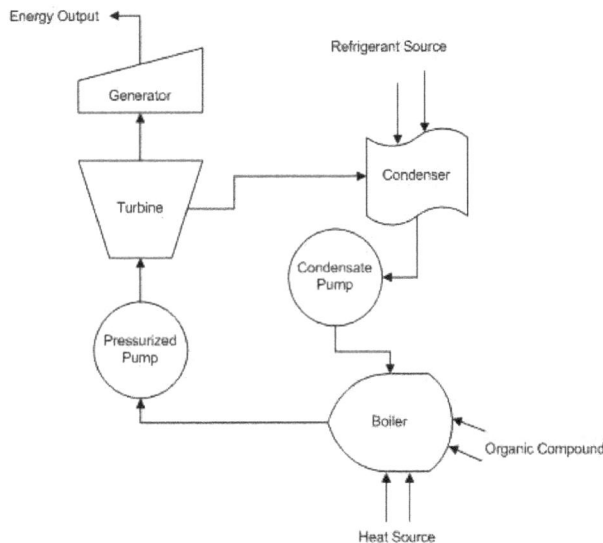

Figure 1. General Processing Flow of ORC

The complex process of ORC plants has resulted in different development of control strategies by several authors and experts [2, 3]. A predictive application of control is necessary for self-tuning is necessary for an efficient system [3]. A linear quadratic Gaussian (LQG) controller is proposed by Cori and Maffezzoni in [1, 8]. Pellegrinetti and Bentsman designed a robust controller for boilers [1, 9]. Ben- Abdennour and Lee presented loop transfer recovery method [6]. These controllers and models are designed using the mathematical modeling techniques for the efficient control of the ORC plant. Several Other techniques are also implemented in the domain based on controller designs, many artificial intelligence techniques such as dynamic

matrix control [12] fuzzy control algorithms [13], neural network control methodologies [14], genetic algorithm method [15] are applied to thermal power plant control.

The proposed Spiral Predictive Model (SPM) based on step response of the plant, using optimized calculation, relegating the error between output through Response Collector (RC) and Feedback Check (FC) modules of Spiral Predictive Model (SPM). It does need prior response of the process structure to avoid the complex process disruptions identification. In this paper, Spiral Model Controllers can adapt with the concurrently varying conditions like: stability and adaptability [1]. As Spiral Model using the optimization technique to compute a sequence of inputs, responses and feedback to predict the future outputs within the designed range; as addressed to be done in several studies [16, 17], and the whole process is repeated at each modular interval.

Therefore, it has good tracking of feedback and performance to compensate a dynamic recital output. Due to the dynamics of boiler-turbine system [17], single-loop control can't achieve desired performance [1].

3. Proposed Spiral Predictive Model

This paper elaborate the SPM as predictive and robust responsive process model to control all the functional module of an ORC plant as it include a lot working conditions to be checked and corrected to be lied within the desired range. This model consists of the nested-loop strategy to cascade the sudden disruptions and irregularities occurring at the turbine rolling or the run time correction. Each loop consists of a Spiral Model-Response Collector (SPM-RC) and a Spiral Model-Feedback Correction (SPM-FC). Both of these modules in accordance to the other act elegantly to analyze the signals and their trend and correct the disruptions occurring in the turbine. SPM up to high extent provides the solution to the compromising issues of earlier represented control models.

Hence for highly non stationary signals this analysis insufficient. But on the flipside benefit of Fourier Transformation; the correlation can be found between the time and frequency domain of a signal. Still the problem occurs for short term, short range signals with varying frequency. In recent years there has been an uplifting in the research of signal analysis concerning the time-frequency domain.

The time-frequency analyses should be able to analyze a non stationary signal with not only signal frequencies, but also the time range when these frequencies occur. In principle there are two basic approaches to analyze a non stationary vibration signal in time and frequency domain simultaneously. One approach is to split a non stationary vibration signal at first into segments in time domain by proper selection of a window function and then to carry out a Fourier transform on each of these segments separately. This is the basic idea for the calculation of the Short Time Fourier Transforms (STFT). Gang Zhao [7] and Wang [18] present Short Time Fourier Transform (STFT) to be a powerful tool in detecting distortion in the signal at an early state. The short-time Fourier transform (STFT) is used to determine the sinusoidal frequency and phase content of signal as it changes over time. Simply, in the continuous-time case, mathematically, this is written as:

As shown in the figure-2:

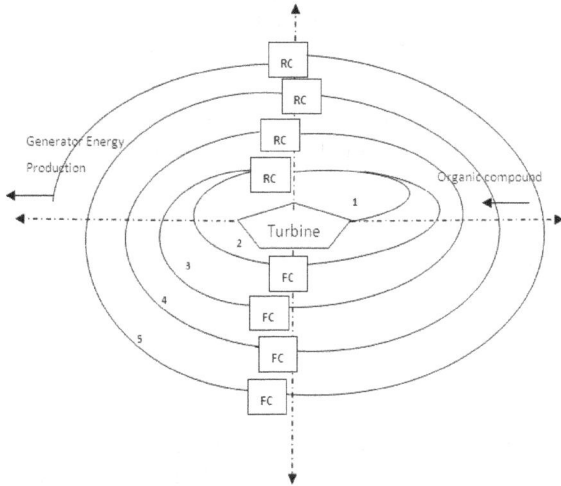

Figure 2. *Spiral Predictive Model*

$$\{x(t)(\tau, \omega)\} = \int_{-\infty}^{\infty} (t)(t - \tau)e^{-j\omega t} \, dt \qquad (1)$$

Where w(t) is a window function and x(t) is the signal to be transformed, x(t)w(t-τ), a complex function representing the phase and magnitude of the signal over time and frequency. In the discrete time case, the data to be transformed could be broken up into chunks or frames and then each chunk is Fourier transformed and iteratively summed up into the earlier values. For discrete signals the mathematical representation can be expressed as:

$$\{x[n]\}(\tau, \omega) = \int_{-\infty}^{\infty} x[n]\omega[n - m]e^{-j\omega n} \qquad (2)$$

In this case, m is discrete and ω is continuous, but in most typical applications the STFT is performed on a computer using the Fast Fourier Transform, so both variables are discrete and quantized. STFT is static as it cannot be altered whine one been selected for a specific window. The other approach is the wavelet transform (WT) for non stationary vibration signal analysis. The soul of this technique is filtration into different frequency bands split into segments in time domain. Wavelet Transform (WT), eliminating the drawbacks of the Fourier Transform, uses a dynamic windowing technique. Wavelet Transform provides the dual benefit; more precise low-frequency information and shorter regions high-frequency information. The definition of the continuous wavelet transform (CWT) is written by Gang Zhao in [7] as following in mathematical notation:

$$CWT(a, b) = \frac{1}{\sqrt{a}} \int s(t)\psi * conj.(\frac{t - b}{a}) \, dt \qquad (3)$$

Wavelet transform can be used to detect the fault of fans and the signal component indicative of a fault can be identified from the sound signals [19]. These signal analysis strategies can provide the robust solution to the disruptions occurring in signal analysis and turbine flow if they are used in the increments for continuous improvements.

4. Turbine Disruptions

Power generation in distributed environment normally contain the high use of steam turbines. These units have increased considerably due to a restructuring of the energy sector worldwide [10]. Steam Turbines use the process of converting thermal energy into electrical energy. Steam turbine does have a balanced construction, high efficiency, easy maintenance, and availability in large sizes these features lead it to the best of the turbines in general [11]. There are so many disruptions occur in the process flow of the turbine which are necessary to be discussed.

For disruption detection the vibration diagnosis has a wide scope as a tool as with the vibration analysis it is possible to detect a disruption in any process in any interval of time which reduces production overhead and loss of time [12]. The vibration diagnosis is normally carried out in the following main steps: signal measurement, signal analysis, diagnosis and strategic decision, where the signal analysis plays a key role and has the task of extracting useful information, filtering noise from a measured vibration signal and finding the fault feature and its developing trend. Traditional spectral analysis techniques, based on the Fourier transformation provide a good description of stationary and pseudo stationary signals [7], [12].

Several faults of steam turbine are simulated and discussed in the paper. The most commonly fault of steam turbine is unbalance. By applying Fourier Transformation the time-frequency graphs shows the similar results for stationary signal as mentioned by Gang Zhao in [7]. The peak frequency and the power distribution in the time frequency are almost the same. The next fault of steam turbine is component loss such as blades and wings of the turbine. It will cause damage of turbine and performance decline. Hence early diagnosis is very necessary for risk mitigation. These two faults are eminent in practice of steam turbine and they have the similar symptom by using Fourier Transformation. The third fault of steam turbine is misalignment. When allay of a coupled shafts do not coincide with respect to the time domain. Parallel misalignment occurs when shafts are centered align are parallel but not coincident and the vice versa is the angular misalignment. The fourth fault of steam turbine is component laxity. Component Laxity between a machine and its component will lead to the looseness which ultimately increase the vibration disruption in the direction if the least stiffness. This is usually the horizontal direction, but it depends on the physical layout of the machine. Low-order harmonics are also commonly produced if the laxity is severe. Component flexibility or looseness can be caused by loose bolts, corrosion, or cracking of mounting hardware.

Another common problem in newly rebuilt, modified or customized turbine rotors is a slight rubbing condition if the turbine rotors are initially operated. Rotor rubs never operates over an extended period of time and long delay; they usually increase the clearances until the rub has been cleared or, if not corrected, they will wear away the internal clearances until the machine cannot be operated. The rub fault of steam turbine can be detected in the early stages to cascade the failure ratio.

5. Response Check and Feedback Correction

Structure and representation of the Response Control Thread and Feedback Corrective Control Thread is mentioned in the figure-3.

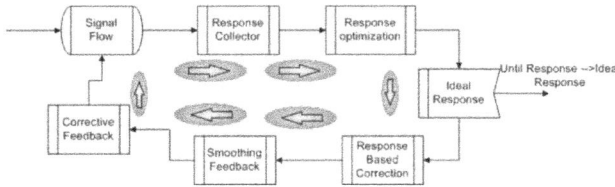

Figure 3. *SPM-RC and SPM-FC Loop*

Response Check of the Spiral Model (SPM-RC) utilizes the sequences of the feedbacks and responses in the spiral and looping scenario.

SPM-RC can be represented with the following mathematical equation:

$$W(n) = [\omega(n + i) \mid i \in N] \wedge i \leq N]^T \qquad (4)$$

$$\Delta u(n) = c^T[A^T Q(j)A + R(j)I]^{-1} A^T Q(\omega(k) - y(k)) \quad (5)$$

Where $\Delta u(n)$ is the control increment, $Q(j)$, $R(j)$ are the output error coefficient and $W(n)$ is the response. Response optimization is also necessary because responses are normally generated on the basis of the ideal situations so it need to be optimized. The optimization strategy can be repressed as:

$$J(k) = \sum_{i=0}^{N} Q(j)[y(k + j) - \omega(j + j)]^2$$

$$+ \sum_{i=0}^{M} R(j)[y(k + j) - \omega(j + j)]^2$$

Minimum and Optimized value of $J(k)$ can be achieved by $\frac{\partial j}{\partial \Delta u}$ using equation (5). Feedback correction module of Spiral Predictive Model (SPM-FC) receives and corrected the output on the basis of response been collected earlier by the SPM-RC module. Mathe-

matically it can be represented as:

$$Y_n = \sum_{i=0}^{n} Y_i(k) + A * \Delta U_M(k) \qquad (6)$$

Her in equation (6), $Y_i(k)$ is the previous output, A is the Dynamic Matrix and $\Delta U_M(k)$ is the ideal output.

This will yield the expected output but this estimated output may contain the errors which will eventually lead towards the disruption again. As SPM provides the best Risk Mitigation Plan hence it is necessary to address the smooth error free expected output. The following equation is used for the smooth output:

$$\omega(k + j) = \sum_{j=0}^{M} \beta^j y(k) + (i - \beta^j)y_s \qquad (7)$$

$$e(k + 1) = y(k + 1) - Y_n(k + 1) \qquad (8)$$

$$\yen_{CF} = Y_n + C_v * e(k + 1) \qquad (9)$$

Equation 8 represents the error by removing the run time response to the ideal response. The finding will lead to the corrective feedback formulation as errors have been found and now no risks are left as this formulation leads to the dual and flip risk mitigation as correcting the corrective output. Equation (9) represents the corrective feedback. $\yen_{CF}$ is Corrective Feedback Y_n is the previously formulated feedback on the basis of prior responses and C_v is the Correction Vector.

6. Conclusion

This paper analyzes the ORC plant management, control strategy, risk mitigation by proposing the best suiting controlling scheme named as Spiral Predictive Model (SPM). SPM strategy is an effective method to control rankine system nonlinearity, parameters uncertainty and long delay problems by using its incremental approach. Response Check controller and module can change the feedback with the continuously changing operating condition, prediction with the help of Feedback Correction Controller and Module. The ideal output is found and compensated with the smooth error free output with good dynamic and static performance. Additionally, because of the spiral intrinsic behavior of SPM, inhibits the disruptions occurring in the turbine flow by sensing and analyzing the signal. The design of the strategy of SPM-RC and SPM-FC is easy and simple to be implemented and it is well defined in the paper. So the proposed predictive control strategy can be deployed in practical industrial process in order to achieve robustness and elegancy in operation of organic rankine cycle. As everything comes up with its pros and cons so this strategy also includes a limitation that for long-range predictions it gets complex. Hence for successful implication needs a deep cautiousness while implementing Response Collector and Feedback Correction Module of SPM.

References

[1] Main-Steam Temperature Control for Ultra- Supercritical Unit Using Multi-Model Predictive Strategy, Ding Li, Hong Zhou, 978-1-4673-4584-December 2012 IEEE

[2] V. Maizza*, A. Maizza, Unconventional working Fluids in organic Rankine-cycles for waste energy recovery systems (2000)

[3] Samuel Sami, A concept of power generator using wind turbine hydrodynamic retarder, and organic Rankine cycle drive, Journal Of Renewable And Sustainable Energy 5, 023123 (2013)

[4] B. Boukhezzar, H. Siguerdidjane, and M. Maureenhand, "Nonlinear control of variable-speed wind turbines for generator torque limiting and power optimization," J. Sol. Energy Eng. 128(4), 516–530 (2006).

[5] I. Daubechies, Ten Lectures on Wavelets (Society for Industrial and Applied Mathematics, Philadelphia, PA, 1992).

[6] A. Kusiak, Z. Song, and H. Zheng, "Anticipatory control of wind turbines with data-driven predictive models," IEEE Trans. Energy Convers. 24(3), 766–774 (2009).

[7] Gang Zhao, Dongxiang Jiang, Jinghui Diao, Lijun Qian, "Application of wavelet time-frequency analysis On fault diagnosis for steam turbine, Surveillance 5CETIM Senlis2004

[8] R. Cori and C. Maffezzoni, "Practical optimal control of a drum boiler power plant," Automatica, vol. 20, pp.163-173, 1984.

[9] G. Pellegrinetti and J. Bentsman, "H Controller design for boilers, Int. J. Robust Nonlinear Control," vol. 4, pp.645-671, 1994.

[10] B. W. Hogg and N. M. Ei-Rabaie, "Multivariable generalized predictive control of a boiler system," IEEE Trans. Energy Convers, vol. 6, pp. 282-288, June. 1991.

[11] Un-Chul Moon and Kwang Y. Lee, "An Adaptive Dynamic Matrix Control With Fuzzy-Interpolated Step-Response Model for a Drum- Type Boiler-Turbine System," IEEE Trans. Energy Convers, vol. 26, pp. 393-401, June. 2011.

[12] Xiangjie Liu, Xuewei Tu, Guolian Hou and Jihong Wang, "The Dynamic Neural Network Model of a Ultra Super-critical Steam Boiler Unit," American Control Conference on O'Farrell Street, San Francisco CA, USA, June 29-June 01,2011.

[13] Adarsha Swarnakar, Horacio Jose Marquez and Tongwen Chen, "New Scheme on Robust Observer-Based Control Design for Interconnected Systems With Application to an Industrial Utility Boiler," IEEE Transactions on control systems technology, vol.16, pp. 539-547, May.2008.

[14] Shaoyuan Li, Hongbo Liu, Wen-Jian Cai, Yeng-Chai Soh and Li-Hua Xie, "New Coordinated Control Strategy for Boiler-Turbine System of Coal-Fired Power Plant," IEEE Transactions on control systems technology, vol.13, PP. 943-954, Nov.2005.

[15] Kang Y. Lee, Joel H. Van Sickel, Jason A. Hoffman, Won-Hee Jung and Sung-Ho Kim, "Controller Design for a Large-Scale Ultrasupercritical Once-Through Boiler Power Plant," IEEE Trans. Energy Convers, vol. 25, pp. 1063-1070, Dec 2010.

[16] Ajay Gautam, Yun-Chung Chu and Yeng Chai Soh, "Optimization Dynamic Policy for Receding Horizon Control of Linear Time-Varying Systems with Bounded Disturbances," IEEE Trans. Autom. Control, vol. 57, PP. 973-988, April 2012.

[17] Y.I.Lee and B.Kouvaritakis, "Receding horizon output feedback control for linear systems with input saturation," IEE. Proc.-Control Theory Appl., Vol. 148, pp. 109-115, March 2001.

[18] Wang, W.J., McFadden, P.D., "Early Detection of Gear Failure by Vibration Analysis–I. Calculation of the Time-Frequency Distribution", Journal of Mechanical Systems and Signal Processing, 1993,7(3):193-203.

[19] Katshuhiko Shibata, Atsushi Takkahashi, Takuya Shira "Fault Diagnosis of Rotating Machinery Through Visualization of Sound Signals", journal of Mechanical Systems and Signal Processing, 2000 14(2): 229-241.

28

Transient stability of 11-bus system using SVC and improvement of voltage profile in transmission line using series compensator

Ramlal Das, D. K. Tanti

Electrical Engineering Department, Bit Sindri

Email address:

ramlal0326@gmail.com (R. Das),dktanti66@gmail.com (D. K. Tanti)

Abstract: Power system stability is defined as the ability of power system to preserve its steady stability or recover the initial steady state after any deviation of the system's operation. Present time power systems are being operated nearer to their stability limits due to economic and environmental reasons. Maintaining a stable and secure operation of a power system is therefore a very important and challenging issue. Transient stability has been given much attention by power system researchers and planners in recent years, and is being regarded as one of the major sources of power system insecurity. Shunt FACTS devices play an important role in improving the transient stability, increasing transmission capacity and damping low frequency oscillations. In this work 11-bus power system network has been modeled using MATLAB SIMULINK software. The power system network under study consist of three units of power plant each producing 20 KV and step up by two winding transformer to 230 KV. For parallel operation of two different power plants, the frequency and the terminal voltage has been kept constant to avoid circulating current in the existing network. A Static VAR Compensator and a series compensator have been used in the considered network for improving the transient stability and to increase the transmission capacity of the system.

Keywords: Power System Stability, Transient Stability, SVC, Series Compensator

1. Introduction

Power systems generally consist of three stages: generation, transmission, and distribution. In the first stage, generation, the electric power is generated mostly by using synchronous generators. Then the voltage level is raised by transformers before the power is transmitted in order to reduce the line currents which consequently reduce the power transmission losses. After the transmission, the voltage is stepped down using transformers in order to be distributed accordingly. Power systems are designed to provide continuous power supply that maintains voltage stability. However, due to undesired events, such as lightning, accidents or any other unpredictable events, short circuits between the phase wires of the transmission lines or between a phase wire and the ground which may occur is called a *fault*. Due to occurring of a fault, one or more generators may be severely disturbed causing an imbalance between generation and demand. If the fault persists and is not cleared in a pre-specified time frame, it may cause severe damages to the equipment's which in turn may lead to a power loss and power outage. Therefore, protective equipment's are installed to detect faults and clear/isolate faulted parts of the power system as quickly as possible before the fault energy is propagated to the rest of the system.

2. Literature Review

For the purpose of this review, a literature survey has been carried out including two of the most important and common databases, namely, the IEEE/IEE electronic library and Science Direct electronic databases. The survey spans over the last 15 years from 1990 to 2004. For convenience, this period has been divided to three sub-periods; 1990–1994, 1995–1999, and 2000–2004. The number of publications discussing FACTS applications to different power system studies has been recorded. The results of the survey are shown in Figure 1. It is clear that the applications of FACTS to different power system studies have been drastically increased in last five years. This observation is more pronounced with the second generation devices as the interest is almost tripled. This shows more interest for the VSC-based FACTS applications. The results also show a decreasing interest in TCPS while the interest in SVC and TCSC slightly increase. Generally, both generations of FACTS have been applied to different areas in power system studies including optimal power flow [5–9], economic power dispatch [20], voltage stability [11; 22], power system security [3], and power quality [7].

Applications of FACTS to power system stability in particular have been carried out using same databases. The results of this survey are shown in Figure 7. It was found that the ratio of FACTS applications to the stability study with respect to other power system studies is more than 60% in general.

3. Classification of Stability

Figure 1.1 provides a comprehensive categorization of power system stability. As Depicted by Figure 1, there are two main classes of stability: rotor angle stability and voltage stability. Rotor angle stability has two main subclasses: small-disturbance angle (steady-state) stability and transient stability. A power system is considered to be steady-state stable if, after any small disturbance, it reaches a steady state operating condition which is identical or close to the pre disturbance operating condition. A power system is transient stable for a large disturbance or sequence of disturbances if, following that disturbances it reaches an acceptable steady-state operating condition. Unlike steady-state stability which is a function only of the operating condition, transient stability is more complicated since it is a function of both operating condition and the disturbance [2]. Voltage stability also has two main subclasses: large disturbance voltage stability and small disturbance voltage stability.

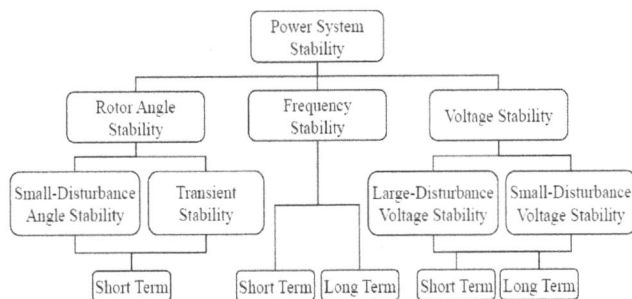

Figure 1. Classification of power system stability

4. Transient Stability

Transient stability is the ability of the power system to maintain synchronism when subjected to sever transient disturbance. The response to this type of disturbance involves large excursions of rotor angles and is influenced by nonlinear power-angle relationship. Stability depends on the initial operating state of the system and the severity of the disturbance. The system usually altered after the disturbance which may cause the system to operate in a different steady-state status from that prior the disturbance. Power systems are designed to be stable for a selected set of contingencies. The contingencies usually considered are short-circuits of different types: phase-to-ground, phase-to phase- to-ground, or three-phase. They are usually assumed to occur on the transmission lines, but occasionally bus or transformer faults are also considered. Figure 2.1 illustrates the behavior of a synchronous machine for stable and unstable situations. In Case 1, the rotor angle increases to a maximum, then decreases and oscillates with decreasing amplitude until it reaches a steady state. This case is considered transient stable. In Case 2, the rotor angle continues to increase steadily until synchronism is lost. This type on transient instability is referred to as first-swing instability. In Case 3, the system is stable in the first swing but becomes unstable as a result of growing oscillations as the end state is approached. This form of instability occurs when the post fault steady-state condition is itself is small-signal unstable. In transient stability studies, the study period of interest is usually limited to 3 to 5 seconds following the disturbance, although it may extend to about ten seconds for very large systems with dominant inter area modes of oscillation.

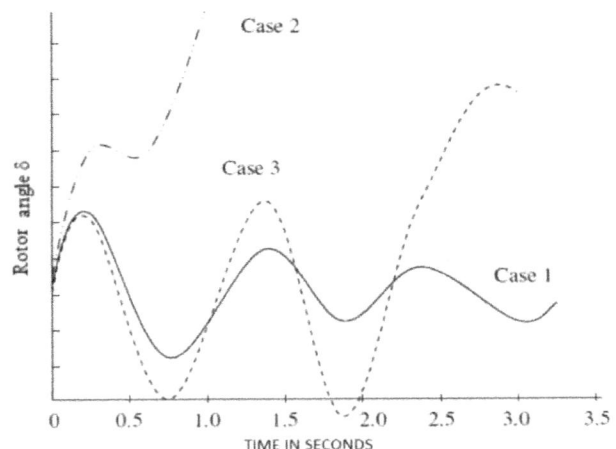

Figure 2. Rotor angle response to a transient disturbance.

Methods of transient stability analysis
1.   Swing equation
2.   Equal area criterion
3.   Numerical Integration Methods
4.   Direct Methods Transient Stability Analysis

5. Static Var Compensator (SVC)

The Static Var Compensator (SVC) is a shunt device of the Flexible AC Transmission Systems (FACTS) family using power electronics to control power flow and improve transient stability on power grids [1]. The SVC regulates voltage at its terminals by controlling the amount of reactive power injected into or absorbed from the power system. When system voltage is low, the SVC generates reactive power (SVC capacitive). When system voltage is high, it absorbs reactive power (SVC inductive). The variation of reactive power is performed by switching three-phase capacitor banks and inductor banks connected on the secondary side of a coupling transformer. Each capacitor bank is switched on and off by three thyristor switches (Thyristor Switched Capacitor or TSC). Reactors are either switched on-off (Thyristor Switched Reactor or TSR) or phase-controlled (Thyristor Controlled Reactor or TCR).

It is a variable impedance device where the current through a reactor is controlled using back to back connected thyristor valves. The application of thyristor valve technology to SVC is an offshoot of the developments in HVDC technology. The major difference is that thyristor valves used in SVC are rated for lower voltages as the SVC is connected to an EHV line through a step down transformer or connected to the tertiary winding of a power transformer.

The application of SVC was initially for load compensation of fast changing loads such as steel mills and arc furnaces. Here the objective is to provide dynamic power factor improvement and also balance the currents on the source side whenever required. The application for transmission line compensators commenced in the late seventies. Here the objectives are:

1. Increase power transfer in long lines
2. Improve stability with fast acting voltage regulation
3. Damp low frequency oscillations due to swing (rotor) modes
4. Damp sub synchronous frequency oscillations due to torsional modes
5. Control dynamic over voltages

5.1. Configuration of SVC

There are two types of SVC:
1. Fixed Capacitor-Thyristor Controlled Reactor (FC-TCR)
2. Thyristor Switched Capacitor (TSC-TCR).

The second type is more flexible than the first one and requires smaller rating of the reactor and consequently generates fewer harmonic. The schematic diagram of a TSC-TCR type SVC is shown in Fig.3 this shows that the TCR and TSC are connected on the secondary side of a step-down transformer. Tuned and high pass filters are also connected in parallel which provide capacitive reactive power at fundamental frequency. The voltage signal is taken from the high voltage SVC bus using a potential transformer.

The TSC is switched in using two thyristor switches (connected back to back) at the instant in a cycle when the voltage across valve is minimum and positive. This results in minimum switching transients. In steady state, TSC does not generate any harmonics. To switch off a TSC, the gate pulses are blocked and the thyristors turns off when the current through them fall below the holding currents. It is to be noted that several pairs of thyristors are connected in series as the voltage rating of a thyristor is not adequate for the voltage level required. However the voltage ratings of valves for a SVC are much less than the voltage ratings of a HVDC valve as a step down transformer is used in the case of SVC. To limit $\frac{di}{dt}$ in a TSC it is necessary to provide a small reactor in series with the capacitor.

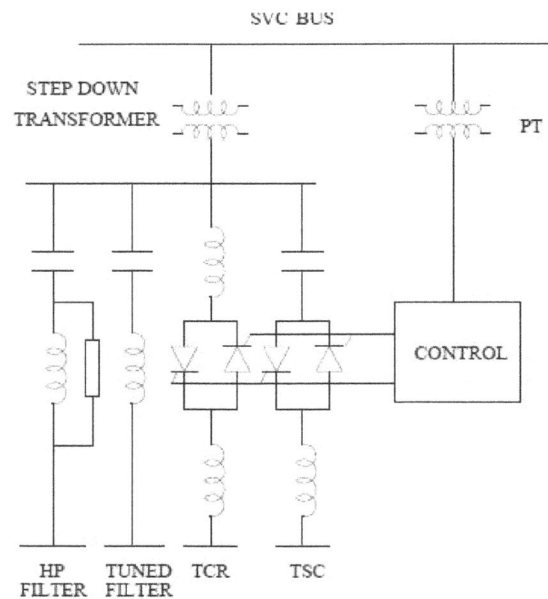

Figure 3. *A Typical SVC (TSC-TCR) Configuration*

5.2. SVC Controller

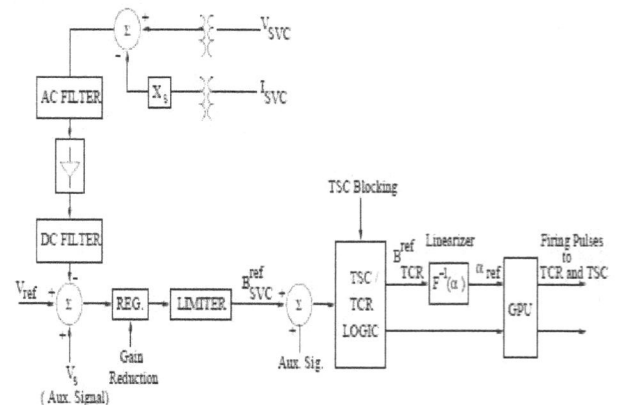

Figure 4. *SVC Controller*

The block diagram of basic SVC Controller incorporating voltage regulator is shown in Fig.4 This shows that both voltage (V_{SVC}) and current (I_{SVC}) signals are obtained from potential and current transformers and then rectified. The AC filter is basically a notch filter to eliminate the signal component of frequency corresponding to the parallel resonance in the system viewed from the SVC bus. The line capacitance (in parallel with SVC capacitance) can result in parallel resonance with the line inductance. The SVC voltage regulator has a tendency to destabilize this resonant mode of oscillation and the notch filter is aimed at overcoming this problem. As a matter of fact, any parallel resonance mode (of frequency below second harmonic) can have adverse interaction with SVC voltage regulator. If series capacitors are used along with SVC, then they can cause parallel resonance with a neighboring shunt reactor. If the second (parallel resonance) mode has a lower frequency (say below 20 Hz), a high pass filter in addition to the notch filter has been suggested

The rectified signal is filtered. The DC side filters include both a low pass filter (to remove the ripple content) and notch filters tuned to the fundamental and second harmonic components. The notch filters are provided to avoid the adverse interactions of SVC caused by second harmonic positive sequence and fundamental frequency negative sequence voltages on the SVC bus. For example, second harmonic positive sequence voltages at the SVC bus cause a fundamental frequency component in the rectified signal that results in the modulation of SVC susceptance at the same frequency. This in turn (due to amplitude modulation) results in two components at side band frequencies (0,2f) in the SVC current. The dc component can result in unsymmetric saturation of the SVC transformer and consequent increase in the magnetization current containing even harmonics. It has been observed that this adverse harmonic interaction between the SVC and the network can result in large distortion of the SVC bus voltage and impaired operation of SVC (termed as second harmonic instability).

The auxiliary signals mentioned in Fig.3.2 are outputs from the Susceptance (or reactive power) Regulator (SR) and Supplementary Modulation Controller (SMC). The Susceptance Regulator is aimed at regulating the output of SVC in steady state such that the full dynamic range is available during transient disturbances. The output of Susceptance Regulator modifies the voltage reference V_{ref} in steady state. However its operation is deliberately made slow such that it does not affect the voltage regulator function during transients.

6. Series Compensator

Series capacitors are connected in series with the line conductor to compensate for inductive reactance of the line. This reduces the transfer reactance between the buses to which the line is connected, increases maximum power that can be transmitted, and reduces the effective reactive power (XI^2) loss. Although series capacitors are not usually installed for voltage control as such, they contribute to improved voltage control and reactive power balance. The reactive power produced by a series capacitor increases

with increasing power transfer; a series capacitor is self-regulating in this regard.

6.1. Series Compensation Subsystem

The three-phase module consists of three identical subsystems, one for each phase. The transmission line is 40% series compensated by a 62.8 µF capacitor. The capacitor is protected by the MOV block. A gap is also connected in parallel with the MOV block. The gap is fired when the energy absorbed by the surge arrester exceeds a critical value of 30 MJ. To limit the rate of rise of capacitor current when the gap is fired, a damping RL circuit is connected in series. Open the Energy & Gap firing subsystem. It shows how you calculate the energy dissipated in the MOV by integrating the power (product of the MOV voltage and current). When the energy exceeds the 30 MJ thresholds, a closing order is sent to the Breaker block simulating the gap.

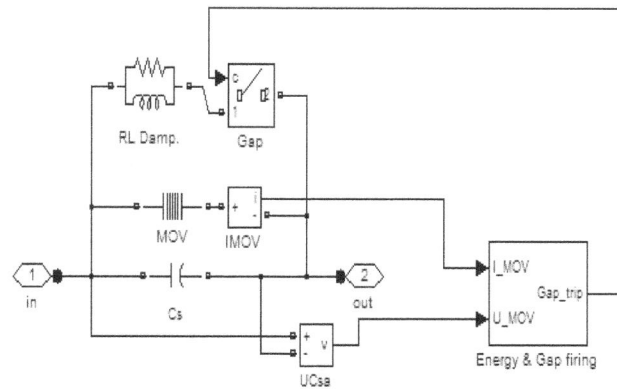

Figure 5. Series Compensation of Subsystem

7. Simulation and Results

7.1. Single Line Diagram of 11-Bus System

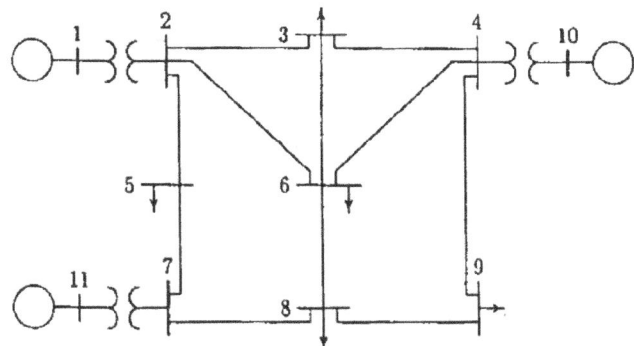

Figure 6. Single Line diagram of 11-bus power system

In this paper three plants has been considered of 900 MW each generating 20 KV and then fed to two winding transformer where voltage is step up to 230 KV and then connected to the transmission system which consist of 11-buses. There are five loads connected in the buses 3,5,6,8 and 9 respectively. Simulation model has been developed using MATLAB SIMULINK software. A three phase fault has been created near bus no 10 for time 0.01 to 0.03 sec and its impact has been taken down with the help of different bus scopes connected in the model

In the same system I have introduced a series compensator which is 55 % compensated near bus no. 6 to reduce the overshoot caused due to fault and see the amount of overvoltage and over current reduced by the use of it.

7.2. Simulation Model of 11-Bus System with SVC

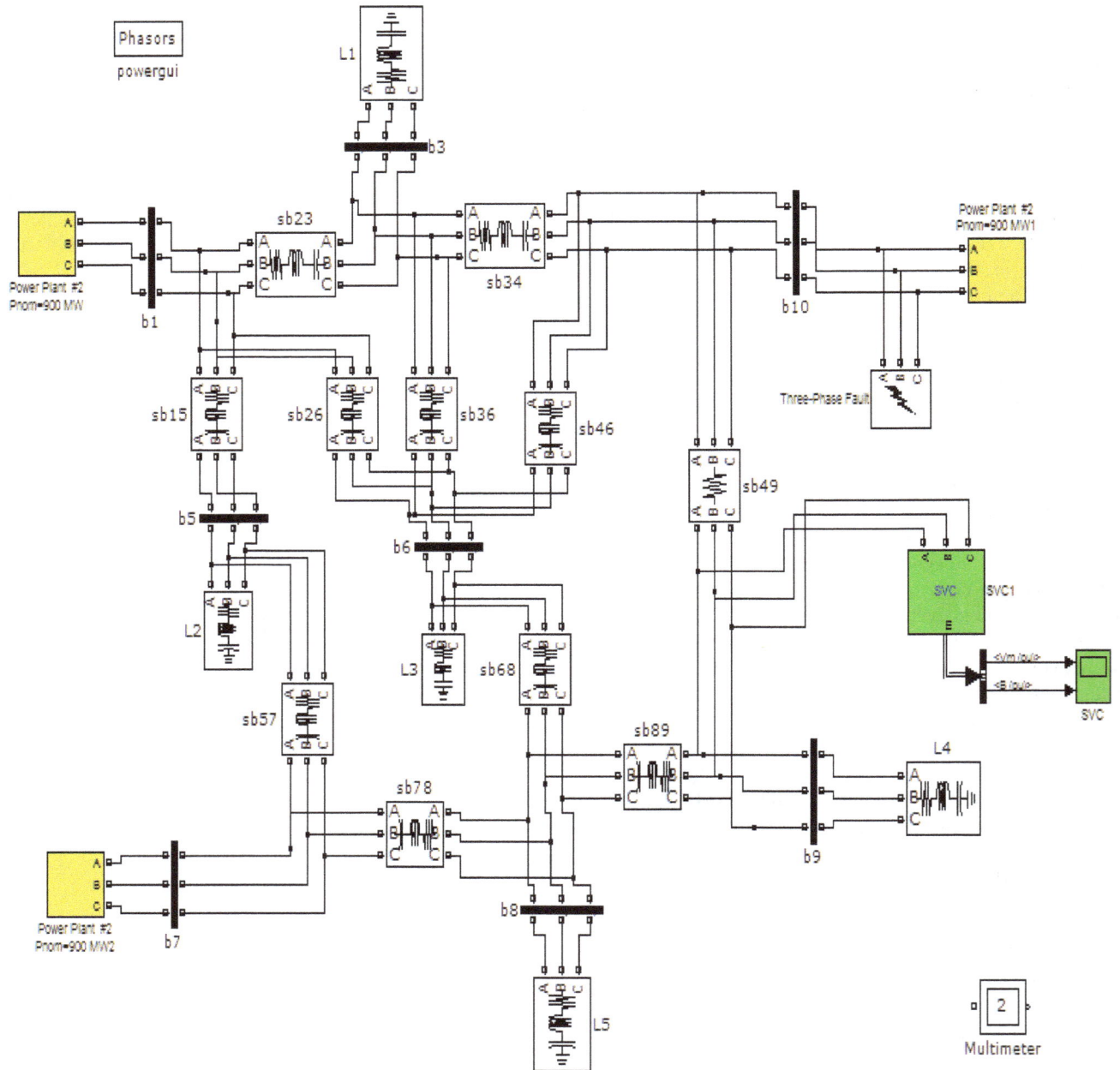

Figure 7. Simulation model of 11-bus system with SVC

7.3. Simulation Model of 11-Bus System with Series Compensator

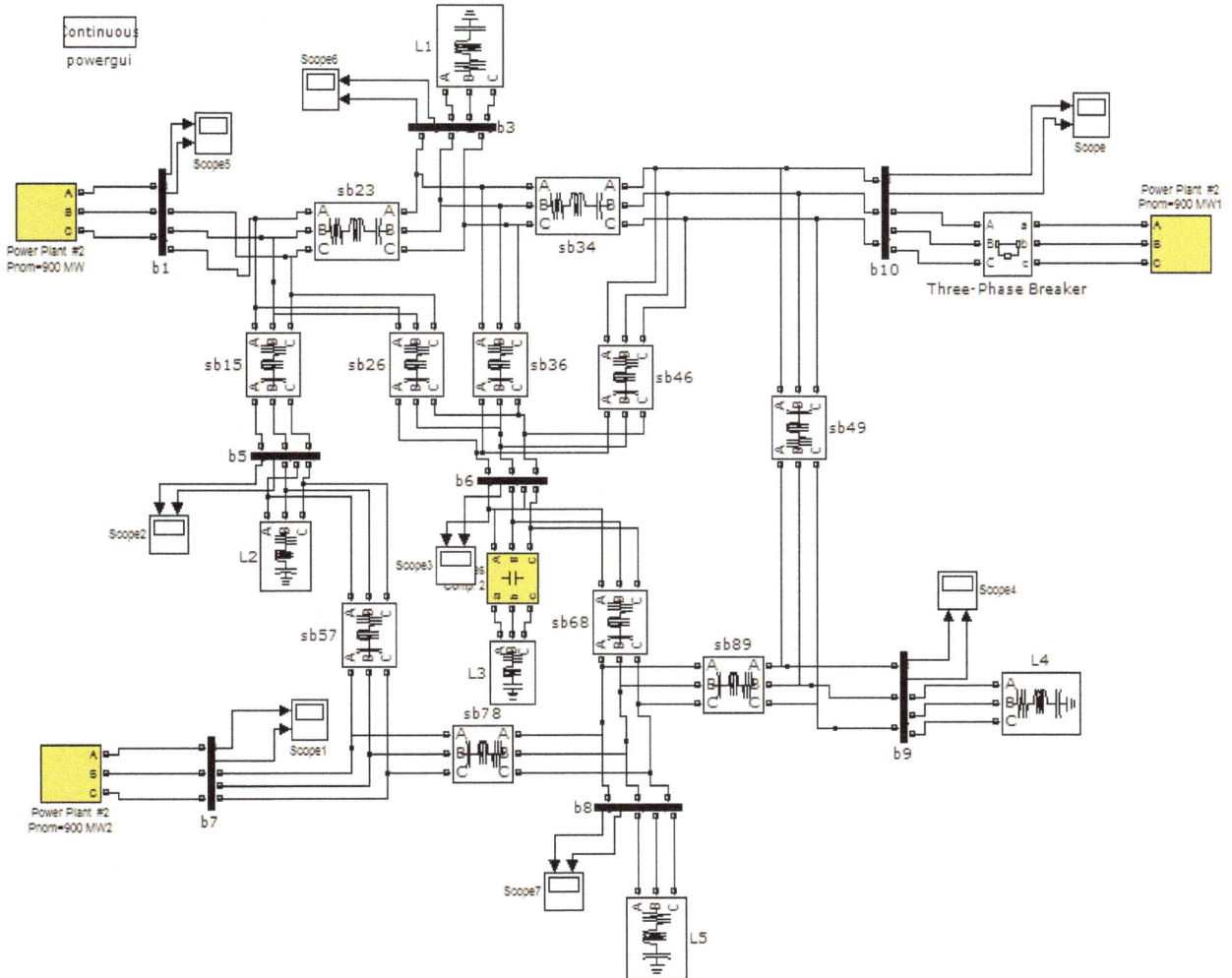

Figure 8. Simulation model of 11 bus system with series compensator

8. Results

8.1. Voltage Waveform of Bus 10 without SVC

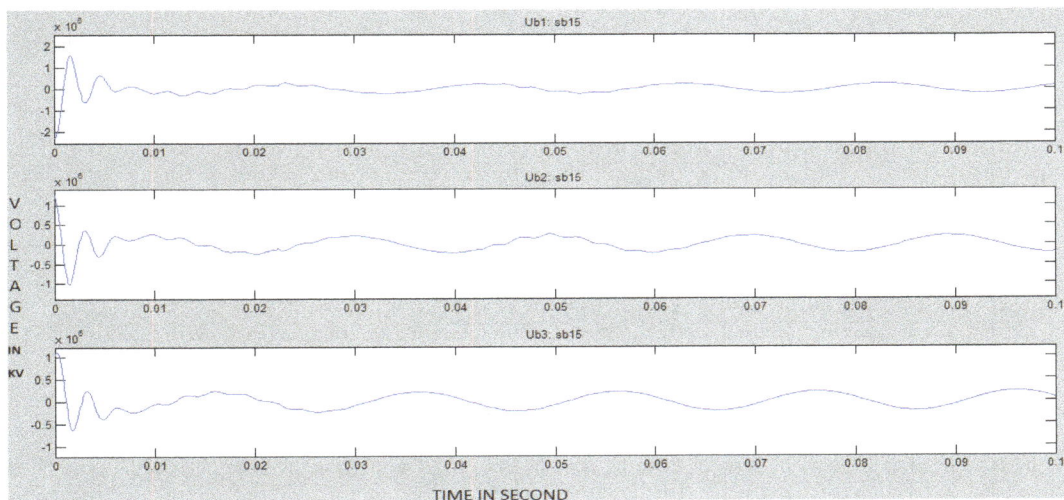

Figure 9. Voltage Waveform of Bus 10 without SVC

8.2. Voltage Waveform of Bus 10 with SVC

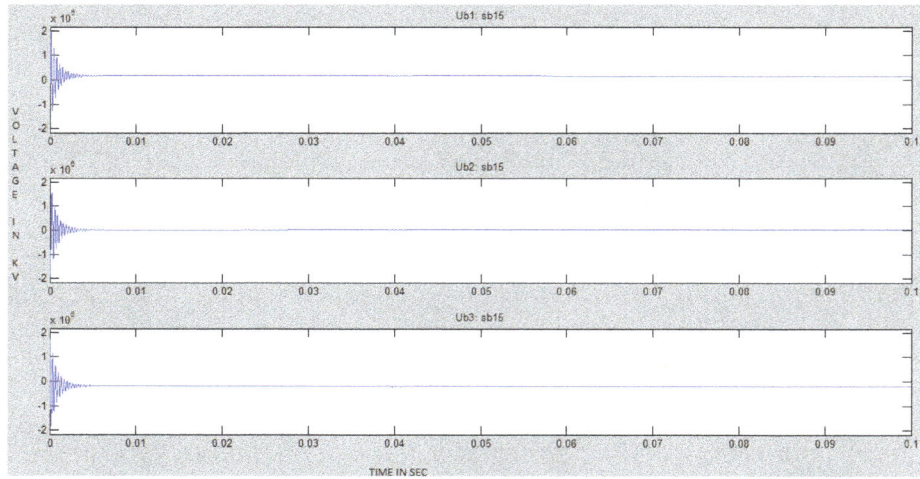

Figure 10. *Voltage Waveform of Bus 10 with SVC*

8.3. Current Waveform of Bus 10 without SVC

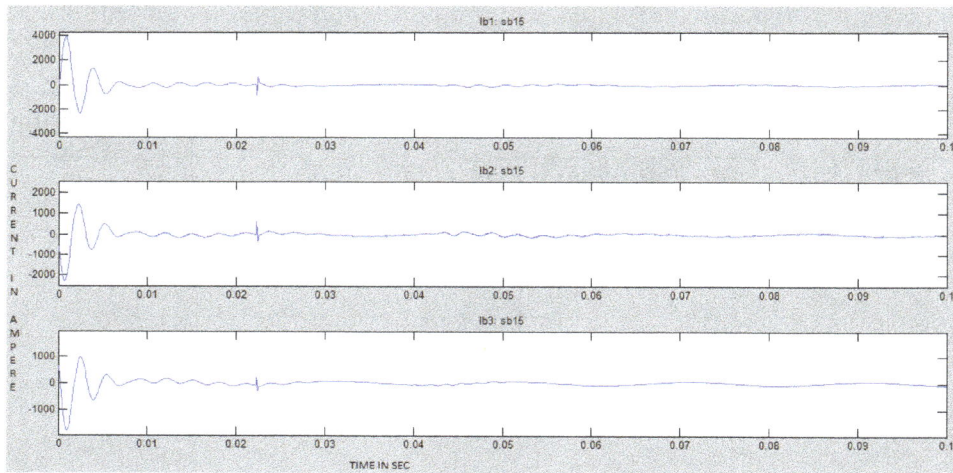

Figure 11. *Current Waveform of Bus 10 without SVC*

8.4. Current Waveform of Bus 10 with SVC

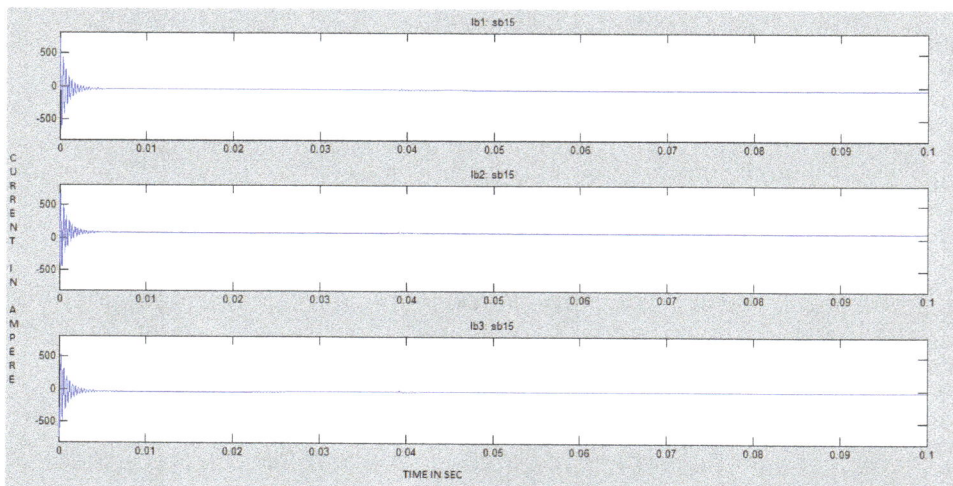

Figure 12. *Current Waveform of Bus 10 with SVC*

8.5. Voltage Waveform of Bus 1 without Compensator

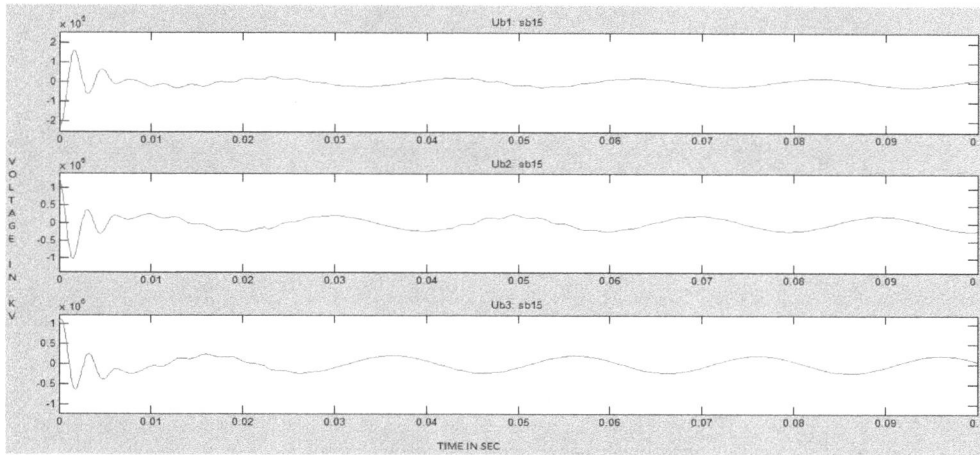

Figure 13. *Voltage Waveform of Bus 1 without Compensator*

8.6. Voltage Waveform of Bus 1 with Compensator

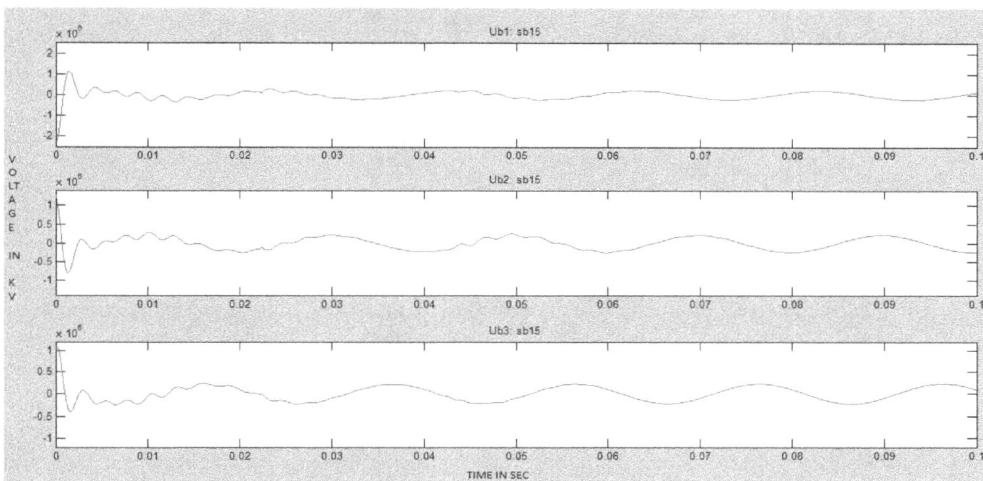

Figure 14. *Voltage Waveform of Bus 1 with Compensator*

8.7. Current Waveform of Bus 1 without Series Compensator

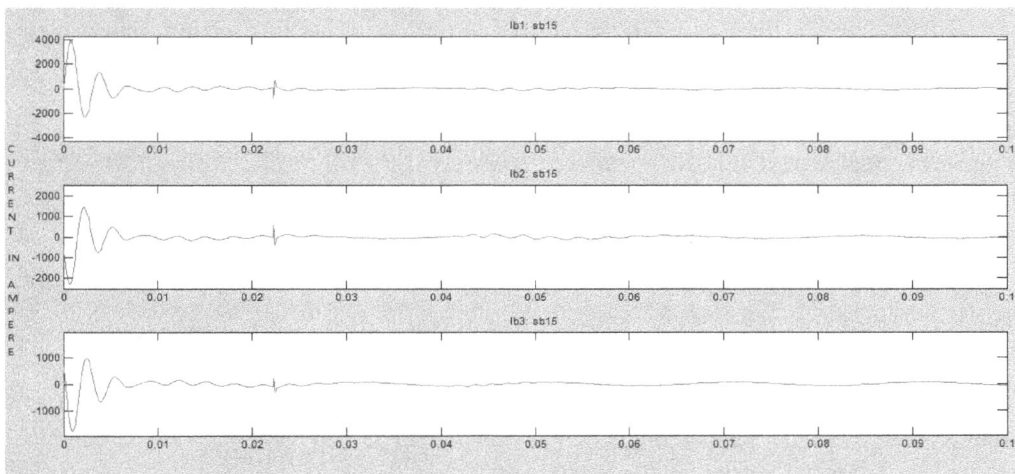

Figure 15. *Current Waveform of Bus 1 without Series Compensator*

8.8. *Current Waveform of Bus 1 with Series Compensator*

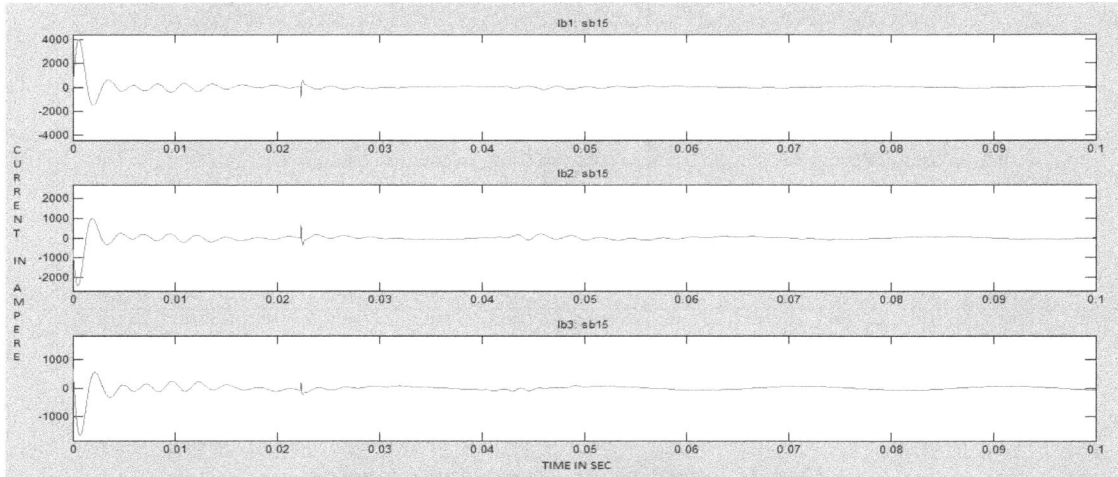

Figure 16. *Current Waveform of Bus 1 with Series Compensator*

Table 1. *Line and transformer data for 11-bus system*

From bus	To bus	R pu	X pu	½ B pu
1	2	0.00	0.06	0.0000
2	3	0.08	0.30	0.0004
2	5	0.04	0.15	0.0002
2	6	0.12	0.45	0.0005
3	4	0.10	0.40	0.0005
3	6	0.04	0.40	0.0005
4	6	0.15	0.60	0.0008
4	9	0.18	0.70	0.0009
4	10	0.00	0.08	0.0000
5	7	0.05	0.43	0.0003
6	8	0.06	0.48	0.0000
7	8	0.06	0.35	0.0004
7	11	0.00	0.10	0.0000
8	9	0.052	0.48	0.0000

9. Conclusion

The system is modeled without and with SVC to observe the phenomena of transient oscillation and subsequently how it is damped out by the SVC. It is found that the SVC can effectively damp out the resulting oscillating. In the result part we have compared the outputs of two models with and without SVC. Thus it has been clearly observed that the transient oscillations sustain up to 0.03 sec without SVC and with the use of SVC the transient oscillation vanishes before 0.01 sec. Further it has been observed that with use of series compensator the fault amplitude for current and voltages at different buses reduces and transmission capacity increases thus improving the power transfer capability.

References

[1]   L. Gyugyi, \ Reactive Power Generation and Control by Thyristor Circuits" IEEE Trans., v. IA-15, n.5, 1979, pp.521-531

[2]   A.Olwegard et al, \Improvement of Transmission Capacity by Thyristor Controlled Reactive Power", IEEE Trans .v. PAS-100, 1981, pp. .3930-3939.

[3]   R.M. Mathur, Editor, Static Compensators for Reactive Power Control, Canadian Electrical Association, Cantext Publications, Winnipeg, 1984

[4]   A. E. Hammad, "Analysis of power system stability enhancement by static var compensators", IEEE Trans. On Power Systems, vol. 1, No. 4, pp. 222-227, 1986.

[5]   M. O'Brien and G.Ledwich, \ Static Reactive Power Compensator Controls for Improved System Stability", IEEE Proc., v.134, Pt.c, n.1, 1987, pp.38-42

[6]   L.Gyugyi, \Power Electronics in Electric Utilities: Static Var Compensators", Proc. IEEE, v. 76, n.4, 1988, pp.483-494

[7]   L. Gyugyi, "Power Electronics in Electric Utilities : Static Var Compensators" in Proc. IEEE' 76, paper 4, p. 483–494, 1988.

[8]   K. R. Padiyar and R.K.Varma, \Concepts of Static Var System Control for Enhancing Power Transfer in Long Transmission Lines", Electric machines and Power Systems, v.18, 1990, pp.337-358

[9]   K. R. Padiyar, and R. K. Verma, "Concepts Of Static VAR System Control For Enhancing Power Transfer In Long Transmission Lines", Electric Machines and Power Systems, vol. 18, p. 337-358, 1990.

[10]   K. R. Padiyar and R.K. Varma, \ Damping Torque Analysis of Static Var System controller ", IEEE Trans., Power Systems, v.6, n.2, 1991, pp.458-465

[11]   P. Kundur, *Power System Stability and Control*, EPRI Power System Engineering Series, New York, McGraw-Hill Inc, 1994.

[12] V. Rajkumar and R. R. Mohler , \ Nonlinear control methods for power systems : A comparison", IEEE Trans. on Control Systems Technology, v. 3, n. 2, 1995, pp. 231-237.

[13] V. Venkatasubramanian , K.W. Schneider and C.W. Taylor, \ Improving Pacific intertie stability using existing static VAR compensators and Thyristor Controlled Series Compensation", Bulk Power System Dynamics and Control IV { Restructuring, Santorini, Greece, August

[14] N. G. Hingorani, and L. Gyugyi, Understanding *FACTS, Concept and Technology of Flexible AC Transmission Systems*, New York, Wiley Publishers, 2000.

[15] H. Saadat, *Power System Analysis*, Tata McGraw-Hill, 2002.

[16] G. Sybille and P. Giroux, " Simulation of FACTS Controllers using the MATLAB Power System Blockset and Hypersim Real-Time Simulation", IEEE PES, Panel Session Digital Simulation of FACTS and Custom-Power Controllers Winter Meeting, New York, p. 488– 49, 2002.

[17] M. H. Hague, "Improvement of first stability limit by utilizing full benefit of shunt FACTS devices", IEEE Transactions On Power Systems, vol. 19, no.4, pp. 1894 – 1902, 2004.

[18] IEEE TASK FORCE: " Proposed Terms and Definitions for Flexible AC Transmission, vol.12, No.4, Systems (FACTS)", IEEE Trans. On Power Delivery 2005.

[19] S. Panda, and Ramnarayan M. Patel, "Improving Power System Transient Stability with an. Off–Centre Location of Shunt Facts Devices ", Journal of Electrical Engineering, vol. 57, No. 6, 2006

[20] K. R. Padiyar, *FACTS Controllers in Power Transmission and Distribution*, New Age International Publishers, 2007.

[21] A. Ghosh, D. Chatterjee, "Transient Stability Assessment of Power Systems Containing and Shunt Compensators", Power Systems, IEEE Transactions on Power Delivery, vol. 22, no.3, p. Series 1210-1220, Aug. 2007.

[22] A. A. Edris, R. Aapa, M. H. Baker, L. Bohman and K. Clark, "Proposed Terms and Definitions for Flexible Ac Transmission Systems (FACTS)", IEEE Trans. On Power

[23] V. Mahajan, "Power System Stability Improvement with Flexible A .C. Transmission System (FACTs) Controller," Power System Technology an IEEE Power India Conference, POWERCON 2008.

29

Experimental identification of the equivalent conductive resistance of a thermal elementary model of an induction machine

R. Khaldi, N. Benamrouche, M. Bouheraoua

LATAGE Laboratory, Department of Electrical Engineering, Mouloud Mammeri University, Tizi-Ouzou, Algeria

Email address:

khaldi_rabah@yahoo.fr (R. Khaldi), benamrouchen@yahoo.com (N. Benamrouche), Bouheraoua@hotmail.com (M. Bouheraoua)

Abstract: This paper proposes a basic thermal model to estimate the temperature in different points of an induction motor, totally enclosed with external ventilation, for different loads at steady state. This basic model consists simply of a conductive thermal resistance for each point considered in the machine. Thereafter, the intermediates thermal resistances of conduction of the model are deduced. This approach is very easy to implement, requiring no geometrical data, or thermo-physical coefficients, or complex methods of implementation of a thermal model. Indeed, by knowledge of total losses in the machine, the temperature of the carcass, and the temperature of any point inside of the latter allows to deduct the equivalent thermal resistance of conduction of the different points and so the corresponding temperature.

Keywords: Induction Motor, Temperature, Heating, Conductive Resistance, Thermal Model

1. Introduction

The thermal performance of an electrical machine are crucial to guarantee its lifetime and thus to establish periods of maintenance. The insulation system of the motor and the lubricant deteriorate, and duration of life is reduced by half for each increase of temperature of 8 to 12°C [1].

The large number of works that are performed to study the thermal behavior of electric machines by detailed thermal models [2-9, 19] and simplified [10-13, 17, 18], shows the complexity of thermal phenomena come into play in these machines. These models are based on a nodal network of various dimensional interconnect by convective, conductive and radiative thermal resistances. Often to determine the latter, using semi-empirical relationships involving dimensionless numbers, especially for convection coefficients (particularly at the air gap). The difficulty also lies in the determination of contact resistance stator iron-frame requiring experimental tests for calibration [14]. All these models require a perfect knowledge of the geometry and physical properties of all elements constituting the machine.

On the other hand, the distribution of losses must be accurately known in the various constituent elements heat sources of the thermal problem.

Nowadays, iron and additional losses are always news about their assessment accurately [15, 16] and [9].

In this paper, one frees oneself of all the data needed, using a thermal model elementary of the motor, simply represented by a single equivalent conductive resistance. The total losses constitute the only source of heat in the machine, and are imposed at the point considered.

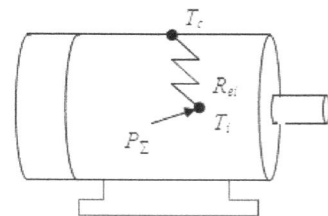

Figure 1. *Circuit diagram of the elementary thermal model of an asynchronous machine*

Figure 1 shows a diagram of the elementary thermal model where:

T_c : carcass temperature

T_i : Temperature of the point considered inside of the machine.

R_{ei} : Equivalent conductive resistance

P_Σ : Total losses in the machine.

Generally the equation of heat in the thermal network is represented by the thermal bilan:

$$P_i = C_i \frac{dT_i}{dt} + \sum_{i=1}^{N} \frac{(T_i - T_j)}{R_{ij}} \qquad (1)$$

where:

P_i : Losses in the i part machine

C_i : Thermal capacitie

T_j : Temperature of the adjacent point considered

R_{ij} : Conductive resistance between the node i and j

The equation of heat to the thermal mode established is expressed by (1) when $j=c$ and $R_{ij} = R_{ei}$:

$$P_\Sigma = \frac{T_i - T_c}{R_{ei}} \qquad (2)$$

The conduction resistance equivalent in function of the load of a given point of the machine can be deduced from two tests: nominal load and empty (or half load). The other point's just one test. Subsequently, the conductive resistance between points (nodes) of the model are easily deduced.

2. Experimental Procedure

The basic test rig consists of a three-phase cage induction motor of 2.2 kW, 380 V, 5.2 A, Δ connection, 1420 rpm, mechanically coupled to a dc machine with separate excitation. A variable voltage three-phase 50Hz is used to supply the motor test.

The material used in the tests is shown in Fig.2.

Figure 2. *Test rig.*

Several tests are carried out on the machine studied. The temperature measurement at various points in the machine tested, it is obtained using eight temperature sensors (thermocouples) placed in strategic locations on the motor.

The machine is loaded until thermal equilibrium is reached. Then, the final temperature, torque, current, voltage, the power absorbed and speed are recorded.

2.1. Location of Sensors

At the fixed part of the machine the sensors are placed more or less easily. From the moment that the machine is already realized, the sensor slot stator is placed in the space between the conductors and the insulating spacer closing.

The end winding, iron and of the tooth on the front are easily installed. At the air gap, we took advantage of the void left by the slots for the introduction, since the thickness of air gap is very small.

By cons, in the rotor, the investment is less easy as long as this part is rotating. The thermocouple is placed in a corresponding hole on the short-circuit ring. The signal issued by the latter is achieved through a system of rings and brushes made. The voltage drop across the sensor, resulting from the use of this method is systematically corrected. This drop is easily determined by a test drive of the motor under test, not powered by a DC motor.

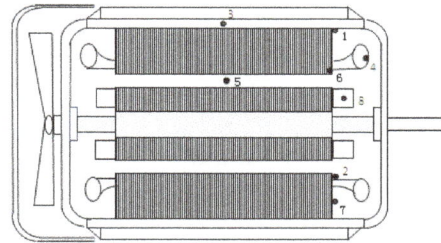

1: Iron-frame, 2: stator tooth, 3: Frame, 4: end winding, 5: air gap, 6: slot winding, 7: stator yoke, 8: short-circuit ring rotor.

Figure 3. *Location of sensors.*

3. Estimation of Equivalent Conductive Resistance

Total losses (P_Σ) of the motor can be easily deduced from the difference between power input (P_{in}) and power output (P_{out}).

$$P_\Sigma = P_{in} - P_{out} \qquad (3)$$

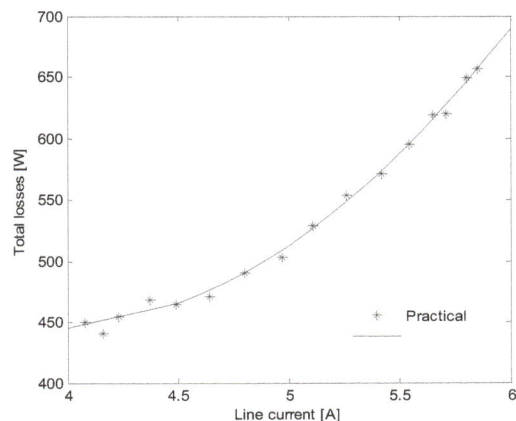

Figure 4. *Total losses as a function of load current.*

The Fig. 4 above shows the total losses according to the load current of the motor. The experimental points are extrapolated by a polynomial function cut the ordinate axis by a point-coordinated (no load current, total losses to empty) Fig. 4.

The temperature difference between the frame and a point considered inside the machine is shown in Fig. 5. The experimental points are extrapolated, for a given point, by a polynomial function of second order.

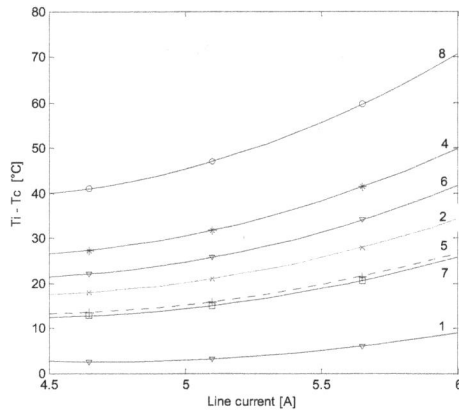

Figure 5. *(Ti –Tc) as a function of load current.*

On Figure 6 and 7 are shown the equivalent thermal resistance of the stator winding based on total losses and current load, respectively, where the experimental points are extrapolated by a polynomial function.

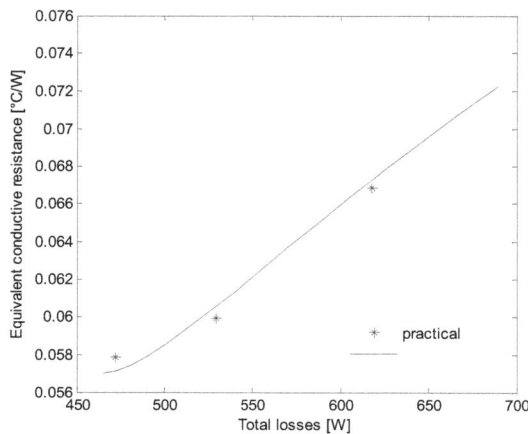

Figure 6. *The equivalent conductive resistance of the end winding according to the total losses.*

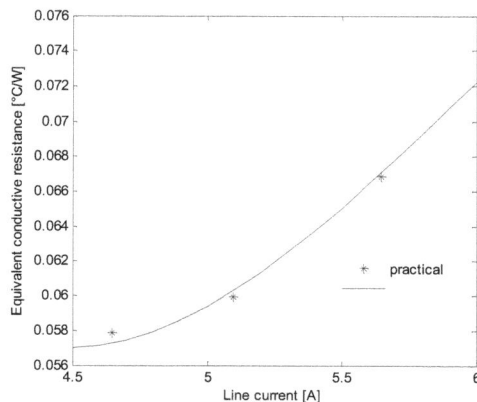

Figure 7. *The equivalent conductive resistance of the end winding according to the load current.*

The conductive resistance equivalent to the other points

are deducted by a single heating test (half load test for example). This allows us to have the gap conductive resistance between points and considered the reference point in thermal regime established. These differences in absolute values are respectively:

Table 1. *Difference Between Equivalent Conduction Resistance of Different Points in the machine compared of that End winding.*

Point considered	Difference [°C/W]	Equivalent conductive resistance [°C/W]
Rotor (8)	0.0287	0.0858
Stator yoke (7)	0.0305	0.0266
Iron-frame (1)	0.0518	0.0053
Air gap (5)	0.0289	0.0283
End winding (4), (reference)	0.0000	0.0571
Slot winding (6)	0.0111	0.0461
Stator tooth (2)	0.0194	0.0377

The equivalent of the conduction resistance of experimental point of reference (end winding (4)) is:

$R_{ei} = 0.0571 \,°\, C \,/\, W$, the motor operating half load.

It can be seen in Table I, the equivalent conduction resistance of the rotor is the most important. This corresponds to reality where the rotor temperature is the most important in a cage induction machine (Fig. 5). By cons, that of the stator-frame interface is lowest (Fig. 5). This is certainly due to the proximity of the interface of the stator frame.

Are then obtained parallel lines (Figure 8) to the line shown in Figure 6 with the differences in Table 1.

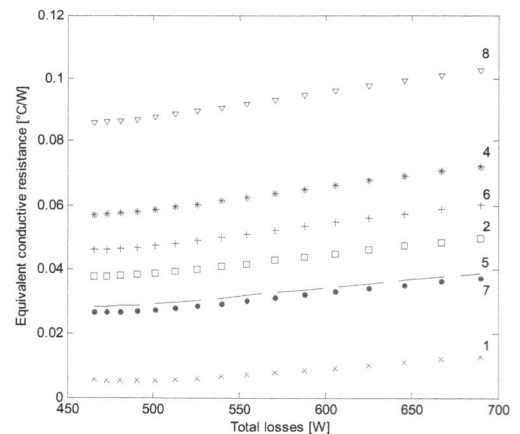

Figure 8. *Equivalent conductive resistance according to total losses.*

Knowing the conduction resistance of the equivalent points (1, 2, 4, 5, 6, 7, 8) Figure 8, from the point situated on the carcass, one can obtain the equivalent resistance of conduction between the different points on using the following equations:

$$R_{73} = R_{71} + R_{13} \qquad (4)$$

$$R_{63} = R_{67} + R_{71} + R_{13} \qquad (5)$$

$$R_{23}=R_{27}+R_{71}+R_{13} \qquad (6)$$

$$R_{53}=R_{56}+R_{67}+R_{71}+R_{13} \qquad (7)$$

or

$$R_{53}=R_{52}+R_{27}+R_{71}+R_{13} \qquad (8)$$

$$R_{43}=R_{46}+R_{67}+R_{71}+R_{13} \qquad (9)$$

$$R_{83}=R_{85}+R_{56}+R_{67}+R_{71}+R_{13} \qquad (10)$$

or

$$R_{83}=R_{85}+R_{52}+R_{27}+R_{71}+R_{13} \qquad (11)$$

Where:

R_{73}, R_{63}, R_{23}, R_{53}, R_{43}, R_{83}: Equivalent conductive resistance between considered node and the frame.

R_{71}, R_{13}, R_{27}, R_{56}, R_{67}, R_{52}, R_{46}, R_{85}: the intermediate equivalent conductive resistance.

The reference point (end windings) R_{43} is fully known. Knowing the gap of thermal conduction resistances equivalent of the various points (Table I), relative to that of the end winding, it is deduced the value of resistance easily intermediate heat.

Then we deduce the overall thermal elementary model with the thermal conductive resistances of interconnection between different nodes shown in Figure 9.

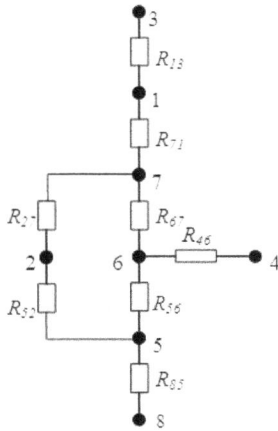

Figure 9. *An thermal elementary model decomposed of cage induction machine.*

The value of these resistances is given in the following table 2.

All of these intermediate thermal resistances are constant, except R_{13} is given by the following equation:

$$R_{13}=R_{43}+\Delta R_{13} \qquad (12)$$

Where:

ΔR_{13}: represents the difference of the value of the thermal resistance of the interface (iron-carcass) to the end winding.
In our case: $\Delta R_{13} = 0.0518$ °C/W.

$$R_{43}=a*P_\Sigma+b \qquad (13)$$

with:

a, b: are constants (in our case:
$a = 7.100813E-05$ °C/W^2, $b = 0.023265036$ °C/W) and,
P_Σ: total losses in induction machine.

Table 2. *The Intermediate Thermal Resistances of Conduction of The Thermal Model*

The intermediate thermal resistances of conduction	Values in [°C/W]
R_{71}	0.0213
R_{67}	0.0195
R_{27}	0.0111
R_{56}	-0.0178
R_{52}	-0.0094
R_{46}	0.0111
R_{85}	0.0576

4. Results

Figures 10, 11 and 12 represent the temperature of the carcass, at the thermal regime established for different loads, depending on the temperature of the point considered, respectively of the end winding , of the stator yoke, the rotor, the iron-frame, of the air gap, the stator slot and finally the stator tooth. In these figures, are shown the theoretical curves and experimental points with an offset for certain points not exceeding 5 ° C, as shown in Table III. The accuracy of the results depends mainly on test conditions and measurements. Specifically, measurement of total losses and temperatures.

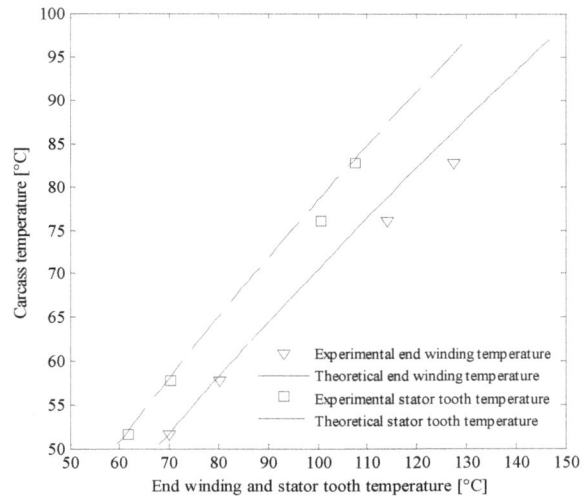

Figure 10. *The frame temperature according to the end winding and the stator tooth temperature.*

The ambient temperature is taken into account by the theoretical model as a given. In this case it is equal to 30 ° C.

The following table shows the comparison in results obtained by test and those by the thermal elementary model.

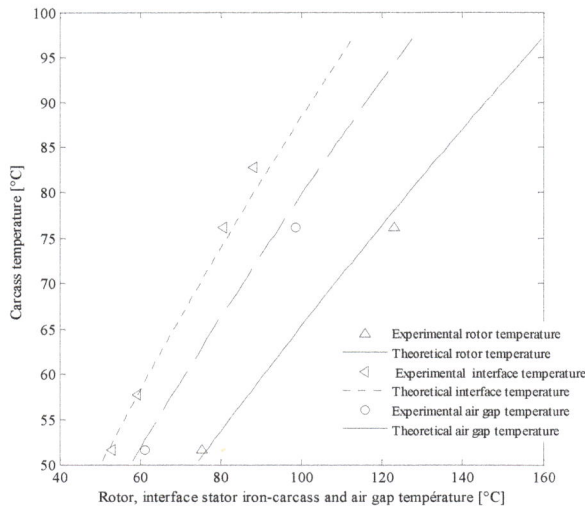

Figure 11. *The frame temperature according to the rotor, interface iron-carcass and air gap temperature.*

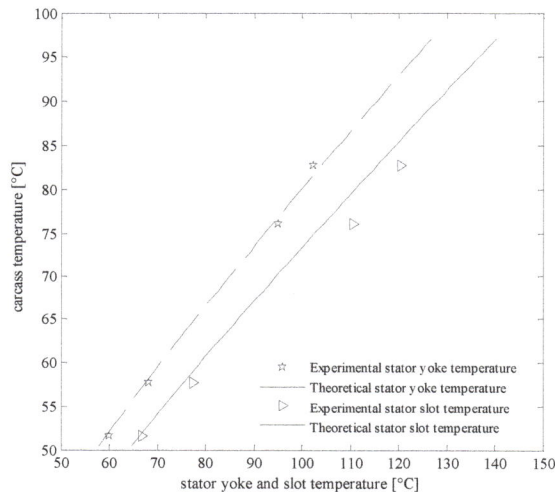

Figure 12. *The frame Temperature according to the slot and the stator yoke temperature.*

Table 3. *Comparison of Theoretical and Practical Temperatures at No-load and Full load at Steady State*

Temperature of the point considered Point considered	Practice [°C]		Theoretical[°C]	
	No-load	Full load	No-load	Full load
Rotor (8)	75.23	122.9	75.85	120.86
Stator yoke (7)	60.10	95.10	58.90	95.07
Iron-frame (1)	53.00	80.60	51.60	83.94
Air gap (5)	61.15	98.70	60.23	97.09
End winding (4)	70.00	114	69.20	110.74
Slot winding (6)	66.90	110	65.88	105.69
Stator tooth (2)	61.70	100.60	60.57	97.60
Frame (3)	51.60	76.10	51.25	76.85

These results show an acceptable agreement between the points of experimental measurements and theoretical curves, as shown in Table III, in number, for the heating test at no-load and full load at steady state.

Therefore the thermal elementary model presented in this paper can be validated and applied to the permanent thermal regime for different charge of the induction machine studied.

The model thermal, shown on Figure 9, shows that for an asynchronous machine of given materials construction, the contact conductive resistance is a very important factor in heat transfer.

Finally, the application of this thermal elementary model allows us to deduce the temperature of the internal parts of the machine with sufficient accuracy, by the knowledge of the carcass temperature, load current (total losses) and ambient temperature.

The temperature of the carcass can also be easily obtained by an infrared camera Fig. 13.

Figure 13. *The frame Temperature obtained by an infrared camera.*

5. Conclusion

In this paper, a thermal elementary model of an asynchronous machine with self-ventilated cage is presented.

This model is characterized by single equivalent conduction resistance and whose determination is very easy. Indeed two heating runs suffice for identifying for one a given point.

This method allows us to overcome the knowledge of the loss distribution in the different materials. This is because the total losses are introduced in a concentrated manner at the point considered. It also does not require geometric data and thermo-physical factors, the various constituents' material of machine. It is shown in this paper that for a given material of construction for an asynchronous machine, the contact resistance iron-carcass is a major factor in heat transfer.

The results show a maximum of about 5 ° C difference for some points, between the theoretical curves and experimental points. This difference can certainly be reduced by more accurate measurements of total losses and temperatures.

Finally, this model can be easily applied to a series of asynchronous machine to provide the temperature inside of the machine at the points considered.

References

[1]   M. Kostenko et L. Piotrovski, "Machines Electriques," Tome II, Machines à Courant Alternatif, 3ème édition, Mir-Moscou, 765 p., 1979.

[2] A. Boglietti, A. Cavagnino, and D. Staton, M. Shanel, M. Mueller and C. Mejuto, "Evolution and Modern Approaches for Thermal Analysis of electrical machines," IEEE Trans. On indust. Electronics, vol. 56, No. 3, pp. 871-882, March 2009.

[3] J. F. Trigeol, Y. bertin, and P. Lagonotte "Thermal Modeling of an Induction Machine through the Association of Two Numerical Approches," IEEE Trans. On Energy conversion, vol. 21, No. 2, pp. 314-323, June 2006.

[4] D. Staton, A. Boglietti, and A. Cavagnino, "Solving the More Difficult Aspects of Electric Motor Thermal Analysis in Small and medium Size Industrial Induction Motors," IEEE Trans. On Energy conversion, vol. 20, No. 3, pp. 620-628, September 2005.

[5] O. I. Okoro, "Steady and Transient states thermal Analysis of Induction Machine at Blocked rotor Operation," European trans. On Electr. Power, pp. 109-120, October 2005.

[6] G. Kylander, "Thermal Modelling of small Cage Induction Motors," Int. symp., 1995, On Elect. Power Engin., pp. 235-240.

[7] P. H. Mellor, D. Roberts, and D. R. Turner, " Lumped parameter thermal model for electrical machines of TEFC design," in Proc., Sep. 1991 IEE, vol. 138, No. 5, pp. 205-218.

[8] R. Glises, A. Miraoui, and M. Kauffmann, "Thermal modelling for induction motor," J. Phys. III France 3, September 1993, pp. 1849-1859.

[9] A. Bousbaine, W. F. Low, M. McCormick and N. Benamrouche, "Thermal modelling of induction motors based on accurate loss density measurements," ICEM, Sept. 1992, pp. 953-957.

[10] C. A. Cezário and H. P. Silva, " Electric Motor Winding Temperature Prediction Using a Simple Two-Resistance Thermal Circuit," Proc. of the 2008 International Conference on Electrical Machines, IEEE, Paper ID 1383, pp.1-3.

[11] N. Jaljal, J. F. Trigeol, and P. Lagonotte "Reduced Thermal Model of an induction Machine for real-time thermal Monitoring," IEEE Trans. On Indust. Electronics, vol. 55, No. 10, pp. 3535-3542, October 2008.

[12] A. L. Shenkman, and M. Cherktov, "Experimental Method for Synthesis of generalized Thermal Circuit of Polyphase induction Motors," IEEE Trans. On Energy conversion, vol. 15, No. 3, pp. 264-268, September 2000.

[13] J. T. Boys, M. J. Miles, "Empirical thermal model for inverter-driven cage induction machines," IEE Proc-Electr. Power Appl., vol. 141, No.6, pp. 360-372, Nov. 1994.

[14] J. F. Moreno, F. P. Hidalgo and M. D. Martinez, "Realisation of the tests to determine the parameters of the thermal model of an induction machine," in Sept. 2001 IEE, Proc.-Electr. Power App., vol. 148, No.5, pp. 393-397.

[15] A. Bousbaine, W. F. Low, and M. McCormick, "Novel approach to measurements of iron and stray load losses in induction motors," in Proc. 1996 IEE, vol. 143, No. 1, pp. 78-86.

[16] E. Olivier, R. Perret and J. Perard, "Localization of the losses in an induction machine supplied by an inverter," Electric Machines and power Systems, No. 9, pp. 401-412, 1984.

[17] M. Aníbal Valenzuela, and Pablo Reyes, "Simple and reliable Model for the Thermal Protection of Variable-Speed Self-Ventilated Induction MotorDrives," IEEE Transactions of Industriy Applications, Vol. 46, No. 2, pp. 770-778, March/April 2010.

[18] M.J. Picazo-Ródenas, R. Royo, J. Antonino-Daviu, J. Roger-Folch, "Use of the infrared data for heating curve computation induction motors Application to fault diagnosis" Engineering Failure Analysis, Vol. 35, pp. 178-192, 15 December 2013.

[19] Kai Li, Shaoping Wang, John P. Sullivan, "A novel thermal network for the maximum temperature-rise of hollow cylinder," Applied Thermal Engineering, Vol. 52, Issue 1, pp. 198-208, 5 April 2013.

30

Scenario of electricity trading in South Asia: Perspective and feasibility of trading between India and Bangladesh

Anmona Shabnam Pranti[1], Arif Mohammad Shahed Iqubal[2], Mohammad Shawkut Ali Khan[2], Mohammad Kayesar Ahmmed[3]

[1]Department of Electrical and Electronics Engineering, International University of Business Agriculture and Technology, Dhaka, Bangladesh
[2]Department of Mechanical Engineering, International University of Business Agriculture and Technology, Dhaka, Bangladesh
[3]Greenland Builders Limited, Dhaka, Bangladesh

Email address:

Anmona.s.p@iubat.edu (A. S. Pranti), iqubal.shahed@iubat.edu (A. M. S. Iqubal), shawkut.ali@iubat.edu (M. S. Ali Khan), maks_eee_cuet@yahoo.com (M. K. Ahmmed)

Abstract: Uninterrupted electricity supply is the precondition of economic development of a country. Bangladesh has huge electrical power deficiency and to minimize this lagging, Government is importing 250 MW electrical power from India. India is itself an electricity deficiency country and has already relationship of trading with two south Asian countries, but the policies of those trading are not beneficial for those respective countries. On the other hand Bangladesh is richer than many other south Asian countries in respect to primary energy reserve. Proper management of this primary energy and electricity sector, to produce and use electricity properly could be a good solution of running electricity scarcity problem of Bangladesh. In some context inter-countries trading is important in south Asia to have a sustainable economic growth when it is beneficial and preserve the right for both countries. Purchasing power from India cannot be a fruitful solution because it will increase our dependency on them. By this trading we may reduce the electricity crisis instantly but could be a cause of long term negative effect on our overall economy. Moreover most of the energy specialists and learned people have negative opinion about this electricity trading. This trading cannot be the good and permanent solution to solve power scarcity problem of this country. Bangladesh should try to be self sufficient in electrical power by proper management and utilization of its own resources instead of spending money for purchasing power from India.

Keywords: Electricity Trading, South Asia, India-Bangladesh

1. Introduction

Electricity is an essential element of modern living and key of Economic and social development in an industrial country. Our personal and professional life activities like communication, transportation, basic and luxurious equipments entirely depend upon sufficient supply of electricity at cheap rate. So uninterrupted electricity supply at considerable rate is the precondition of national development and Government's constitutional duty which has not been proved yet. Though electricity is considered as a primary right, the people of Bangladesh are far away from this because of the destruction of electricity sector. The industries are the heart of economy, sometimes it becomes difficult for industrial and developing countries like Bangladesh to supply sufficient electricity by setting power plant as it needs time, huge investment, and fuel reserve. As Bangladesh is trying to become a middle income country within 2022 [68], Government is taking different strategies to minimize the difference between demand and supply of electricity and importing electricity from India is one of them. The regional cooperation is necessary for development and to strengthen South Asian union, country and their motto of cooperation, long term economic analysis, political assessment and public opinion are important as people are the owner of all resource. People may have different thoughts and argument on 25 MWh electricity import from India from August 2013 [68]. This paper has shown feasibility of this electricity import especially from India on the basis of regional energy and energy exchange analysis and survey on people opinion.

2. Literature Review

An agreement of power transmission for 35years between India and Bangladesh for importing 250 MW of electricity to Bangladesh was made in July 2010 by the Bangladesh Power Development Board (BPDB) and Power Grid Company of India Ltd. (PGCIL)[73,67,72]. BPDB will pay power and transmission tariff according to the decision of India's Central Electricity Regulatory Commission (CERC)[73,67,72]. Power Secretary of Bangladesh has declared that the government of Bangladesh is going to import 250 MW of electricity from India at 4 taka per Kwh (without transmission line cost) and the power secretary of India has confirmed the completion of transmission line installation within this July [73,67,72]. As Bangladesh wants to be a middle income country by 2022, Government is importing electricity from India since August 2013. For this purpose grid connection between Bheramare (Bangladesh) and Behrampur (India) has been constructed according to the meeting of Ministers Dipu Moni and Salman Khurshid in Dhaka[74,67]. Bangladesh has to buy this 250 MWh/day electricity from India [74]. Moreover, a joint coal based power plant (Bangladesh-India) of 1320 MW is being constructed at Rampal, Bangladesh and will start in 2017 under the supervision of Bangladesh Power Development Board and National Thermal Power Company of India according to the agreement of August, 2010 at 50% ownership for each[74,69,67,70].

Bangladesh government has constructed 400 KV high voltage transmission line for importing 500MW of electricity according to the decision of Prime Minister Level meeting [69]. On the other hand, the government of India now has committed to supply only 250 MW of electricity from 'Unallocated Resource' at low cost and said that Bangladesh can import another 250 MW of electricity in future from 'Power Pool' of India [69,72]. The length of the transmission line between Bheramara and Behrampur is 125km of which 40km is inside Bangladesh area and transmission voltage level are 230 KV and 400 kV respectively with a capacity of 500 MW [72]. The total construction cost of this transmission line is 82.2 billion INR of which India paid only 17.8 billion INR and Bangladesh had to pay 64.4 billion INR [72].

Most of the existing transmission line of Bangladesh is of 132KV. For supplying the imported power at 230KV voltage level, new transmission line has to be built and a tender was invited by Power Grid Company Bangladesh to construct transmission line of 30km (400Kv and 230 Kv, Ishurdi-Khulna); 165 km (230 kV, Bibyana-Comilla) and a HVDC line at Bheramara [72]. $100 million loan was approved by Asian Development Bank (ADB) for cross-border transmission line construction [72]. But the loan had not been got because of the problem in power

purchase agreement between BPDB and the NTPC Limited [72]. Moreover, Bangladesh's Planning Commission had to pros pond the project because India did not guarantee the power supply [72]. Power Grid Company Bangladesh is giving 1 billion BDT to a Spanish company to construct the transmission line of Bangladesh portion and $170 million to Siemens to construct a converter station at Bheramara to convert 400Kv from India to 230Kv to Bangladesh [72]. For greater trade with India, PGCB has constructed another 500 MW substation at Bheramara [72].

Figure 1. *proposed construction line from Behrampur to Bheramara*

According to the MoU signed by the Government of Bangladesh and Myanmar, Bangladesh will be able to import 500MW of electricity from Lemro river of Rakhain state, Myanmar by 2018 [69].

Nepal is always willing to share of its' hydro electricity resource of 40000 MW with the neighbors for the benefit of this region and Nepalese Ambassador to Bangladesh said that Nepal is interested to export hydro power in Bangladesh [71]. He also said that though Nepal has huge hydro electric potential, Bangladesh cannot use this because of a gap in trade relationship and both countries have to move forward to minimize the gap by establishing physical connectivity [71].

3. Energy Security, Electricity and Fuel Relationship for Economic Development of Bangladesh

The education rate, per capita income and expenses, GDP increment rate, standard of living are the main determinants of human development which entirely depend upon per capita electricity use of a country [1,2]. Only 42% people of Bangladesh get electricity with a rate of only 188 KWh/person [3]. On the other hand, Pakistan (developing country) and USA (developed country) use electricity of 456 KWh and 13640 KWh per capita respectively [3]. 80% rural people of our country depend on agricultural residue for meeting their energy need [76]. Human life security and sustainable development of a country cannot be earned by unrestricted use of national energy reserve [4]. Though electricity is the best form of energy, it cannot be preserve for future use. Moreover, uninterrupted electricity supply depends on secure transmission and distribution system. Current demand and the generation of electricity are 7500MW and 5063 MW respectively [75,68,74]. The government plans to increase the overall capacity up-to 20000 MW by 2021[9]. Current GDP growth rate of Bangladesh is 6.3% [66]. 7.5% GDP growth rate is necessary to reduce half of the poverty of Bangladesh and energy use increment rate should be 2.5 times [6]. Electricity scarcity is the main obstructer of the development of Bangladesh. From 1973 to 2006, increment of GDP growth rate is lower than increment of per capita electricity use rate in Bangladesh [7]. So it is clear that electricity was used for living only rather than using in production purpose and it is seemed that the trend will continue. The imported electricity will use for increasing living standard only.

4. Comparative Energy Situation and Economics in South-East Asia

The countries of South East Asia have been developed rapidly in last ten years. But most of these countries cannot meet energy demand for sustaining this development growth. Table below represents the condition of primary and secondary energy for different countries of this region.

975 million people of 1330 million of China are electrified [12,23]. GDP growth rate is 11.4% [15]. China had 25000MW of electricity deficiency in 2005-06 [22] and imported 5392 million KWh [25] and exported 11270

million KWh [26] in that year. Generally China import electricity from Russia [27] and Export to Vietnam [29]. Now China has taken different strategy of electricity production such as hydro power by creating Dam in Himalayan Rivers and setting their own plant in Kazakhstan, Kirgizstan, Cambodia and Mongolia [28].

Though Bhutan has surplus hydro electricity reserve, it imported 20million KWh electricity from India in 2006 [25]. On the other hand it exported electricity of 1500 million KWh to India in the same year [30]. The electrification rate is only 60% [32] and the country has to import Oil and petroleum to meet energy demand [31].

Nepal is also endowed with hydro electricity around 83000 MW [10]. Only 40% population is electrified [10] and electricity consumption rate is 66kWh/cap. Instead of having enormous hydro resource, to meet the energy demand it has to import oil (87%), coal and electricity [31]. It imported electricity of 266 million KWh [25] from India and exported electricity of 101 million KWh [26] to India in 2006. Though there is serious scarcity of electricity, Nepal export electricity and it seems that electricity export revenue could not change the economic condition of the country.

Though Sri Lanka does not have sufficient primary energy and hydro potential reserve, 78.1% people are electrified [35]. Sri Lanka does not import electricity; only import primary energy to produce electricity to obtain $1195/capita GDP [14] which is the second highest in South Asia.

Table 1. Primary and Secondary Energy reserve in South Asian Countries

Country	Population			Primary Energy							Electricity (million KWh)			Per capita consumption		Per capita GDP	
	Population in million [12]	Increment rate % [13]	Oil [17] bbl	Gas (tcf)	Coal (MT)	Nuclear Energy (ton)	toe/cap	Hydro MW	Hydro W/cap	Consumption [76]	Import [25]	Export [26]	KWh	Kg oe	US $ [14]	growth rate [15]
China	1330.04	0.63	12800	66.54 [42]	114500 [18]	68000 [19]	374.73	165000 [24]	124	2859000	5390	11270	2150	1075[33]	3174	11.4
India	1148.00	1.58	5848	37.26 [42]	92445 [18]	73000 [19]	448.41	143311 [32]	130	488500	1764	286	426	351 [34]	1078	8.5
Pakistan	167.76	1.81	289.20	30.02 [42]	3300 [18]	6000 [21]	240.32	17369 [32]	248	67060	0	0	400	337 [35]	931	6.3
Bangladesh	153.55	2.02	28	13.77 [42]	2221 [44]	-	10.86	555 [10]	3.6	19490	0	0	127	117 [36]	541	6.0
Afghanistan	32.74	2.63	-	1.678 [16]	100 [39]	-	3.04	183.35 [38]	6	800	100	0	24	12 [37]	393	7.5
Sri Lanka	21.13	0.94	-	-	-	-	-	2000 [10]	95	7070	0	0	335	195 [38]	1195	6.0
Nepal	29.52	2.10	-	-	-	-	-	83290 [10]	2776	1960	266	101	66	45 [39]	428	2.5
Bhutan	0.68	1.30	-	-	-	-	-	30000 [10]	44118	380	20	1500	559	63 [40]	2005	8.8
Maldives	0.38	2.69	-	-	-	-	-	-	-	160	0	0	421	808 [41]	3358	5.5
Myanmar	47.77	0.80	1963	9.59 [16]	258.11 [20]	-	13.51	108000 [32]	2260	3740	0	0	78	82 [42]	287	5.5

Afghanistan imports electricity of 4 MW from Iran, 5 MW from Kazakhstan, 150 MW from Uzbekistan and 8 MW from Turkmenistan [22]. It imported 96 MW of electricity in 2007 [39].

Instead of having no primary energy reserve and hydro potential, the per capita electricity consumption of Mal dives is 421 KWh. Moreover, they have planned to generate 232.5 MW wind electricity and to set half kilometer solar panel [40]. So Maldives is developing rapidly without importing any electricity.

Pakistan has sufficient primary energy reserve and the coal reserve is the 6th highest in the world along with nuclear energy reserve [31]. 36.8% produced gas is used in electricity generation and 33.4% generated electricity comes from hydro power plant [31]. Instead of having 40316.08 Mtoe primary energy and 41722 MW hydro electricity reserves [32], Pakistan is going to import electricity from Iran, Tajikistan and Uzbekistan in near future [43].

88.45% electricity generation of Bangladesh is dependent on Natural gas [44] and 0.5 million ton coals is used to produced 250 MW electricity each year [31]. Though Bangladesh has 2221 million ton high grade coal reserve, it cannot be used properly due to lack of technological support [44]. Electricity production from oil and hydro potential is negligible and the government has not taken proper steps to enhance electricity sector yet.

India can meet its 70.73% energy demand by its' own energy and it has to import Gas, Coal, Oil and Hydro electricity for other 29.27% [31]. India imported 1764 million KWh and exported 286 million KWh of electricity in 2006-07 [25]. There was electricity deficiency. The above information shows that, India is still not solvent in electricity production for serving its own population.

Myanmar has 10.8 million MW hydro potential reserve and they are producing 35% (3145 MW) and 55% of electricity from hydro potential and gas respectively [45, 30]. Myanmar is going to export electricity of 360 MW to Thailand and 1200 MW to India in near future [22].

4.1. Energy Exchange in South Asian Region

- Different international monetary organizations think energy cooperation in South Asia will help people to get energy at a cheap rate because of open border energy market which will result technological and economical development of this region [10]. The Energy ministers of SAARC countries decided to create "Energy Ring" for interchanging Gas and Electricity in 2009 in Colombo, Sri Lanka [11]. The decision of energy exchange between South Asian countries has been taken on the basis of following reasons:[11]

- Though some of the South Asian Countries have sufficient energy reserve and there is a possibility of exchanging energy, the countries are amongst the lowest income countries of the world due to low per capita energy use.

- To obtain sustainable development by using these huge amounts of energy resource, internal relation, cooperation and trust is needed among the governments.

- World Bank stated that it is possible to earn 6.6-11% GDP growth rate in SAARC region in 15-20 years if "Cross Border Energy Trade" will complete properly [11].

4.2. Scenario of Existing Cross-Border Electricity Exchange in South Asia

4.2.1 Exchange between India and Nepal

Table 2. Electricity exchange between Nepal and India (1990-2006) [46].

Million KWh	'90	'91	'92	'93	'94	'95	'96	'97	'98	'99	'00	'01	'02	'03	'04	'05	'06	Total
Export KWh	23	81	85	46	51	40	87	100	67	64	95	126	134	192	140	113	101	676
Import	61	34	55	82	103	104	37	154	210	232	232	227	238	150	127	241	266	1381
difference	38	-47	38	36	52	64	-50	54	143	168	137	100	104	-42	-13	128	165	704.9

From the above statistics we can see that, Nepal had to import more electricity from India than export to India in each years and the trend was uprising. Nepal imported electricity of 1381 million KWh, where as exported only 676 million KWh electricity in 17 years. In this trade, Nepal spent 70438.51 million INR for importing and earned 3605.82 million INR for exporting [46]. In 1996, electricity exchange rate was 1.67 INR at 8.5% increment rate at 230KV level [46]. According to this, the rate would be 2.96 INR in 2003, 3.10 INR in 2004 and 3.77 INR in 2008 and 3.96 INR in 2009 [46].

Cross border exchange between Nepal and India was started in 1954 after the sign of Kasi MoU with the installation of a 20 MW hydro-electric power plant [46]. According to the MoU, Nepal can buy 50% (10 MW) of generated electricity at predefined rate. India can get rest of 50% electricity by only minimum royalty [46]. In 2005-06, Nepal bought electricity of 30 million KWh at a rate 2.38

INR from its own available hydro potential [46].

According to Mahakali MoU, India set a 120 MW hydro electric power plant at Nepal which can produce 484.4 million KWh/year among which Nepal can get only 70 million KWh/year of electricity [30]. India will get rest of all [29]. However, Nepal has never got electricity of 70 million KWh from 1990; only get 52.83 million KWh/year of electricity [29].

India agreed to export 150 MW of electricity to Nepal in 2001 but Nepal has to pay extra charge for electricity of higher than 50 MW [46]. Nepal has to lose $1.44 million for this trade [46]. Moreover, Snowy Mountain Engineering Company has set a 750 MW hydro electric power plant under West-Seti project and all electricity of this project will be exported to India, Nepal will get 90% royalty money instead of electricity [47]. According to MOU of that project, India imports electricity of 3 billion KWh/year at 0.0496$/KWh [46].

4.2.2. Exchange Program between India and Bhutan

Table 3. Electricity generation, use and export from Bhutan to India in between 2000-08

year	2000	2001	2002	2003	2004	2005	2006	2007	2008	Total
Generation [49]	1788	1856	1876	1896	1896	2001	1882	2050	2000	17245
Use [54]	345	191	381	389	380	313	250	527	380	3146
excess	1443	1665	1495	1517	1517	1688	1639	1524	1620	14101
Export [77]	1339	1550	1385	1400	1400	1560	1510	1400	1500	13044

Bhutan generated 17245 million KWh, Use 3145 million KWh and exported 14101 million KWh of electricity in 9 years (2000-08). This is a one way trade in which India only imported electricity.

Chukha hydro electric power plant (336 MW) was built by the fund of Indian Government in 1988 [50]. 70% of 1320 million KWh/year of generated electricity is exported to India [50]. Until 2004, the electricity rate was 1.50 INR and after 2005 it became 2.00 INR [50].

Kurichu hydro electric power plant (60 MW) was built by Indian Government Fund in 2002 and the rate of electricity is 1.5 INR/Kwh [48].

1020 MW Tala Hydro electric power plant, capacity 3962 million KWh/year, was funded by Indian Government at 9% interest rate and electricity rate is 1.6 INR [51].

On the other hand, Basochu hydro electric power plant (64 MW) was built by Austrian Government fund without any interest and all electricity is used only in Bhutan [52].

4.2.3. Electricity Exchange between Myanmar and Bangladesh

Myanmar's electricity generation capacity has increased to 1335 MW from 706.82 MW between 1988 and 2005 and in 2008 it generated 6.154 billion KWh of electricity [45, 53]. Myanmar uses only 3.744 billion KWh of electricity annually, so half of the generated electricity remains unused [54]. Myanmar also planned to construct 16 new hydro electric power plant, 5 of which has already completed and 11 are in under construction [30]. At the completion of those power plants, 3445 MW power will be added to the total generation capacity of Myanmar [30]. In 2008, Bangladesh gave proposal to set a power plant in Myanmar with its' own fund [61]. According to the proposal, Bangladesh would get 70% of generated electricity from that plant and Myanmar accepted it at that time [55].

From the Above information it is clear that, India has created electricity trade with Nepal and Bhutan in the name of electricity exchange. According to above statistics, India exported more electricity to Nepal than imported electricity from Nepal. Currently Nepal is facing serious electricity scarcity problem and it is considered as the poorest country of this region having GDP growth rate of 4.6% [77]. Though Nepal has huge hydro electric potential, this electricity could not be helpful in economical development because of not having proper ownership or royalty. Nepal export its' generated electricity. On the other hand it has to import electricity at high rate from India which has a negative effect on economy. Almost all generated hydro electricity of Bhutan, 14101 million KWh of 17245 million KWh, is exported to India.

4.3. India's Policy of Cooperation

All most all main rivers of Himalayan region are international; such as Shindu, Started from Tibet and flows through Pakistan and India, Mohakali, Started from Nepal and flows through India, Ganges, Started from Himalaya and flows through India and Bangladesh, Iraboti, Stars from Himalaya and flows through China and Myanmar and Yamuna flows through Tibet, India and Bangladesh [56]. These are the main source of human living in this region and about 200000 million people depend on these rivers [56]. Bangladesh has 54 and 3 common river with India and Myanmar respectively. The problem with water flow through India has not been solved yet [56]. India and china's thrust for electricity is increasing continuously and they have already started generating electricity by creating dam at Himalayan Rivers which creates serious water scarcity in other downstream countries [32]. India is planning to make huge money from this generated power

[56]. The Government of Arunachal province (India) has said that, Arabian countries are floating on Petro Dollar and their country will float on hydro dollar [32]. 200000 million downstream people's life will be in danger for those electricity generation dams of Himalayan Rivers. That is a question of preserving human primary right, the water. Indian Supreme court stated in a suit that if ecology is hampered for creating a suitable environment of people living, consolation should be given [57]. Four main rivers of Bangladesh, Padma, Brahmaputra, Meghan and Karnafuli come from Himalaya through India. So Bangladesh will be the main victim of Indian's plan of hydro electricity generation and Bangladesh is not getting any consolation for this. India is going to export electricity in near future and the main source of this electricity are those hydro power plants which reflected in 11-15th electricity plan of India [58].

5. Survey on the Opinion of Electricity Trading between India and Bangladesh

Bangladesh government has taken different policy to overcome the barrier of insufficient electricity supply to the people and electricity trade from India is one of the policies. People of Bangladesh have different opinion regarding this matter. From one point of view, it has positive effect on total electricity condition of Bangladesh. On the other hand it may have negative economical effect in far future. However a survey has been done for people opinion about different issue related to this electricity trading. A questioner has been prepared and distributed among 100 teachers of different discipline such as Economics, English, Engineering, Management, Accounting, Marketing, Tourism & Hospitality Management of IUBAT—Internal University of Business Agriculture and Technology for giving their opinion. From that survey we have got the following result. The first question was, "Bangladesh Government will import 250MW electricity from India by August, 2013 which will cost TK 4/Unit excluding transmission expenses. Do you think that it is a right decision?" 58% teachers said no and only 9% teachers said yes. 33% teachers did not give any opinion.

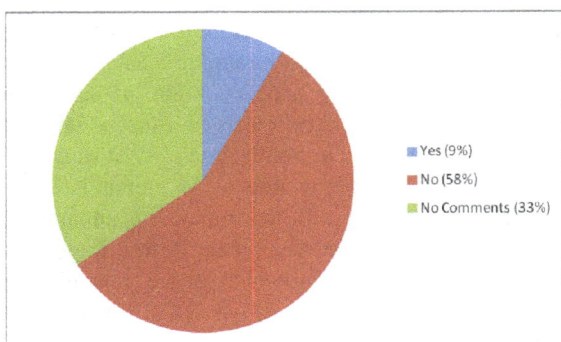

Figure 2. *pi chart showing the opinion in % on question 1*

40% teachers have given positive answer and 31% teachers have given negative answer to the question 2 which was "Including the transmission and maintenance cost, the total purchase cost of electricity from India will increase far more. Do you think that it will increase the cost per unit of electricity?"

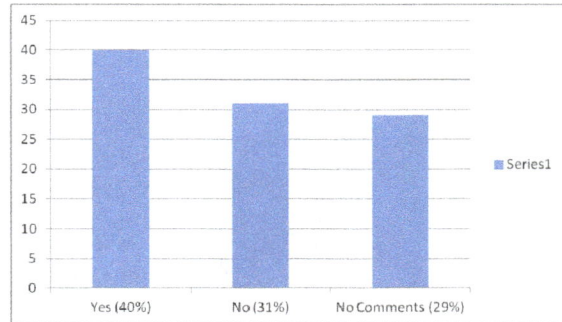

Figure 3. *graph showing the opinion in % on question 2*

Only 18% teachers agreed and 59% teachers disagreed on question 3, "Do you think importing only 500 MW is enough to solve the load shading problem of our country?"

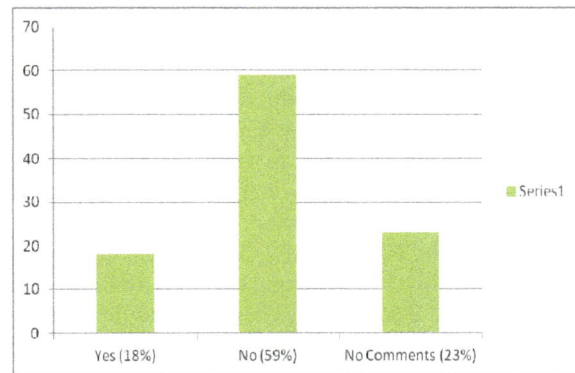

Figure 4. *graph showing the opinion in % on question 3*

The 4th question was "At present, the government cannot supply electricity to many industries, residential and commercial buildings due to lack of electricity. Do you think that this problem can be solved by this imported extra power from India?" only 35% of the teachers said yes and 43% said no.

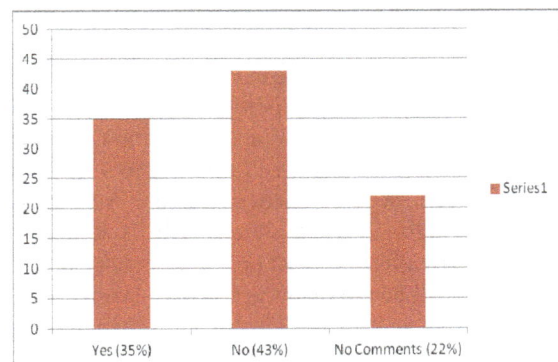

Figure 5. *graph showing the opinion in % on question 4*

In question 5 we asked that "What can be the reason for purchasing electricity from India. 17.02% teachers think that we have lack of man power to produce enough electricity. 42.55% teachers said that we don't have sufficient money to produce electricity and 40.43% teachers think Bangladesh has limited primary fuel reserve to produce necessary electricity.

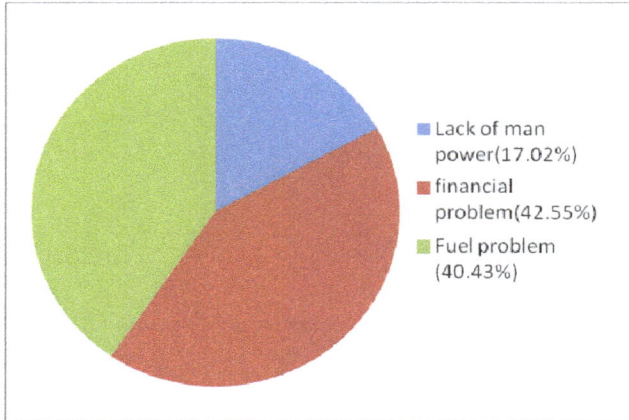

Figure 6. graph showing the opinion in % on question 5

6. Discussion & Recommendation

The main reason of electricity scarcity in Bangladesh is lack of gas supply and 770 MW electricity generations was stopped on 22 October 2008 for this reason [5, 59, 60]. 3115 MW of Electricity supply can be ensured by ensuring gas supply [60]. Total generation of our country will be 7000 MW after the addition of 2130 MW gas power plant in this year (2013) [59]. Bangladesh has sufficient coal reserve also and 37% electricity is generating by coal [59]. Only half of the reserved commercial energy is used to produce electricity in Bangladesh and contribution of renewable energy is only 22 MW [59, 63]. Current minimum electricity rate is 3.33 Taka per KWh [78].

From the survey it is clear that most of the people think electricity import from India cannot solve our electricity problem properly and it has some disadvantages. Including distribution cost the tariff of imported power will be at least approximately Tk 6.41[79] which is far more from our current domestic rate. The history of electricity trade of different south Asian countries with India is not satisfactory and in all cases other countries were dominated. India is non cooperative in case of sharing common resources like water with other countries. Moreover, in 2007-08, India's peak demand was 108886 MW and generation was 90973 MW and in 2027 the demand will raise up to 700000 MW [56]. India is a power deficiency country. So export of electricity by India is quite illogical. It is a matter of question that how effective will be the electricity trade and which country will be the main beneficiary. From the above information it can be said that electricity exchange between two countries is a good idea to solve running electricity problem of Bangladesh. But the trading should be effective,

clear, at the cheapest rate and beneficial to both countries and we should be careful in case of choosing trading partner. However, self sufficiency is the best policy for achieving economic development. Improper policy is the main obstructer in proper electricity generation. The government has not taken effective plan to generate electricity of its own. The government should reform the electricity sector. Bangladesh has mainly one coal based power plant of 250MW and 30 million ton coal remain surplus after the generation of electricity each year [3]. Huge amount of power is lost due to mismanagement of the system and it is considered as system loss where the loss in the transmission system cannot be greater than 10% as Bangladesh has moderate transmission system [64]. In 2008, total system loss was considered 21.25% where transmission loss was only 5.62% and it is possible to save 400 MW of electricity by limiting system loss at 10% [65]. Though the coal mine was identified in Bangladesh before the liberation war, no initiative has been taken yet by the Government for improving this sector [9]. Electricity crisis can be minimized by proper management of power generation and distribution sector of Bangladesh.

7. Conclusion

Though electricity connection is the precondition of sustainable development, 80% of world's un-electrified people live in South Asia [62] and 60% of Bangladeshi people are not connected to national grid [59]. Bangladesh has to earn electricity consumption rate of 383 KWh/capita to be a middle income country [61]. Some experts thing that electricity problem of Bangladesh can be solved by exchanging electricity with neighbor countries. But the exchange condition should be logical and favorable for both countries. But Bangladesh is not enough economically solvent to purchase electricity at high rate to support the people. If the country needs to import electricity badly, it can import from Myanmar at a cheap rate. One of the duties of the government is to secure energy and electricity demand and supply at affordable rate and one steps of the government is to import electricity from India. But most of the learned people have negative thoughts about this energy trading. They think that it is not a suitable way to solve our electricity problem which will affect our economy. The problem can be solved by reconstructing energy sector management system and infrastructure to reduce the system loss. Though mutual cooperation among SAARC countries is advantageous in many aspects, right of all country involved in trading should be reserved equally. History of cooperating in such trade in South Asia are not satisfactory, on the other hand electricity rate of Myanmar is low and Bangladesh has a border with Myanmar, it is advantageous to import electricity from Myanmar. Power import from India is not only the solution to solve our electricity problem and may create a negative effect on our overall economy.

References

[1] M.S. Alam, A. Roychowdhury, K.K. Islam and A.M.Z. Huq, A Revisited Model for the Physical Quality of Life (PQL) as a Function of of Electrical Energy Consumption. ENERGY, UK. 28 (1998), pp.-791-801.

[2] M.S. Alam, B.K. Bala, A.M.Z. Huq and M.A. Matin, A Model for Quality of life as a Function of Electrical Energy Consumption. ENERGY, UK. 16 (1991), pp.-739-745.

[3] M Samsul Alam, Vulnerable Power Sector and National Interest, Main Article present in round table dialogue, National Committee of Oil-Gas-Mine and Electricity-Port Saving, 04 January 2008. (based on the information getting by research of IEP, CUET) [in Bangla]

[4] M. Nurul Islam, Sustainability & Energy Security for Bangladesh, Prepared for A Symposium on Bangladesh Energy Sector: The Way Forward, Organized by Ministry of Energy, Power & Mineral Resources and GTZ-German Technical Cooperation, Dhaka, 12April 2007.

[5] BPBD (Bangladesh Power Development Board), Summary of the Daily Electricity Production, 22 November, 2008

[6] M Shamsul Alam, Suvanir, Energy Security and right of information, weekly 2000, Vol-29, November 2007 [in Bangla]

[7] B.D. Rahmatullah, Vision Bangladesh-2021: Energy and Power Crisis – Reform Attempt For Sustainable Development (Power point), 21 January, 2008.

[8] Khatib, H., and M. Munasinghe, "Electricity, the Environment and Sustainable World Development", Paper presented at the 15th Congress of the World Energy Council, Madrid, September, 1992.

[9] M. Jamaluddin, SAARC Regional Trade Study: Country Report – Bangladesh, February, 2008.

[10] Mahendra P Lama, Mohan Man Sainju, and Q K Ahmad, "Reforms and Power Sector in South Asia: Scope and Challenges for Cross-Border Trade", Economic Development in South Asia (edited by Mohsin Khan), Tata McGraw-Hill, New-Delhi, 2005.

[11] Saleque Sufi, "The Rationale of SAARC Energy Ring", Vol. 6 (17), pp. 1-5, Power & Energy, Dhaka, February, 2009.

[12] Country Comparison>Population, Source: CIA World Fact book, 1 January, 2008, www.indexmundi.com/g/r.aspx?t=0&v=21&l=en

[13] Country Comparison> Population growth rate, Source: CIA World Factbook, 1 January, 2008, www.indexmundi.com/g/r.aspx?t=0&v=24&l=en

[14] List of countries by GDP (nominal) per capita, source: The World Fact book, Central Intelligence Agency for the year 2008, Wikipedia, the free encyclopedia. en.wikipedia.org/wiki/List_of_countries_by_GDP_(nominal)_per_capita

[15] Country Comparison > GDP - real growth rate, Source: CIA World Factbook, 1 January, 2008, www.indexmundi.com/g/r.aspx?t=0&v=66&l=en

[16] Country Comparison > Natural gas – proved reserves, Source: CIA World Fact book, 1 January, 2008, www.indexmundi.com/g/r.aspx?t=0&v=98&l=en

[17] Oil – proved reserves Country Comparison, Source: CIA World Fact book, 1 January, 2008, www.indexmundi.com/g/r.aspx?t=0&v=97&l=en

[18] Proved recoverable coal reserves at end-2006, Coal, Wikipedia, the free encyclopedia, Accesses March 2009, en.wikipedia.org/wiki/Coal#Production_trends

[19] Country Uranium reserves in tones, Uranium reserves, Wikipedia, the free encyclopedia, Accessed March 2009, en.wikipedia.org/wiki/Uranium_reserves

[20] Myanmar to produce an additional 260 million tons of coal, Published in ELEVEN no. 1 News Media in Myanmar, 02 July 2013, http://elevenmyanmar.com/national/2654-myanmar-to-produce-an-additional-260-million-tons-of-coal

[21] Rediff India Abroad, Pakistan has 1000 uranium reserve, 14 April, 2007, http://www.rediff.com/news/2007/apr/14pak.htm

[22] Future Projects, Export oriented projects, www.aseanenergy.org/energy/sector/electricity/myanmar

[23] The hkskyline, Honkong, "South Asia Plagued by Electricity Shortages", 31 May 2006, www.skyscrapercity.com/archive/index.php/t-356380.html

[24] The Peoples Daily Online, "China has huge potential in hydroelectric generation", 24 October 2005, http://webcache.googleusercontent.com/search?q=cache:http://english.peopledaily.com.cn/200510/24/eng20051024_216366.html

[25] Country Comparison > Electricity – imports, Source: CIA World Factbook, 1 January, 2008, www.indexmundi.com/g/r.aspx?t=0&v=83&l=en

[26] Country Comparison > Electricity – exports, Source: CIA World Factbook, January 1, 2008, www.indexmundi.com/g/r.aspx?t=0&v=82&l=en

[27] Reuters, "UES to Increase Electricity Export to China this Year", The Russia Journal, 17 May 1999. www.russiajournal.com/node/807.

[28] The CHINA.ORG.CN, "China to Import Electric Power from Bordering Countries", 19 June, 2006. www.china.org.cn/english/BAT/171947.htm

[29] The People's Daily Online, "Vietnam to import electricity from China till 2010", 22 August, 2006 english.peopledaily.com.cn/200608/17/eng20060817_293849.html.

[30] Juliet Shwe Gaung, Hydro leads way in power generation, Myanmar times special feature, July 2008, www.mmtimes.com/feature/energy08/eng005.htm

[31] Samira Kalam, "Energy 2050: Commercial Energy Flow in South Asia", A paper submitted for partial fulfillment of the degree of M-Phill in Institute of Energy Technology, CUET, Chittagong.

[32] Shripad Dharmadhikary, "Mountains of concrete, Dam building in the Himalayas", Published by International Rivers- a non-governmental organization, December 2008, www.internationalrivers.org/files/attached-files/ir_himalayas_rev.pdf

[33] Bhutan , UN data a world of information (Source: World Statistics Pocketbook | United Nations Statistics Division), Accessed March 2009, data.un.org/CountryProfile.aspx?crName=Bhutan

[34] Nepal, UN data a world of information (Source: World Statistics Pocketbook | United Nations Statistics Division). Accessed March 2009, data.un.org/CountryProfile.aspx?crName=Nepal

[35] Ceylon Electricity Board, Colombo, Sri Lanka, "Statistical Digest, 2006".

[36] Sri Lanka, UN data a world of information (Source: World Statistics Pocketbook | United Nations Statistics Division). Accessed March 2009, data.un.org/CountryProfile.aspx?crName=Sri Lanka

[37] Afghanistan, UN data a world of information (Source: World Statistics Pocketbook | United Nations Statistics Division). Accessed March 2009, data.un.org/CountryProfile.aspx?crName=Afghanistan_

[38] Ministry of Energy & Water, Islamic Republic of Afghanistan, "Power Sector Strategy for the Afghanistan National Development Strategy (with focus on prioritization)", April 2007.

[39] Islamic Republic of Afghanistan Ministry of Energy & Water," Power sector Strategy for The Afghanistan National Development Strategy", 2007

[40] President of Maldives, "Maldives is the first carbon free country of Bangladeshi", report of the Daily Prothom Alo (Bangla daily), pp 6, 24 March, 2009.

[41] Maldives, UN data a world of information (Source: World Statistics Pocketbook | United Nations Statistics Division) Accessed March 2009, data.un.org/CountryProfile.aspx?crName=Maldives

[42] British Petroleum, "Natural gas reserves table: Natural gas reserves", 2008, www.bp.com/liveassets/bp_internet/globalbp/globalbp_uk_e nglish/reports_and_publications/statistical_energy_review_2 008/STAGING/local_assets/downloads/pdf/gas_table_of_pr oved_natural_gas_reserves_2008.pdf

[43] Background paper, Central Asia/South Asia Electricity Trade Conference, Islamabad, Pakistan, May 8-9, 2006.

[44] The Bangladesh News.Com (Online news paper), "Coal can change the economy of Bangladesh", March 2009, www.bangladeshnews.com/bd

[45] The People's Daily Online, "Myanmar raises electricity charge prices", May 2006, english.peopledaily.com.cn/200605/22/eng20060522_26760 4.html

[46] Dwarika N. Dhungel, "Energy, Reforms and Cross-border Cooperation between Nepal and India: A Professional Perspective", A paper submitted for a dialogue programme on economic reforms and development dynamics: A cross borders perspective between India and Nepal, Organized by University of Sikkim, Gangtoc, April 18-20, 2008.

[47] Rabin Subedi & Ratan Bhandari, February 2008, w4pn.org/index.php?option=com_content&task=view&id=5 3&Itemid=30

[48] The Bhutan News Online, "KURICHHU HYDRO-ELECTRIC PROJECT", 06 April, 2005. www.bhutannewsonline.com/kurichhu _hydroproject.html

[49] Bhutan - Electricity - production (Source: CIA World Factbook, 1 January, 2008, www.indexmundi.com/g/g.aspx?c=bt&v=79

[50] The Bhutan News Online, "Chukha Hydro Power Corporation (CHPC)", 6 April, 2005. www.bhutannewsonline.com/chukha_hydro.html

[51] The Bhutan News Online, "Tala Hydroelectric Project Authority (THPA), 06 April, 2005, www.bhutannewsonline.com/tala_hydroproject.html

[52] The Bhutan News Online, "Basochu Hydro Power Project", 06 April, 2005. www.bhutannewsonline.com/basochu_hydro.html

[53] Burma - Electricity - production (Source: CIA World Factbook, 1 January, 2008, www.indexmundi.com/g/g.aspx?c=bm&v=79

[54] Burma - Electricity - consumption (Source: CIA World Factbook, 1 January, 2008, www.indexmundi.com/g/g.aspx?c=bm&v=81

[55] Asean Affairs, "Hydro Power Project: Myanmar agrees to push Bangladeshi's proposal", 9 October, 2008, www.aseanaffairs.com/hydro_power_project_myanmar_agr ees_to_push_bangladeshi_s_proposal

[56] Weekly (weekly magazine) 552 dam of India Pakistan Nepal Bhutan and hydro electricity politics of India, pp 30-42, 19 February, 2009.

[57] M Shamsul Alam, Coal: Environment in improving minr, the Daily Somokal 21 October, 2007 [in Bangla]

[58] Interim Report of the Independent People's Tribunal (IPT) on Dams in Arunachal Pradesh, India. 3 February 2008. [http://www.iptindia.org]

[59] PDB, "An open discussion on planning of power generation increase, Demand side management & system loss reduction", A power point presentation in BERC meeting, Dhaka, 16 April, 2009.

[60] M. Fouozul Kabir Khan, Energy sector: Challenges of Adding New Capacity, A paper presented in the conference on Development With Equity and Justice, organized by Centre for Policy Dialogue (CPD), Dhaka, 28-29 March 2009.

[61] Planning Commission, Government of Bangladesh, Development Parameters of Bangladesh Present Vs. MI and UMI References for Background Papers, Outline Participatory Perspective Plan (OPPP) Project 2007-2021, 2008.

[62] Electricity for all: Targets, timetables, instruments, An initiative to make electricity available, accessible and affordable to all: Proposal for a global debate, GENI (Global Energy Network Institute), October 2002, www.geni.org/globalenergy/library/media_coverage/electricite- de-france/electricity-for-all--targets-timetables-instuments.shtml

[63] B.D. Rahmatullah, Study on Critical Factors in Determining Success of Renewable Energy Projects in South Asia, A Country Report-Bangladesh, November 2008.

[64] M.S. Alam, E.Kabir, M.M. Rahman, M.A.K. Chowdhury, "Power sector reform in Bangladesh: Electricity distribution system", ENERGY, UK. 29 (2004), pp.-1773-1783.

[65] B. D. Rahmatullah, Vulnerability of Power Sector, way of solving problem and vision 2021, Dr. Golam Mohiuddin memorial speech, Bangladesh Udichi Shilpi Gosti, 24 April, 2009. [in Bangla]

[66] Trading Economics, Babgladesh GDP Grouth Rate, www.tradingeconomics.com/bangladesh/gdp-growth,2013

[67] REUTER, "Bangladesh to import 250 MWh power/day from India, 18 February 2013, in.reuters.com/article/2013/02/18/bangladesh-power-india-i dINDEE91H0AS20130218,

[68] Power and Energy Sector Road Map : An Update, Finance Division, Ministry of Finance Government of People's Republic of Bangladesh, Accessed December 2012, www.mof.gov.bd/en/budget/11_12/power/power_energy_en. pdf, 2011

[69] The EnergyBangla (online news paper), 13 November 2013, www.energybangla.com/2013/04/17/2626.html#.UXY1glJ4r CM,

[70] Global Transmission Report, "Project Updates", gust 22, 2013, www.globa ltransmission.info/archive.php?id=3976, Au

[71] Bangla News 24.com, "Power Import from India in July",

January 31, 2013, www.banglanews24.com/English/detailsnews.php?nssl=21e 7a3199904eacbfc122e0e681c6aba&nttl=2013013163262,

[72] Business Recorder, "Bangladesh to import 250 megawatts-hours a day from India", February 19, 2013, http://www.brecorder.com/fuel-a-energy/193/1155498/

[73] Power Grid Company of Bangladesh Ltd, National Load Dispatch Center, Daily Report, www.bpdb.gov.bd/bpdb/pdb_utility/maxgen/dailygen_archi ve/2019page1.pdf

[74] AHM. Mustain Billah, Energy and Environment: Demand for Wood Energy in Bangladesh, SEMP-SDN Project, Bangladesh Institute of Development Studies (BIDS), www.sdnbd.org/sdi/issues/energy/publications/energy_and_ environment.htm

[75] International Monetary Fund, IMF Country Report No.12/326, Nepal, 2012 ARTICLE IV CONSULTATION, December 2012, www.imf.org/external/pubs/ft/scr/2012/cr12326.pdf,

[76] Country Comparison > Electricity – consumption, Source: CIA World Factbook, 1 January, 2008, www.indexmundi.com/g/r.aspx?c=pk&v=81

[77] Bhutan - Electricity – exports, Source: CIA World Factbook, January 1, 2008, www.indexmundi.com/g/g.aspx?v=82&c=bt&l=en

Permissions

All chapters in this book were first published in EPES, by Science Publishing Group; hereby published with permission under the Creative Commons Attribution License or equivalent. Every chapter published in this book has been scrutinized by our experts. Their significance has been extensively debated. The topics covered herein carry significant findings which will fuel the growth of the discipline. They may even be implemented as practical applications or may be referred to as a beginning point for another development.

The contributors of this book come from diverse backgrounds, making this book a truly international effort. This book will bring forth new frontiers with its revolutionizing research information and detailed analysis of the nascent developments around the world.

We would like to thank all the contributing authors for lending their expertise to make the book truly unique. They have played a crucial role in the development of this book. Without their invaluable contributions this book wouldn't have been possible. They have made vital efforts to compile up to date information on the varied aspects of this subject to make this book a valuable addition to the collection of many professionals and students.

This book was conceptualized with the vision of imparting up-to-date information and advanced data in this field. To ensure the same, a matchless editorial board was set up. Every individual on the board went through rigorous rounds of assessment to prove their worth. After which they invested a large part of their time researching and compiling the most relevant data for our readers.

The editorial board has been involved in producing this book since its inception. They have spent rigorous hours researching and exploring the diverse topics which have resulted in the successful publishing of this book. They have passed on their knowledge of decades through this book. To expedite this challenging task, the publisher supported the team at every step. A small team of assistant editors was also appointed to further simplify the editing procedure and attain best results for the readers.

Apart from the editorial board, the designing team has also invested a significant amount of their time in understanding the subject and creating the most relevant covers. They scrutinized every image to scout for the most suitable representation of the subject and create an appropriate cover for the book.

The publishing team has been an ardent support to the editorial, designing and production team. Their endless efforts to recruit the best for this project, has resulted in the accomplishment of this book. They are a veteran in the field of academics and their pool of knowledge is as vast as their experience in printing. Their expertise and guidance has proved useful at every step. Their uncompromising quality standards have made this book an exceptional effort. Their encouragement from time to time has been an inspiration for everyone.

The publisher and the editorial board hope that this book will prove to be a valuable piece of knowledge for researchers, students, practitioners and scholars across the globe.

List of Contributors

Funso K. Ariyo
Department of Electronic and Electrical Engineering, Ile-Ife, Nigeria
Obafemi Awolowo University, Ile-Ife, Nigeria

A. S. M. Monjurul Hasan and Md. Habibullah
Department of Electrical & Electronic Engineering, IUT, Dhaka, Bangladesh

A. S. M. Muhaiminul Hasan
Department of Electrical & Electronic Engineering, AUST, Dhaka, Bangladesh

Bouhadouza Boubekeur, Ahmed Gherbi and Hacene Mellah
Department of Electrical Engineering, Sétif-1 University, Algeria

A. M Bouzid
Departmentof Electrical and Computer Engineering, University of Quebec at TroisRivieres UQTR, QC, CANADA
Department of electrical engineering, University USTO MB, Oran, ALGERIA

A. Cheriti
Departmentof Electrical and Computer Engineering, University of Quebec at TroisRivieres UQTR, QC, CANADA

M. Bouhamida and M. Benghanem
Department of electrical engineering, University USTO MB, Oran, ALGERIA

M. Kavitha
Dept of Computational Engineering, APIIIT, RGUKT, Nuzvid, A.P,India

T. Chandrasekhar
Dept of CSE, APIIIT, RGUKT, Nuzvid, A.P,India

D. Mohan Reddy
Dept of EEE, SVIET,Pedana (M), Krishna (Dt),A.P,India

Hamed H. H. Aly and M. E. El-Hawary
Department of Electrical and Computer Engineering, Dalhousie University, Halifax, Nova Scotia, Canada, B3H 4R2

Dhiya A. Al-Nimma, Majid S. M. Al-Hafidh and Saad Enad Mohamed
Electrical Engineering Department, University of Mosul, Mosul- IRAQ

Benheniche Abdelhak
Electrotechnical Department, Mouloud Mammeri Tizi-Ouzou University
Electronic Department, Badji Mokhtar Annaba University

Bensaker Bachir
Electrotechnical Department, Mouloud Mammeri Tizi-Ouzou University
Electronic Department, Badji Mokhtar Annaba University

Adepoju Gafari Abiola
Electronic and Electrical Engineering Department, LAUTECH, Ogbomoso, Nigeria

Tijani Muhammed Adekilekun
Electrical and Electronics Engineering Department, Federal Polytechnic, Ede, Nigeria

E. A. Kamba, A. U. Itodo and E. Ogah
Department of Chemical Sciences, Federal University Wukari, Nigeria

Funso K. Ariyo
Department of Electronic and Electrical Engineering, Ile-Ife, Nigeria
Obafemi Awolowo University, Ile-Ife, Nigeria

Obodeh, O.
Mechanical Engineering Department Ambrose Alli University, Ekpoma, Edo State, Nigeria

Ugwuoke, P. E.
Mechanical Engineering Department Petroleum Training Institute, Effurun, Delta State, Nigeria

Jyothilal Nayak Bharothu, K Lalitha
Sri Vasavi Institute of Engineering & Technology, Nandamuru, A.P.; India

Mohd Tariq
M.Tech Student, Department of Electrical Engineering, Indian Institute of Technology, Kharagpur, India

Sagar Bhardwaj
B.Tech Student, Department of Electrical Engineering, Aligarh Muslim University, Aligarh, India

Mohd Rashid
B.Tech Student, Department of Computer Science, Jamia Hamdard University, New Delhi, India

Mohd. Tariq
M.Tech Student, Department of Electrical Engineering,
Indian Institute of Technology, Kharagpur, India

Khyzer Shamsi
Protection design engineer, Al Jazirah Engineers and
consultants, Al-khobar, Kingdom of Saudi Arabia

Tabrez Akhtar
Assistant Manager, Delhi Metro Rail Corporation Ltd.,
Delhi, India

Mohammad Saad Alam
Lead Systems Engineer- MagnetiMarelli Holdings of
North America Inc (FIAT GROUP SpA)

**Saravanan Kandasamy, Dawit Leykuen Berhanu and
Getnet Zewde Somanu**
Department of Electrical and computer Engineering,
JIT, Jimma University, Jimma, Ethiopia

Murali. Matcha and Sharath Kumar. Papani
Department of Electrical Engineering, N.I.T Warangal,
A.P, INDIA-506004

Vijetha. Killamsetti
Department of EEE, GMR Institute of Technology,
Srikakulam, A.P, INDIA

Nitin Mohan Lal and Arvind Kumar Singh
Electrical Engg. Deptt, B.I.T Sindri, Dhanbad, India

Pushpalata Khalkho and Arvind Kumar Singh
Electrical Engg. Deptt. B.I.T Sindri, Dhanbad, Govt. of
Jharkhand-828123, India

**Noha Mahmoud Bastawy, Hossam El-din Talaat and
Amr Mohamed Ibrahim**
Electrical Power & Machines Dept., Faculty of
Engineering, Ain Shams University, Cairo, Egypt

**Osama Tarek Al-Taai and Amani Ibraheem Al-
Tmimi**
Department of Atmospheric Sciences, College of
Science, Al-Mustansiriyah University, Baghdad, Iraq

Qassim Mahdi Wadi
Al-MamonUniversity College, Department of Electrical
Power Technical Engineering, Baghdad, Iraq

Mofreh M. Nassief
Faculty of Engineering Zagazig University Egypt

Nahid Ahmadinia
M. A. Electrical Power Engineering, Science and
Research Branch, Islamic Azad University, Broujerd,
Iran

**Bablesh Kumar Jha, Ramjee Prasad Gupta and
Upendra Prasad**
Electrical Engg. Deptt., B.I.T Sindri, Dhanbad
,Jharkhand, India

**Fareed ud Din, Abdul Rehman Raza and Muhammad
Azam**
Faculty of Engineering and Technology, Lahore,
Pakistan

Ramlal Das and D. K. Tanti
Electrical Engineering Department, Bit Sindri

R. Khaldi, N. Benamrouche and M. Bouheraoua
LATAGE Laboratory, Department of Electrical
Engineering, Mouloud Mammeri University, Tizi-
Ouzou, Algeria

Anmona Shabnam Pranti
Department of Electrical and Electronics Engineering,
International University of Business Agriculture and
Technology, Dhaka, Bangladesh

**Arif Mohammad Shahed Iqubal and Mohammad
Shawkut Ali Khan**
Department of Mechanical Engineering, International
University of Business Agriculture and Technology,
Dhaka, Bangladesh

Mohammad Kayesar Ahmmed
Greenland Builders Limited, Dhaka, Bangladesh

Index

www.ingramcontent.com/pod-product-compliance
Lightning Source LLC
Chambersburg PA
CBHW080512200326
41458CB00012B/4179